VI International Workshop on the Dark Side of the Universe 2010

(DSU 2010)

Journal of Physics: Conference Series Volume 315

Guanajuato, Mexico
1-6 June 2010

ISBN: 978-1-61839-133-9
ISSN: 1742-6588

Printed from e-media with permission by:

Curran Associates, Inc.
57 Morehouse Lane
Red Hook, NY 12571

proceedings.com

Some format issues inherent in the e-media version may also appear in this print version.

Printed by Curran Associates, Inc. (2011)

For permission requests, please contact the Institute of Physics at the address below.

Institute of Physics
Dirac House, Temple Back
Bristol BS1 6BE UK

Phone: 44 1 17 929 7481
Fax: 44 1 17 920 0979

techtracking@iop.org

Additional copies of this publication are available from:

Curran Associates, Inc.
57 Morehouse Lane
Red Hook, NY 12571 USA
Phone: 845-758-0400
Fax: 845-758-2634
Email: curran@proceedings.com
Web: www.proceedings.com

Preface

Papers

List of participants

Abigail Alvarez Olarte	CINVESTAV
Alba Leticia Carrillo Monteverde	DCI-UG
Alberto Carramiñana	INAOE
Aldo Morselli	FERMI
Alejandro Castilla	DCI-UG
Alejandro Ibarra	Technical University of Munich
Alma D Rojas Pacheco	FCFM-BUAP
Alma Xochitl Gonzalez Morales	Instituto de Ciencias Nucleares, UNAM
Andrew Walcott Beckwith	American Institute of Beam Energy Physics
Ariadna Montiel Arenas	Departamento de Física, CINVESTAV
Arnulfo Zepeda	Cinvestav
Arturo Alvarez Cruz	Instituto de Fisica, UNAM
Axel de la Macorra	UNAM, IAC
Azar Mustafayev	University of Minnesota
Benjamin Jaramillo	DCI-UG
Vincent Bertin	CPPM-Marseille
Carlos Alberto Vaquera-Araujo	DCI-UG
Carlos Muñoz	Madrid Autonoma U. & Madrid, IFT
Carmine Pagliarone	INFN, FNAL
Carolina Lujan Peschard	DCI-UG
Christiane Frigerio Martins	Universidade Federal do ABC-São Paulo
Csaba Balazs	Monash University
David Delepine	DCI-UG
David G Cerdeno	Universidad Autonoma de Madrid & Instituto de Fisica Teorica
Debasish Majumdar	Saha Institute of Nuclear Physics, Kolkata, India
Dibyendu Panigrahi	Kandi Raj College, Kandi, Murshidabad, INDIA-742137
Dupret Alberto Santana Bejarano	Universidad de Sonora Departamento de Investigacion en Fisica
Ernest Ma	Riverside U.C.
Esteban Alejandro Reyes Pírez Montañez	Instituto de Física, UNAM
Federico Ortiz Trejo	INSTITUTO DE ASTRONOMÍA - UNAM
Francisco José de Anda Navarro	Universidad de Guadalajara
González Alvarez Francisco Javier	CINVESTAV-Depto. Física
Gustavo Medina Tanco	ICN-UNAM

Hernando Efrain Caicedo Ortiz	Instituto Politecnico Nacional - IPN
J D Vergados	CERN & Ioannina U.
James R Boyce	Jefferson Lab
Jason Steffen	FERMILAB
Javier Montaño Domínguez	DCI-UG
Jeevan Solanki	Mandsaur Institue of Technology MP India
Joe Sato	Saitama University
Jorge Luis Navarro Estrada	UNAM-ICN and Universidad del Atlantico (B/quilla-Col.)
Jose A R Cembranos	University of Minnesota
José Díaz	IFIC
Jose Didino Garcia Aguilar	Depto. de Fisica. Cinvestav
Keith Olive	University of Minnesota
Konstanția Balasi	University of Ioannina, Greece
Lilian Prado Gonzalez	BUAP, FCFM
Lorenzo Díaz Cruz	BUAP Facultad de Ciencias Físico Matemáticas
Luis Rey Díaz Barrón	División de Ciencias e Ingenierías
Luis Urena	Universidad de Guanajuato
Magda Lola	Dept. of Physics, University of Patras, Greece
Mahmoud Wahba	Egyptian Center for Theoretical Physics, MTI
Marcus S Cohen	New Mexico State University
Mario A Acero Ortega	ICN - UNAM
Mario E Gomez	Universidad de Huelva
Mark Pipe	University of Sheffield
Mauro Napsuciale	DCI-UG
Mirco Cannoni	Universidad de Huelva
Mónica Felipa Ramírez Palacios	Universidad de Guadalajara
Murli Manohar Verma	Lucknow university, India
Nassim Bozorgnia	UCLA
Octavio Obregón	
Octavio Valenzuela	IA-UNAM
Oleg Kamaev	University of Minnesota
Osamu Seto	Hokkai-Gakuen University
Pedro F González Díaz	IFF, CSIC, Serrano 121, 28006 Madrid, Spain
Qaisar Shafi	Bartol Research Inst. and Delaware U.
Raul Hennings-Yeomans	Los Alamos National Laboratory

René Ángeles Martínez	Departamento de Fisica, del DCI de la Universidad de Guanajuato
Reyna Xoxocotzi	BUAP, FCFM
Rishi Kumar Tiwari	Govt. Model Science College, Rewa (MP) INDIA
Roberto A Sussman	ICN-UNAM
Selim Gómez Ávila	DCI-UG
Sugai Kenichi	Saitama University
Susana Valdez Alvarado	DCI-UG
TVladimir - 2K Collaboration	Colorado State University
Tonatiuh Matos	CINVESTAV
Valeriy Dvoeglazov	Universidad de Zacatecas
Vannia Gonzalez Macias	DCI-UG
Vladimir Avila-Reese	Instituto de Astronomia, UNAM
Wolfgang Bietenholz	INC, UNAM (Mexico)
Yamanaka Masato	Kyoto Sangyo University
Yann Mambrini	LPT Orsay
Yu-Feng Zhou	Institute of Theotretical Physics, Chinese Academy of Sciences, PR China
Aaron Higuera	DCI-UG
Azarael Yebra Pérez	DCI-UG
César Hernández Aguayo	DCI-UG
Jaime Chagoya Alvarez	DCI-UG
Jonathan Rashid Rosique Campuzano	DCI-UG
José Alfredo Soto Álvarez	DCI-UG
Juan Carlos De Haro Santos	DCI-UG
Luis Eduardo Medina Medrano	DCI-UG
Maria Fatima Rubio Espinoza	DCI-UG
Paulo Alberto Rodriguez Herrera	DCI-UG
Roberto Oziel Gutierrez Cota	DCI-UG
Rocha Moran Maria Paulina	DCI-UG
Xareni Sanchez Monroy	DCI-UG

VI International Workshop on the Dark side of the Universe (DSU 2010)

David Delepine

PREFACE

The main indications we have for physics beyond the Standard Model of fundamental interactions come from cosmological data. At the same time, the information collected from particle colliders as LHC or Tevatron shed light on the first moments of the Universe. This correlation between cosmological and astrophysical data and that from particles is so strong that that it has given birth to a whole new field, called astroparticle physics. This is a very active domain of research. It includes an impressive number of fundamental issues, such as the nature of dark matter and dark energy, the origin of the baryon asymmetry of the universe, neutrino physics, physics beyond the Standard Model, nonstandard cosmology, ultra high energy cosmic rays (UHECR), gamma ray bursts, inflation, the nature of electroweak phase transition, detection of gravitational waves, proton decays, etc. With the advent of a new generation of large scale experiments - dedicated to the search for, e.g., dark matter, UHECR, and gravitational waves - the future of astroparticle physics looks very promising. Combining these results with the expected results from the LHC, it will be possible to strongly improve our knowledge of fundamental interactions beyond what is now called the Standard Model of electroweak and strong interactions. Topics in Astroparticle physics are highly multidisciplinary as they involve not only high energy physicists, but also astronomers, astrophysicists, cosmologists, etc. Thus it is very important to regularly hold forums where all these researchers with distinct, but overlapping, expertise can share their knowledge and problems.

Since 2005, the "Dark side of the Universe" (DSU) International Workshop on Dark Matter has offered this possibility, inviting researchers from all over the world working on topics related to astroparticles to present their results. Since then, the DSU workshop has been organized each year in a new continent: in Seoul (2005), Madrid (2006), Minnesota (2007), Cairo (2008) and Melbourne (2009). In 2010, the DSU workshop was organized at Guanajuato University (campus Leon). Mexico has a long history of participation in international collaborations related to astroparticle topics (see for instance the Mexican participation in the Auger observatory). So it was a pleasure for us to welcome the 2010 edition of the DSU International Workshop.

On behalf of the local and national organizing committee, we want to thank all the participants of the workshop. There were 97 participants from 15 different countries. We hope that all of the participants enjoyed their stay in Leon (Guanajuato).

On behalf of the DSU2010 Local and National Organizing Committees

Journal of Physics: Conference Series **315** (2011) 012001 doi:10.1088/1742-6596/315/1/012001

Higher-dimensional Higgs representations in SGUT models

Alfredo Aranda[1,3], **J L Díaz-Cruz**[2,3] **and Alma D Rojas**[3]
[1] Facultad de Ciencias,CUICBAS, Universidad de Colima, Bernal Díaz del Castillo 340, Colima, Colima, México
[2] C.A. de Partículas, Campos y Relatividad , FCFM-BUAP, Puebla, Pue. México
[3]Dual C-P Institute of High Energy Physics

E-mail: fefo@ucol.mx, ldiaz@sirio.ifuap.buap.mx, alma.drp@gmail.com

Abstract.
Supersymmetric Grand Unified Theories (SGUTs) have achieved some degree of succes, already present in the minimal models (with SU(5) or SO(10)). However, there are open problems that suggest the need to incorporate more elaborate constructions, specifically the use of higher-dimensional representations in the Higgs sector. For example, a **45** representation of SU(5) is often included to obtain correct mass relations for the first and second families of d-type quarks and leptons. When one adds these higher-dimensional Higgs representations one must verify the cancellation of anomalies associated to their fermionic partners. One possible choice, free of anomalies, include both representations to cancel anomalies. We review the necessary conditions for the cancellation of anomalies and discuss the different possibilities for supersymmetric SU(5) models. Alternative anomaly-free combinations of Higgs representations, beyond the usual vectorlike choice, are identified, and it is shown that their corresponding β functions are not equivalent. Although the unification of gauge couplings is not affected, the introduction of multidimensional representations leads to different scenarios for the perturbative validity of the theory up to the Planck scale. We study the effect on the evolution of the gauge coupling up to the Planck scale due to the different sets of fields and representations that can render an anomaly-free model.

1. INTRODUCTION

Anomalies in gauge theories. The need to require anomaly cancellation in any gauge theory stems from the fact that their presence destroys the quantum consistency of the theory. It turns out that all one needs to calculate or identify the anomaly is the triangle diagrams. For a given representation of a gauge group G, the anomaly can be written as

$$A(D)d^{abc} \equiv Tr\left[\left\{T_a^{D_i}, T_b^{D_i}\right\} T_c^{D_i}\right], \tag{1}$$

where $T_a^{D_i}$ are the generators of the gauge group G in the representation D_i, and d^{abc} denotes the anomaly associated to the fundamental representation [1].

For a representation R that is the direct sum or tensor product of two representations the anomaly coefficient satisfy

$$\begin{aligned} A_R = A(R) &= A(R_1 \oplus R_2) = A(R_1) + A(R_2), \\ A_R = A(R) &= A(R_1 \otimes R_2) = D(R_1)A(R_2) + D(R_2)A(R_1), \end{aligned} \tag{2}$$

Journal of Physics: Conference Series **315** (2011) 012001 doi:10.1088/1742-6596/315/1/012001

respectively, with $D(R_i)$ denoting the dimensions of representation R_i.

Anomaly coefficients. The anomaly coefficients $A(D)$ for most common representations are known in the literature [1]. We have extended these results to include higher-dimensional representation, with some of them shown in Table 1.

Table 1. Dimension and anomaly coefficients for higher-dimensional representation of SU(N).

Irrep	$dim(r)$	$A(r)$
	$\frac{N(N-1)(N-2)(N-3)}{24}$	$\frac{(N-4)(N-3)(N-8)}{6}$
	$\frac{N(N+1)(N+2)(N+3)(N+4)}{120}$	$\frac{(N+3)(N+4)(N+5)(N+10)}{24}$
	$\frac{N(N+1)(N+2)(N+3)(N-1)}{30}$	$\frac{(N-2)(N+3)(N+5)^2}{6}$
	$\frac{N^2(N+1)(N+2)(N-1)}{24}$	$\frac{N(N+5)(5N^2-3N-50)}{24}$
	$\frac{N^2(N+1)(N-1)(N-2)}{24}$	$\frac{N(N-5)(5N^2+3N-50)}{24}$
	$\frac{N(N+1)(N+2)(N-1)(N-2)}{20}$	$\frac{(N^4-17N^2+100)}{4}$
	$\frac{(N-3)(N-2)(N-1)N(N+1)}{30}$	$\frac{(N-5)^2(N-3)(N+2)}{6}$
	$\frac{N(N-1)(N-2)(N-3)(N-4)}{120}$	$\frac{(N-5)(N-4)(N-3)(N-10)}{24}$
	$\frac{N(N+1)(N+2)(N+3)(N+4)(N+5)}{720}$	$\frac{(N+3)(N+4)(N+5)(N+6)(N+12)}{120}$
	$\frac{N(N+1)(N+2)(N+3)(N+4)(N-1)}{144}$	$\frac{(N+3)(N+4)(N+6)(N^2+5N-12)}{24}$
	$\frac{(N-1)N^2(N+1)(N+2)(N+3)}{80}$	$\frac{3}{40}(N-3)N(N+3)(N+4)(N+6)$

Table 2. Dimension and anomaly coefficients for different representations of SU(5).

Repr.	*Multiplete*	$dim(r)$	$A(r)$	$2T(r)$
[5]	(0, 0, 0, 0)	1	0	0
[1]	(1, 0, 0, 0)	5	1	1
[2]	(0, 1, 0, 0)	10	1	3
[1, 1]	(2, 0, 0, 0)	15	9	7
Ad		24	0	10
[4, 1]	(1, 0, 0, 1)	24	0	10
[1, 1, 1]	(3, 0, 0, 0)	35	44	28
[2, 1]	(1, 1, 0, 0)	40	16	22
[3, 1]	(1, 0, 1, 0)	45	6	24
[2, 2]	(0, 2, 0, 0)	50	15	35

2. Anomaly cancellation.

There are several ways to construct an anomaly free theory:

- The gauge group itself is always free of anomalies. This happens, for instance, for SO(10) but not for SU(5).

- The gauge group is a subgroup of an anomaly free group, and the representations form a complete representation of the anomaly free group. For instance, this happens in the SU(5) case for the $\mathbf{5} + \overline{\mathbf{10}}$ representations, which together are anomaly free.
- The representations appear in conjugate pairs, i.e., they are vectorlike. This is the most common choice when the Higgs sector of SGUT is extended.

Anomaly cancellation with SU(5) representations. Let us consider an SU(5) SGUT model. There are three copies of $\bar{\mathbf{5}} + \overline{\mathbf{10}}$ representations to accommodate the three families of quarks and leptons. Breaking of the GUT group to the SM is achieved by including a (chiral) Higgs supermultiplet in the adjoint representation (**24**). The minimal Higgs sector needed to break the SM gauge group can be accommodated with a pair of **5** and $\bar{\mathbf{5}}$ representations. Within this minimal model one obtains the mass relations $m_{d_i} = m_{e_i}$, which are predicted by the Higgs sector. One way to solve this problem is to add a **45** representation, which couples to the d-type quarks, but not to the u-type, then one obtains the Georgi-Jarlskog factor [3] needed for the correct mass relation. Most models that obtain these relations within an extended Higgs sector, include the **45** conjugate representation to cancel anomalies [4]. This is however not de only possibility.

Consider the representations of SU(5) in Table 2. Then, for the **45**, we can write down the following anomaly-free combinations:

$$\begin{aligned} A(\mathbf{45}) + A(\bar{\mathbf{45}}) &= 0, \\ A(\mathbf{45}) + nA(\bar{\mathbf{5}}) + n'A(\bar{\mathbf{10}}) &= 0, \\ A(\mathbf{45}) + A(\bar{\mathbf{15}}) + 3A(\mathbf{5}) &= 0, \end{aligned}$$

with n and n' integers ≥ 0 and $n + n' = 6$. These are clearly non-equivalent models, with different physical consequences.

3. Gauge coupling unification.

The 1-loop β functions for a general SUSY theory with gauge group G and matter field appearing in chiral supermultiplets are given by $\beta = \sum_R T_R - 3C_A$, where T_R denotes the index for the representation R, and C_A the quadratic Casimir invariant for the adjoint representation. From MSSM-functions, the 1-loop RGEs for the gauge couplings are

$$\frac{d\alpha_i}{dt} = \beta_i \alpha_i^2, \qquad \beta_i = \begin{pmatrix} 33/5 \\ 1 \\ -3 \end{pmatrix} + \beta^X. \tag{3}$$

where $\beta^X = \sum_\Phi T(\Phi)$ are the contributions of the extensions of the MSSM [5] and the sum is over all SU(5) additional multiplets Φ.

We are interested in evaluating the effect of the different representations in the running from $M_{GUT} \sim 2 \times 10^{16} GeV$ up to the Planck scale, and also in finding which representations are perturbatively valid up to the Planck scale. The unified gauge coupling obeys the 1-loop RGE

$$\mu \frac{d\alpha_5^{-1}}{d\mu} = \frac{-\beta}{2\pi} = \frac{3 - \beta^X}{2\pi}. \tag{4}$$

The one loop β functions for some interesting anomaly-free combinations are found to be:

$$\begin{aligned} \beta^X(\mathbf{45} + \bar{\mathbf{45}}) &= 24, \\ \beta^X(\mathbf{45} + 6(\bar{\mathbf{5}})) &= 15, \\ \beta^X(\mathbf{45} + 6(\bar{\mathbf{10}})) &= 21, \\ \beta^X(\mathbf{45} + \bar{\mathbf{15}} + 2(\mathbf{10}) + \mathbf{5}) &= 19, \\ \beta^X(\mathbf{50} + \bar{\mathbf{40}} + \mathbf{5}) &= 29. \end{aligned} \tag{5}$$

Figure 1. Evolution of the gauge coupling for free anomaly combinations listed above, up to the Planck scale .

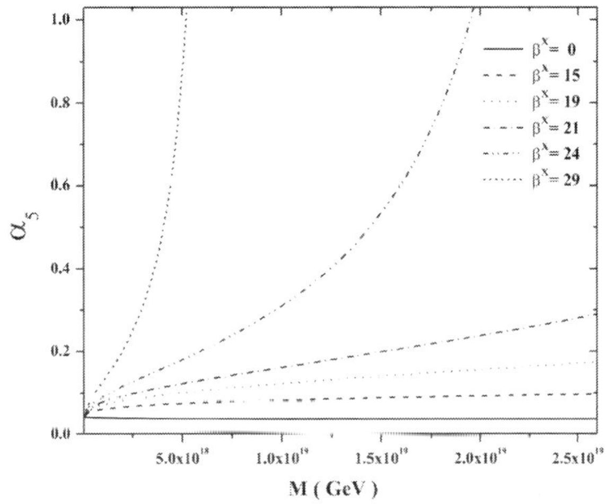

As it is shown in the figure, the model with $\beta^X = 29$ induces a running of the gauge coupling that blows at the scale $M = 6.61 \times 10^{18}$ GeV, while for $\beta^X = 24$ this happens at $M = 2.63 \times 10^{19}$ GeV. Models with $\beta^X = 15, 19, 21$ are found to evolve safely even up to the Planck scale.

4. Yukawa and gauge couplings.

It is also interesting to consider the RGE effect associated with the Yukawa couplings that involve the Higgs representations. In order to do this we shall consider the 2-loop β functions for the gauge coupling, but will keep only the 1-loop RGE for the Yukawa couplings. Thus, we shall consider the following superpotential for the SU(5) SGUT model [2]. This superpotential involves the Higgs representations $H(5)$, $H(5)$ and $\Sigma(24)$, and the matter multiplets $\psi(10)$, $\phi(5)$.

$$\begin{aligned} W &= \tfrac{f}{3}Tr\Sigma^3 + \tfrac{1}{2}fVTr\Sigma^2 + \lambda\bar{H}_\alpha(\Sigma^\alpha_\beta + 3V\delta^\alpha_\beta)H^\beta \\ &\quad + \tfrac{h^{ij}}{4}\varepsilon_{\alpha\beta\gamma\delta\epsilon}\psi_i^{\alpha\beta}\psi_j^{\gamma\delta}H^\epsilon + \sqrt{2}f^{ij}\psi_i^{\alpha\beta}\phi_{j\alpha}\bar{H}_\beta, \end{aligned} \tag{6}$$

Using the RGEs expressions for a general supersymmetric model [6] we obtain the 1-loop RGE for the Yukawa parameters [2]

$$\begin{aligned} \mu\frac{d\lambda}{d\mu} &= \frac{1}{(4\pi)^2}\left(-\frac{98}{5}g_5^2 + \frac{53}{10}\lambda^2 + \frac{21}{40}f^2 + 3(h^t)^2\right)\lambda, \\ \mu\frac{df}{d\mu} &= \frac{1}{(4\pi)^2}\left(-30g_5^2 + \frac{3}{2}\lambda^2 + \frac{63}{40}f^2)^2\right)f, \\ \mu\frac{dh^t}{d\mu} &= \frac{1}{(4\pi)^2}\left(-\frac{96}{5}g_5^2 + \frac{12}{5}\lambda^2 + 6(h^t)^2\right)h^t, \end{aligned} \tag{7}$$

while the 2-loop RGE for gauge coupling that we obtain is:

$$\mu\frac{dg_5}{d\mu} = \frac{1}{(4\pi)^2}(-3g_5^3) + \frac{1}{(4\pi)^4}\frac{794}{5}g_5^5 - \frac{1}{(4\pi)^4}\left\{\frac{49}{5}\lambda^2 + \frac{21}{4}f^2 + 12(h^t)^2\right\}g^3. \tag{8}$$

To solve the RGEs we used values of the coefficients λ, h^t and f that are themselves safe at the Planck scale. The parameters used are $M_{GUT} = 1.28 \times 10^{16}$GeV, $\alpha(M_{GUT}) = 0.040$, $h^t(M_{GUT}) = 0.6572$, $\lambda(MGUT) = 0.6024$, and $f(MGUT) = 1.7210$.

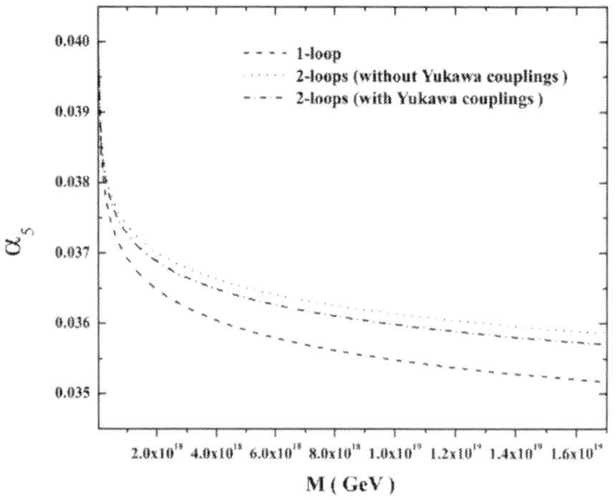

Figure 2. Evolution of the unified gauge coupling for three different cases: i) the 1-loop result, ii) the 2-loop result without including Yukawas, and iii) the 2-loop result including the (1-loop) running of the Yukawas.

5. Conclusions

We have studied the problem of anomalies in SUSY gauge theories, in order to search for alternatives to the usual vectorlike representations used in extended Higgs sector.

The known results have been extended to include higher-dimensional Higgs representations, which in turn have been applied to discuss anomaly cancellation within the context of realistic GUT models of SU(5) type.

We have succeed in identifying ways to replace the $\overline{\mathbf{45}} + \mathbf{45}$ models within SU(5) SGUTs. Then, we have studied the β functions for all the alternatives, and we find that they are not equivalent in terms of their values. These results have important implications for the perturbative validity of the GUT models at scales higher than the unification scale.

We have also considered the RGE effect associated with the Yukawa coupling that involve the additional Higgs representations. We found that there are appreciable differences for the evolution of the gauge coupling when going from the 1 to the 2-loops RGE, but this difference is reduced when one includes the 1-loop Yukawa couplings at the 2-loop level [7, 8].

Acknowledgments

ADRP thanks the Organizing Committee of the Dark Side of the Universe for the fellowship granted to attend this school.

References

[1] J Terning 2006 *Modern Supersymmetry, Dinamics and Duality* (Oxford: Clarendon Press)
U Sarkar 2007 *Int. J. Mod. Phys.* A **22**, 931

[2] J Hisano H. Murayama and T Yanagida 1993 Nucleon decay in the minimal supersymmetric SU(5) grand unification *Nucl. Phys.* B **402** 46 (*Preprint* hep-ph/9207279)
[3] H Georgi and C Jarlskog 1979 *Phys. Lett.* B **86** 297
[4] P Fileviez Perez Unification with and without supersymmetry:adjoint SU(5) *Preprint* arXiv:0710.1321
P Fileviez Perez 2007 *Phys. Rev.* D **76** 071701
[5] R Hempfling 1995 Coupling constant unification on extended SUSY models *Phys. Lett.* B **351** 206 (*Preprint* arXiv:hep-ph/9502201)
[6] S P Martin and M T Vaughn 1994 Two-loop renormalization group equations for soft supersymmetry-breaking couplings *Phys. Rev.* D **50** 2282 (*Preprint* arXiv:hep-ph/9311340) [Erratum-ibid. D 78, 039903 (2008)]
[7] A Aranda, J L Diaz-Cruz and A D Rojas 2009 Anomalies, beta functions, and GUT's AIP Conf.Proc.1116:454-457
[8] A Aranda, J L Diaz-Cruz and Alma D Rojas 2009 Anomalies, beta functions, and supersymmetric unification with multidimensional Higgs representations *Phys. Rev.* D **80** 085027

LIPSS Free-Electron Laser Searches for Dark Matter

J. R. Boyce,[1] A. Afanasev,[2] O.K. Baker,[3] K.B. Beard,[4] G. Biallas,[1] M. Minarni,[5] R. Ramdon,[2] T. Robinson,[1] M. Shinn,[1] P. Slocum[3]

[1]FEL Division, Jefferson Lab, 12000 Jefferson Ave., Newport News, VA 23606.
[2]Department of Physics, Hampton University, Hampton, VA 23668, USA
[3]Department of Physics, Yale University, PO Box 208120, New Haven CT 06520, USA
[4]Muons, Inc., 552 N. Batavia Avenue, Batavia, IL 60510, USA
[5]Department of Physics, Universitas Riau (UNRI) Pekanbaru, Riau 28293, Indonesia

E-mail: boyce@jlab.org

Abstract. A variety of Dark Matter particle candidates have been hypothesized by physics Beyond the Standard Model (BSM) in the very light (10^{-6} – 10^{-3} eV) range. In the past decade several international groups have conducted laboratory experiments designed to either produce such particles or extend the boundaries in parameter space. The **LI**ght **P**seudo-scalar and **S**calar **S**earch (**LIPSS**) Collaboration, using the "Light Shining through a Wall" (LSW) technique, pass the high average power photon beam from Jefferson Lab's Free-Electron Laser through a magnetic field upstream from a mirror and optical beam dump. Light Neutral Bosons (LNBs), generated by coupling of photons with the magnetic field, pass through the mirror ("the Wall") into an identical magnetic field where they revert to detectable photons by the same coupling process. While no evidence of LNBs was evident, new scalar coupling boundaries were established. New constraints were also determined for hypothetical para-photons and for millicharged fermions. We describe our experimental setup and results for LNBs, para-photons, and milli-charged fermions. Plans for chameleon particle searches are underway

1. Introduction

Dark matter and dark energy are postulated to explain astronomical and cosmological observations as well as strong interaction CP violations. However, these postulations are beyond the Standard Model of particle physics, a well tested model of the strong, weak, and electromagnetic interactions. [1-5] Nonetheless, by applying standard theoretical treatments (e.g., string theory) several laboratory experiments have been suggested to search for evidence of WIMPs (weakly interacting massive particles) or WISPs (weakly interacting sub-eV particles). While most of these searches are for WIMPS, a few, such as LIPSS, are exploring WISP parameter space. [6-9]

LIPSS uses the experimental technique "Light Shining through a Wall" (LSW) [10] whereby linearly polarized FEL light passes through and (very weakly) couples to a strong magnetic dipole field thus generating the dark matter candidates: pseudo-scalars if the FEL light polarization is parallel to the dipole field, or scalars if the light polarization is perpendicular to the dipole field. The light is diverted with a mirror (the wall) to a power dump but the generated particles pass on through the mirror and through a second dipole field, identical to the generating dipole field. In the second magnetic field the coupling reconverts particles back to photons that are then detected by a sensitive,

Journal of Physics: Conference Series **315** (2011) 012002 doi:10.1088/1742-6596/315/1/012002

liquid nitrogen cooled, CCD camera located inside a light tight box. Figure 1 depicts the process for (a) Light Neutral Bosons (LNBs) and (b) Para-photons.

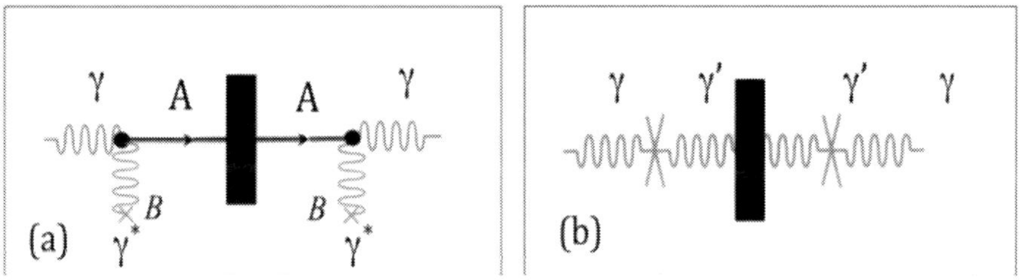

FIGURE 1. The "Light Shining through a Wall" (LSW) Technique. (a) Photons (γ) couple with a virtual photon (γ*) of the dipole's magnetic field (*B*) to produce the Light Neutral Boson (A). The photons are stopped by the wall but the LNBs pass through the wall and couple with the second dipole field thus regenerating photons which are detectable. (b) Photons couple to the hidden sector (U1), predicted by String Theory, pass through the wall and regenerate photons by the same coupling process.

2. Experiment

2.1. Free-Electron Laser

Photons for the experiment are produced by the Jefferson Lab Free-Electron Laser (FEL) located in an underground vault and depicted in Figure 2. An Energy Recovery Linac (ERL) produces a 110 MeV electron beam that is sent through a magnetic wiggler to generate a linearly polarized, tunable infrared (IR), high average power photon beam that is transported to the LIPSS experimental setup in a User Lab. The wavelength for the LIPSS experiment was chosen to be 935 nm since the collection efficiency of the detector and the FEL power output capability were both reasonably high. More details about the FEL facility, including a video demonstrating the tunability of the FEL, are found elsewhere. [11,12]

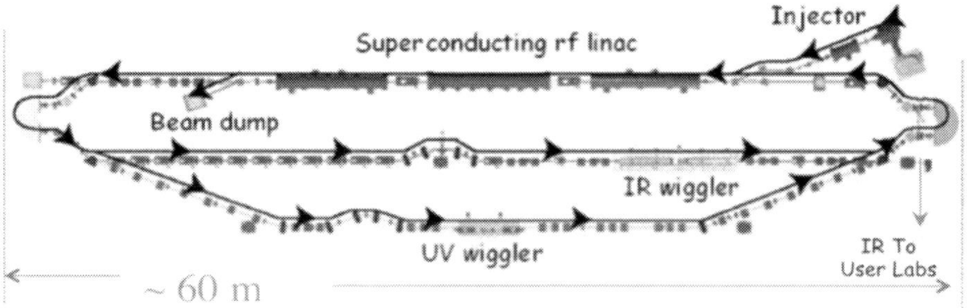

FIGURE 2. Layout of the Jefferson Lab ERL FEL. A beam of 10 MeV electrons is injected into three superconducting accelerators reaching an energy of ~110 MeV, then guided through the IR beam line, through the IR wiggler, bent back around, re-injected 180° out of phase into the SRF accelerators thus decelerating back to 10 MeV and guided to the beam dump.

2.2. Experimental Setup

Once in the User Laboratory, the 935 nm IR FEL beam is collimated to 6 mm diameter beam and sent through the LIPSS experimental setup, depicted in Figure 3, consisting of a vacuum chamber mounted inside a 1.7 T dipole. The chamber has a Brewster window at the entrance and exit. A turning mirror (TM3) deflects the IR beam to a power meter dump. Particles produced in the generating chamber pass through TM3 ("the Wall") and into a light tight regenerating vacuum chamber where the inverse reaction produces photons that enter a Light Tight Box (LTB) where they are directed by TM4 into a liquid nitrogen cooled CCD camera (Princeton Instruments ACTON 10: 400BR/LN). [12]

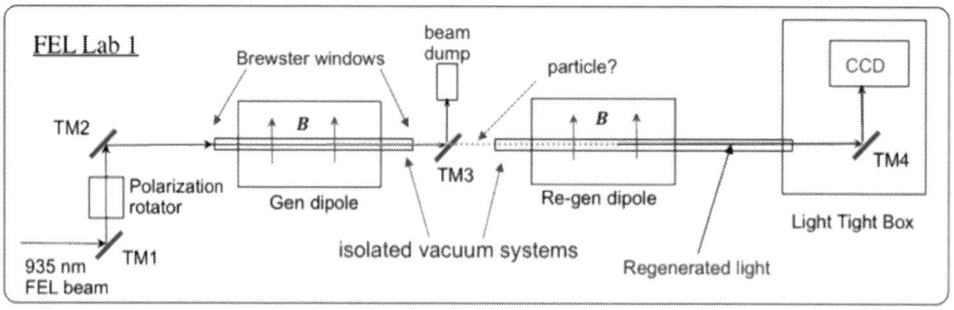

FIGURE 3. Schematic layout of the LIPSS experiment in Lab 1 of the FEL. The 935 nm FEL beam is directed by turning mirror TM1 through a polarization rotator, then by TM2 through the generating vacuum chamber in the 1.7T dipole field, and into the IR beam dump/power meter by TM3 ("the Wall"). Dark matter particles produced in the Gen dipole pass through TM3, through the Re-gen vacuum chamber where the inverse reaction regenerates photons that then enter the Light Tight Box (LTB) and be are directed by TM4 to a LN_2 cooled CCD camera.

The Light Tight Box (LTB) is depicted in Figure 4. On the left is an early photograph of the inside of the LTB. On the right is a schematic of the LTB. Great care was taken to verify that the LTB prevented photons from entering to the level of 1 photon/hour. The inside walls of the LTB were covered with black felt wall paper and the exterior was covered with three layers of Al foil sandwiched between black plastic. Hour long exposures were taken with all combinations of room lights on and off and camera shutter open and closed. With the exception of cosmic rays, there was no statistically significant variation between resulting runs.

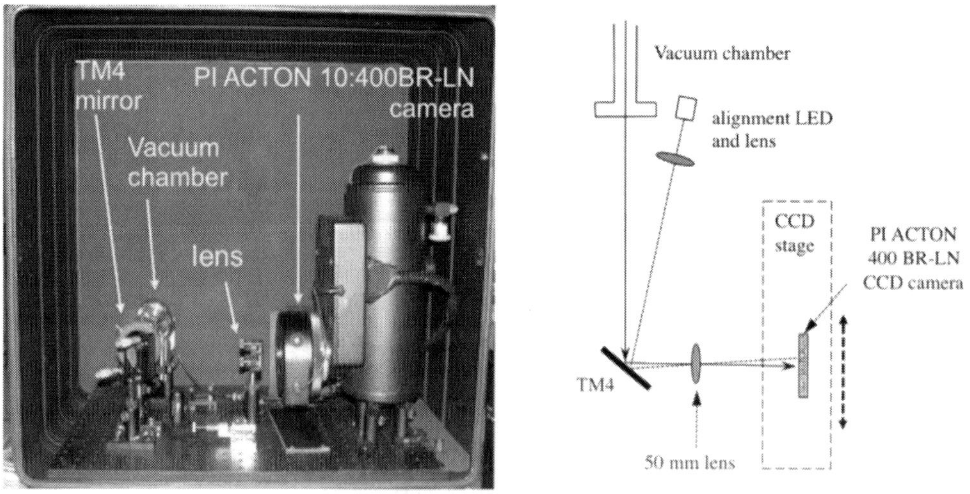

FIGURE 4. Inside layout of the Light Tight Box (LTB) and optics diagram. Regenerated photons enter the LTB through a window at the end of the regeneration vacuum chamber, reflecting off turning mirror 4 (TM4), and finally through a 50 mm lens that focuses down to one pixel in the sweet spot. All optics, window, mirror, and lens were anti-reflection coated. The camera image is 1340 x 400 pixels in size and each pixel is 20 μm x 20 μm in size. The camera was mounted on a translational stage with motion perpendicular to the incoming focused photons. The camera was moved between exposure runs in order not to depend on any one pixel. During the scalar exposure runs, alignment was verified by light from an LED. For the pseudoscalar runs, alignment was verified by pulsed low power FEL light.

A series of exposures were taken with a 0.5 mW red alignment laser to simulate one-hour-long runs with a few counts above background per hour. Neutral density (ND) filters were stacked in front of the entrance to the regeneration vacuum chamber where the solid stainless steel blank-off flange was replaced with a quartz window. Figure 5 shows the attenuation of the laser light as a function of ND filter value. To obtain a photon rate of a few per hour, the rate anticipated based on the PVLAS claim (later withdrawn), the laser had to be attenuated by fifteen (15) orders of magnitude.

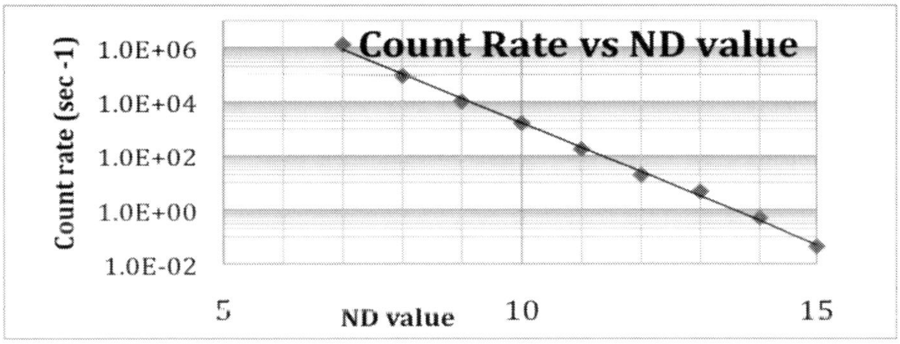

FIGURE 5. System check out and simulation results using a 0.5mW red alignment laser with Neutral Density (ND) filters directly in front of the entrance to the re-generation vacuum chamber. In order to simulate the anticipated WISPS count rate (~ few/hour) the laser needed to be attenuated by fifteen orders of magnitude.

Figure 6 is a typical image resulting from a 1-hour-long simulation using a 0.5 mW red alignment laser and a ND filter value of 15. These data were used not only to illustrate a positive result but also to verify the analysis algorithms were working properly.

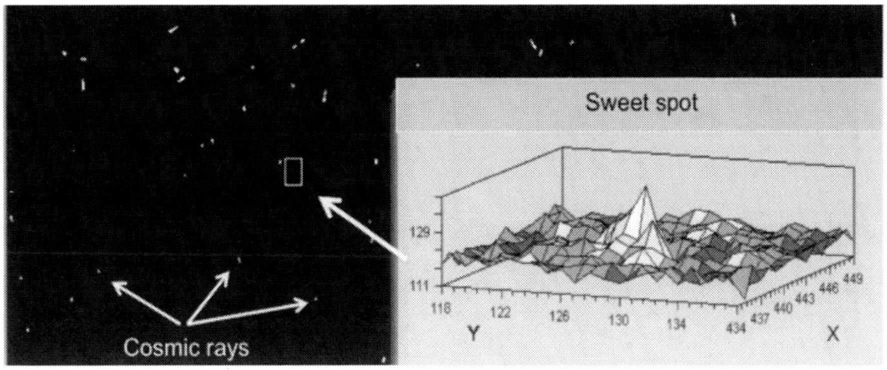

FIGURE 6. Typical 1-hr simulation image from the CCD camera. The white spots are cosmic rays. The "Sweet spot" region is magnified and shown in the inset. Since pixels can "go bad" the camera was moved between each run to remove any dependence on any one pixel. Any image with a cosmic ray near the Sweet spot region was discarded. An algorithm removed cosmic rays from remaining images. The resulting image was examined for deviations from background and readout noise.

3. Results

Table 1 summarizes the relevant parameters for the LIPSS runs with the FEL. Analysis of all the runs did not show any significant deviations from background and readout noise. Thus upper limits were set by these experiments consistent with other LSW results as shown in Figures 7, and 8. [9]. Regions above the curves are excluded by the experimental results.

TABLE 1). LIPSS parameter sets for both scalar and pseudoscalar experiments with the FEL.

Parameter	Scalar Runs (2007)	Pseudoscalar Runs (2009)
B-field	1.7 T	1.7 T
Magnet length	1.0 m	1.0 m
IR FEL power	0.2 kW	0.4 kW
IR FEL wavelength	935 nm (1.3 eV)	935 nm (1.3 eV)
Quantum efficiency	0.4	0.4
Linear polarization	100%	100%
Acceptance	100%	100%
experimental efficiency	~ 90%	~90%
Background rates (all sources)	< 1count/hr/pixel	< 1count/hr/pixel

4. Conclusions

The Jefferson Lab FEL was successfully used to search for three families of dark matter candidate particles: Pseudoscalar and Scalar Light Neutral Bosons, Paraphotons, and Milli-charged fermions. Although no positive evidence was detected, new exclusionary boundaries were established consistent with more recent LSW experiments by other Collaborative Teams. Plans for chameleon particle [14, 15] searches are underway.

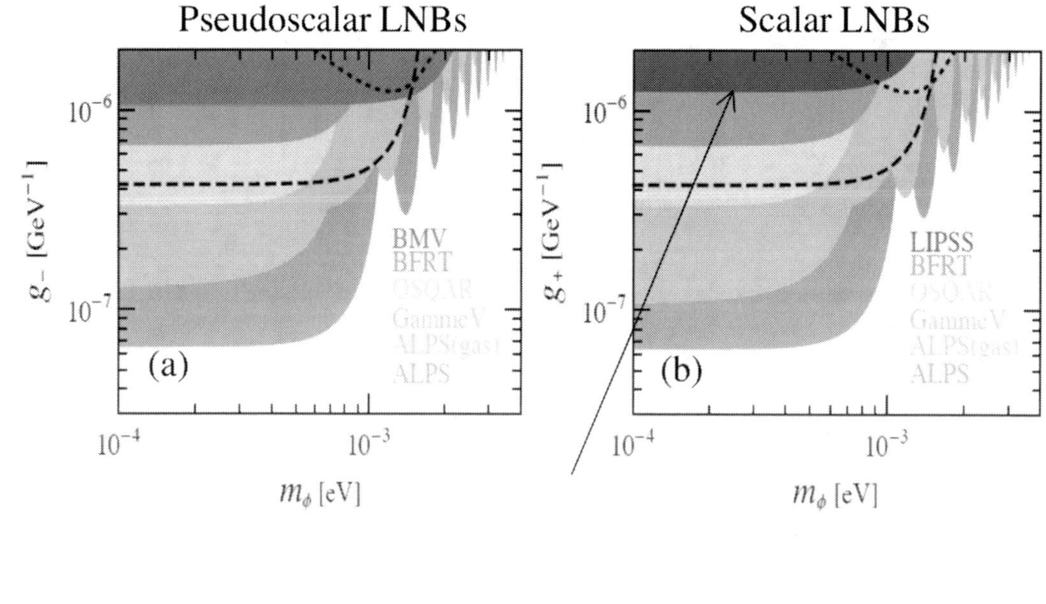

FIGURE 7. LIPSS results compared to results of other LSW experiments. (a) Pseudoscalar Light Neutral Bosons, (b) Scalar Light Neutral Bosons compiled in Reference [9]. Note: these plots from Reference [9] were generated prior to LIPSS pseudoscalar experiment.

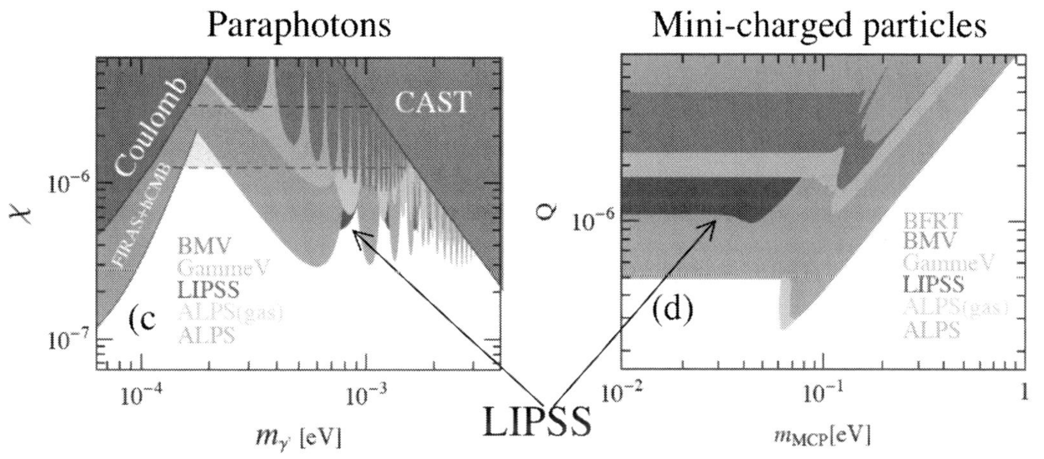

FIGURE 8. LIPSS results compared to results of other LSW experiments: (c) Paraphotons, and (d) Mini-charged particles, compiled in Ref. [9]. Note that the LIPSS Paraphoton data are behind the other results.

Acknowlegements
The authors thank the technical staff of the Jefferson Lab Free-Electron Laser Facility, especially F. Dylla, G. Neil, G. Williams, R. Walker, D. Douglas, S. Benson, K. Jordan, C. Hernandez-Garcia, J. Gubeli, summer 2009 interns Jon Evans, Jennifer Caldwell, and Joe Amma, as well as M.C. Long of Hampton University for their excellent support of the experimental program.

This work was supported by the Office of Naval Research Award N00014-06-1-1168, the Joint Technology Office, the Commonwealth of Virginia, the Air Force Research Laboratory, and by DOE under contract DE-AC05-06OR23177. Notice: Authored by Jefferson Science Associates, LLC under U.S. DOE Contract No. DE-AC05-06OR23177. The U.S. Government retains a non-exclusive, paid-up, irrevocable, world-wide license to publish or reproduce this manuscript for U.S. Government purposes.

References

[1] S. Glashow, Nucl. Phys. **22,** 579 (1961).
[2] S. Weinberg, Phys. Rev. Lett. **19**, 1264 (1967).
[3] A. Salam. In: W. Swatholm (Ed.) *Elementary Particle Theory*, Almquist and Wiksell, Stocholm, 1968.
[4] H.D. Politzer, Phys. Rev. Lett. **30**, 1346 (1973).
[5] D. Gross, F. Wilczek, Phys. Rev. Lett. **30**, 1343 (1973).
[6] E. Zavattini, et al., Phys. Rev. Lett. **96**, 110406 (2006); arXiv:0706.3419 [hep-ex].
[7] A. Afanasev, et al., Phys. Rev. Lett. **101**, 120401 (2008).
[8] A. Chou, et al., Phys. Rev. Lett. **100**, 080402 (2008).
[9] K. Ehret, et al., Phys. Lett. B. **149**, 689 (2010).
[10] K.V. Bibber, et al., Phys. Rev. Lett. **59**, 759 (1987).
[11] S. Benson, et al., "The 4th Generation Light Source at Jefferson Lab", Nucl. Instr. Methods **A582**, 14-17 (2007).
[12] http://www.jlab.org/FEL/
[13] http://www.princetoninstruments.com/products/speccam/spec10/dsheet.aspx
[14] J. Khoury and A. Weltman, Phys. Rev. Lett. **93**, 171104 (2004).
[15] A.S. Chou, et al., arXiv:0806.2438v1 [hep-ex] 15 June (2008).

Channeling Effects in Direct Dark Matter Detectors

Nassim Bozorgnia

Department of Physics and Astronomy, UCLA, 475 Portola Plaza, Los Angeles, CA 90095, USA

E-mail: nassim@physics.ucla.edu

Abstract. The channeling of the ion recoiling after a collision with a WIMP changes the ionization signal in direct detection experiments, producing a larger signal than otherwise expected. We give estimates of the fraction of channeled recoiling ions in NaI (Tl), Si and Ge crystals using analytic models produced since the 1960's and 70's to describe channeling and blocking effects. We find that the channeling fraction of recoiling lattice nuclei is smaller than that of ions that are injected into the crystal and that it is strongly temperature dependent.

1. Introduction

Channeling and blocking effects in crystals refer to the orientation dependence of charged ion penetration in crystals. In the "channeling effect" ions incident upon a crystal along symmetry axes and planes suffer a series of small-angle scatterings that maintain them in the open "channels" between the rows or planes of lattice atoms and thus penetrate much further into the crystal than in other directions. Channeled incoming ions do not get close to lattice sites, where they would be deflected at large angles. The "blocking effect" consists in a reduction of the flux of ions originating in lattice sites along symmetry axes and planes, creating what is called a "blocking dip" in the flux of ions exiting from a thin enough crystal as a function of the angle with respect to a symmetry axis or plane. These effects were first observed in the 1960's and are used in crystallography, in the study of lattice disorder, ion implantation, surfaces, interfaces and epitaxial layers, in measurements of short nuclear lifetimes etc. In particular, avoiding channeling is essential in the manufacturing of semiconductor devices, since ion implantation at a controlled depth is the primary technique.

Ion channeling in NaI (Tl) was first observed in 1973 by Altman *et al.* [1]. They observed that channeled ions produce more scintillation light because they lose most of their energy via electronic stopping rather than nuclear stopping. The potential importance of the channeling effect for direct dark matter detection was first pointed out for NaI (Tl) by Drobyshevski [2] and by the DAMA collaboration [3]. When Na or I ions recoiling after a collision with a dark matter WIMP (Weakly Interacting Massive Particle) move along crystal axes and planes, their quenching factor is approximately $Q = 1$ instead of $Q_{\rm Na} = 0.3$ and $Q_{\rm I} = 0.09$, since they give their energy to electrons.

Ion channeling in crystals could give rise to a daily modulation due to the preferred direction of the dark matter flux arriving on the Earth. Earth's daily rotation naturally changes the direction of the "WIMP wind" with respect to the crystal axes, which produces a daily modulation in the measured recoil energy (equivalent to a modulation of the factor Q). Avignone, Creswick, and Nussinov [4] pointed this out for NaI (Tl) crystals. My collaborators, Graciela Gelmini and Paolo Gondolo, and I [5] present here our analytic calculations of the channeling fraction in NaI (Tl), Si and Ge crystals.

2. Modeling of channeling

Our calculation is based on the classical analytic models developed in the 1960's and 70's, in particular by Lindhard [6, 7, 8, 9, 10, 11, 12, 13]. We use the continuum string and plane model, in which the screened Thomas-Fermi potential is averaged over a direction parallel to a row or a plane. The averaged potential, U is considered to be uniformly smeared along the row or plane of atoms, which is a good approximation if the propagating ion interacts with many lattice atoms in the row or plane by a correlated series of many consecutive glancing collisions. Just one row or plane is considered in this model. Lindhard proved that for an ion propagating with kinetic energy E, and for small angle ϕ between the ion's trajectory and the atomic row (or plane) in the direction perpendicular to the row (or plane), the so called "transverse energy", $E_\perp = E\sin^2\phi + U \simeq E\phi^2 + U$ is conserved.

The conservation of the transverse energy provides a definition of the minimum distance of approach to the string (or plane) of atoms, $\rho_{\rm min}$, at which the trajectory of the ion makes a zero angle with the string (or plane), and also of the angle ψ at which the ion exits from the string (or plane), i.e. far away from it where $U \simeq 0$. In reality the furthest position from a string or plane of atoms is the middle of the channel, whose width we call $d_{\rm ch}$. Thus,

$$E_\perp = U(\rho_{\rm min}) = E\psi^2 + U(d_{\rm ch}/2). \tag{1}$$

Channeling requires that $\rho_{\rm min} > \rho_c(E)$, where $\rho_c(E)$ is the smallest possible minimum distance of approach of the propagating ion with the row (or plane) for a given energy E. Since the potential $U(\rho)$ decreases monotonically with increasing ρ, $U(\rho_{\rm min}) < U(\rho_c(E))$. Using Eq. 1, this can be further translated into an upper bound on $E_\perp$ and on ψ,

$$\psi < \psi_c(E) = \sqrt{\frac{U(\rho_c(E)) - U(d_{\rm ch}/2)}{E}}. \tag{2}$$

$\psi_c(E)$ is the maximum angle the ion can make with the string (or plane) far away from it (i.e. in the middle of the channel) if the ion is channeled. At low enough E, $\rho_c(E)$ becomes close to $d_{\rm ch}/2$, and the critical angle $\psi_c(E)$ goes to zero. This means that there is a minimum energy below which channeling cannot happen, even for ions moving initially in the middle of the channel.

So far we have been considering static strings and planes, but the atoms in a crystal are actually vibrating. We use the Debye model, and take into account thermal effects in the crystal through a modification of the critical distances which was found originally by Morgan and Van Vliet [8] and later by Hobler [13] to provide good agreement with simulations and data. It consists of taking the temperature corrected critical distance $\rho_c(T)$ to be,

$$\rho_c(T) = \sqrt{\rho_c^2(E) + [cu_1(T)]^2}, \tag{3}$$

where u_1 is the one dimensional rms vibration amplitude of the atoms in a crystal, and the factor c is a number between 1 and 2 in different references [8, 9]. As shown in Fig. 1, with this formalism and using $c = 2$ we fit relatively well the critical angles measured at room temperature for B and P ions in a Si crystal in several channels, for energies between 20 keV and 600 keV that Hobler [13] extracted from thermal wave measurements.

As an example, the static ($c = 0$) axial and planar critical distances are presented in Fig. 2(a) for the 100 channel in NaI crystal, together with the amplitude of thermal vibration u_1 at 20 °C, and the Thomas-Fermi screening distances for Na and I ions. Fig. 2(b) shows the temperature corrected axial and planar critical angles at 20 °C (with $c = 1$) for the same channel as functions of energy of the traveling Na and I ions.

3. Channeling of incident particles

The channeling of ions in a crystal depends not only on the angle their initial trajectory makes with strings or planes in the crystal, but also on their initial position. Ions which start their motion close to the center of a channel, far from a string or plane, where they make an angle

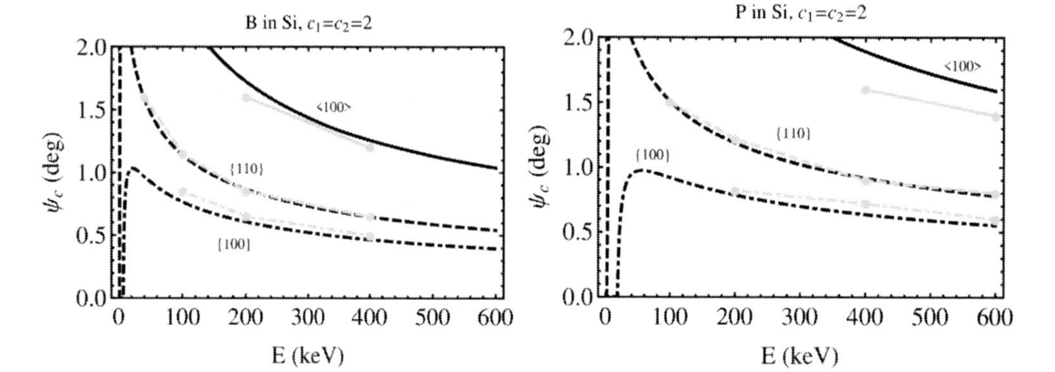

Figure 1. Comparison of theoretical (black lines) temperature corrected critical angles (with $c = 2$) and measured critical angles extracted from thermal wave measurements [13] (green, or gray if color not available, dots, joined by straight lines to guide the eye) as a function of the energy of (a) B ions and (b) P ions propagating in a Si crystal at T=20 °C, for the indicated axial and planar channels.

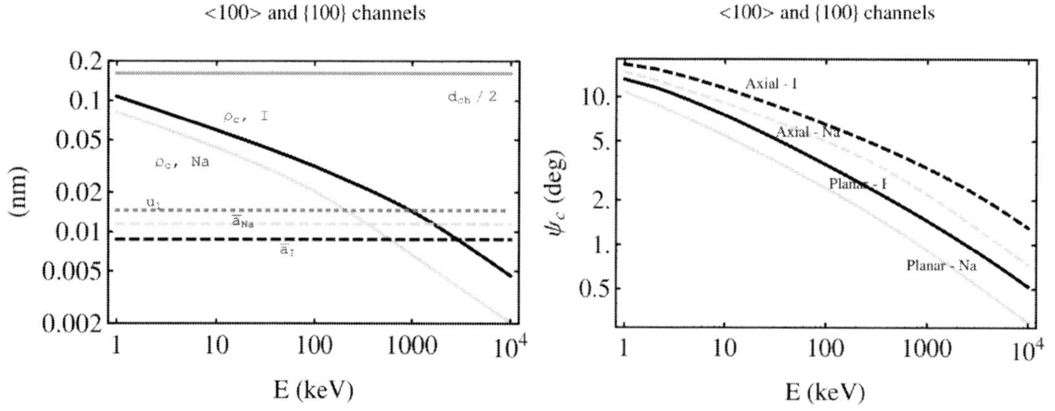

Figure 2. (a) Static critical distances of approach, together with u_1 at 20 °C, and the screening radii $\bar{a}_{\rm Na} = 0.00878$ nm and $\bar{a}_I = 0.0115$ nm [5] and (b) critical channeling angles at 20 °C with $c = 1$ as a function of the energy of propagating Na (green/gray) and I (black) ions in the <100> axial and {100} planar channels.

ψ, are channeled if the angle is smaller than a critical angle (as explain earlier) and are not channeled otherwise. Particles which start their motion in the middle of a channel (as opposed to a lattice site) must be incident upon the crystal.

No data or simulations of Na and I ions propagating in a NaI crystal is available at low energies. We show that to a good approximation we can use analytic calculations and reproduce the channeling fraction in NaI presented in Ref. [3]. Temperature corrections are neglected and we use a static lattice (similar to the approach of Ref. [3]). For an incident angle ψ with respect to each of the channels and an ion energy E, the fraction $\chi_{\rm inc}(E,\psi)$ of channeled incident ions for axial and planar channels is $\chi_{\rm inc} = 1$ if ψ is smaller than the critical angle for the corresponding channel and zero otherwise.

To find the total fraction of channeled incident nuclei, we average $\chi_{\rm inc}$ over the incident direction. The integrals cannot be solved analytically, so we integrate numerically by performing a Riemann sum once the sphere of directions has been divided using the Hierarchical Equal Area iso-Latitude Pixelization (HEALPix) method [14]. The HEALPix method uses an algorithm to

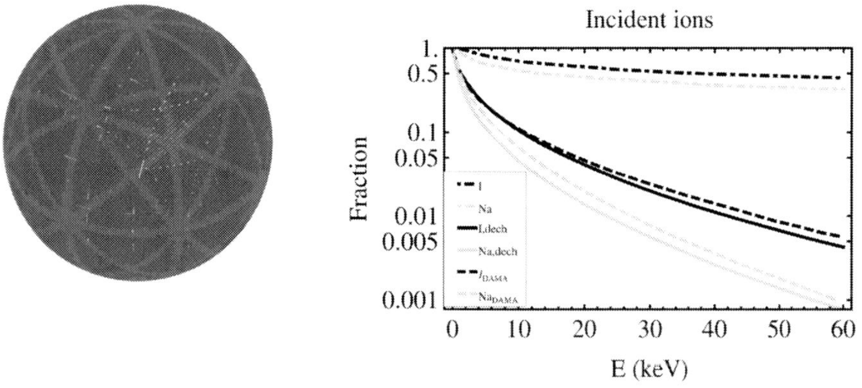

Figure 3. (a) Channeling fraction for a 50 keV Na ion in different directions plotted on a sphere using HEALPix: probability equal to one in red, and probability equal to zero in blue. (b) Fraction of channeled incident I (black) and Na (green/gray) ions as a function of their incident energy E with the static lattice without (dot dashed lines) and with (solid lines) dechanneling due to interactions with Tl impurities. The results of DAMA are also included (dashed lines).

deal with the pixelization of data on a sphere, and it is useful to compute integrals over direction by dividing the surface of a sphere into many pixels, computing the integrand at each pixel (i.e. each direction, see Fig. 3(a)), and finally summing up the values of all pixels over the sphere.

A channeled ion can be pushed out of a channel by an interaction with an impurity such as the atoms of Tl in NaI (Tl). Here we will simply assume that if a channeled ion interacts with a Tl atom it becomes dechanneled and thus it does not contribute to the fully channeled fraction any longer. We thus neglect the possibility that after the interaction the ion may re-enter into a channel, either the same or another. Fig. 3(a) shows the axial and planar channels of the NaI crystal for incoming Na ions with an energy of 50 keV. We include here only the channels with lower crystallographic indices 100, 110 and 111. The fraction of channeled incoming Na and I ions after including dechanneling in the way just described is shown in Fig. 3(b). Our results agree well with those published by the DAMA collaboration [3], also included in the figure.

4. Channeling of recoiling lattice nuclei

The recoiling nuclei start initially from lattice sites (or very close to them), thus blocking effects are important. In fact, as argued originally by Lindhard [6], in a perfect lattice and in the absence of energy-loss processes the probability that a particle starting from a lattice site is channeled would be zero. The argument uses statistical mechanics in which the probability of particle paths related by time-reversal is the same. Thus the probability of an incoming ion to have a particular path within the crystal is the same as the probability of the same ion to move backwards along the same path [15]. This is what Lindhard called the "Rule of Reversibility". Using this rule, since the probability of an incoming channeled ion to get very close to a lattice site is zero, the probability of the same ion to move in the time-reversed path, starting at a nuclear site and ending inside a channel, is zero too. However, any departure of the actual lattice from a perfect lattice, for example due to vibrations of the atoms in the lattice, would violate the conditions of this argument and allow for some of the recoiling lattice nuclei to be channeled, as already understood in the 70's [11, 16]. We now estimate the effect using the formalism presented so far.

In our model, a recoiling ion is channeled if the collision ion-WIMP happens at a distance large enough from the string or plane to which the ion belongs. Namely, channeling happens if the initial position of the recoiling motion is $\rho_i > \rho_{i,\min}$. We define $E_\perp$ in terms of the initial

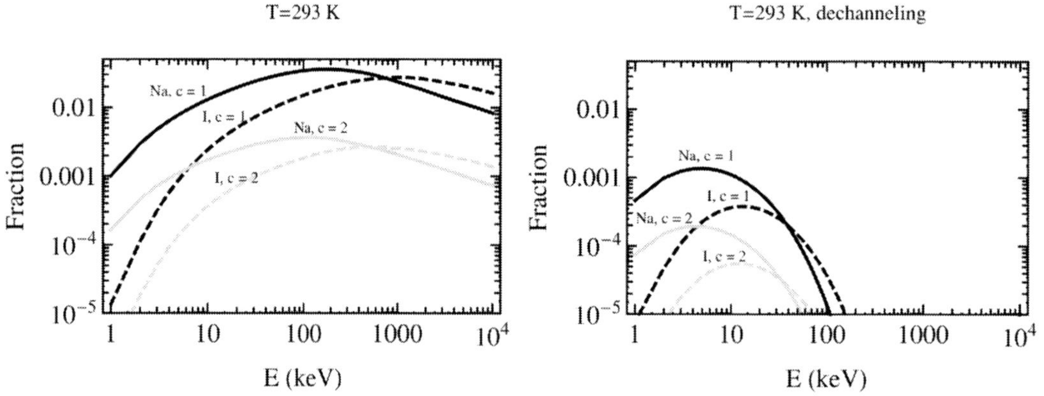

Figure 4. Channeling fractions at T=293 K for Na (solid lines) and I (dashed lines) ions for $c = 1$ (black) and $c = 2$ (green/gray) cases (a) without and (b) with dechanneling included.

recoil energy E of the propagating ion, the angle of the initial recoil momentum with respect to the particular string or plane of atoms ϕ_i, and the initial position ρ_i. The condition

$$E_\perp(E, \phi_i, \rho_i) = U(\rho_{\rm min}) < U(\rho_c), \tag{4}$$

i.e. $E_\perp(E, \phi_i, \rho_{i,\rm min}) = U(\rho_c(E))$ defines the distance $\rho_{i,\rm min}$ in terms of ρ_c.

We take the initial distance distribution, ρ of the colliding atom to be a Gaussian with a one dimensional dispersion u_1, and we obtain the probability of channeling for each individual channel by integrating the Gaussian between the minimum initial distance and infinity (a good approximation to the radius of the channel). The dependence of these probabilities on the critical distances enter in the argument of an exponential (for axial channels) or an erfc function (for planar channels). In order to obtain the total geometric channeling fraction we sum over all the individual channels we consider. The integral over initial directions is computed using HEALPix [14].

Figure 4 shows what we consider to be our main predictions for the range expected as an upper limit to the channeling fraction in NaI (Tl) at 239 K for two different assumptions for the effect of thermal vibrations in the lattice, which depend on the value of the parameter c used in the temperature corrected critical distances of approach. Dechanneling is ignored in Fig. 4(a) and taken into account in Fig. 4(b).

Our results for the total geometric channeling fraction for Si ions propagating in a Si crystal and Ge ions propagating in a Ge crystal at different temperatures are shown in Figs. 5 and 6 for the two cases of $c = 1$ and $c = 2$, respectively. Note that we have not considered the possibility of dechanneling of initially channeled ions, due to imperfections in Si and Ge crystals. Any mechanism of dechanneling will decrease the fractions obtained here for Si and Ge.

5. Conclusions

We have studied the channeling of ions recoiling after collisions with WIMPs within NaI (Tl), Si and Ge crystals. Channeled ions move within the crystal along symmetry axes and planes and suffer a series of small-angle scatterings that maintain them in the open "channels" between the rows or planes of lattice atoms and thus penetrate much further into the crystal than in other directions. Ions which start their motion close to the center of a channel, at an initial angle ψ, are channeled if the initial angle is smaller than the critical angle in Eq. 2, and are not channeled otherwise. We have found that the channeling of lattice ions recoiling after a collision with a WIMP is very different from the case of incident ions, and that the channeling fraction is smaller.

As argued originally by Lindhard [6], in a perfect lattice and in the absence of energy-loss processes, the probability that a particle starting from a lattice site is channeled would be zero.

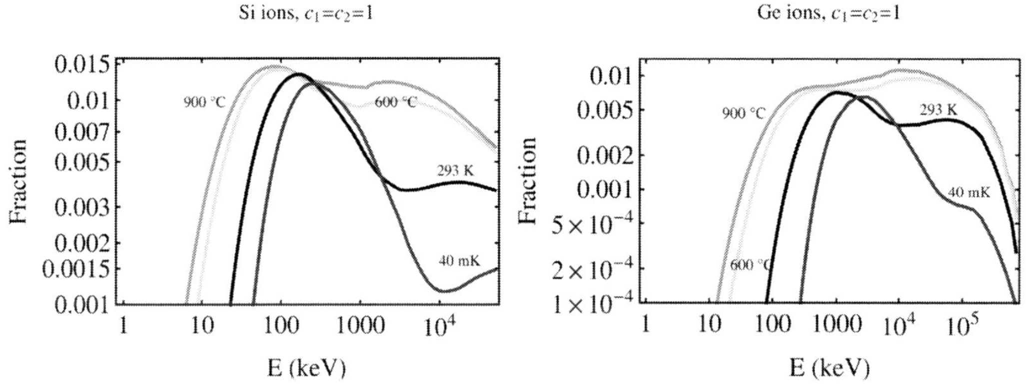

Figure 5. Channeling fractions of (a) Si and (b) Ge recoils in a Si and a Ge crystal respectively, as a function of the ion energy for temperatures T=900 °C (orange or medium gray), T=600 °C (green or light gray), 293 K (black), and 44 mK (blue or dark gray) in the approximation of $c = 1$.

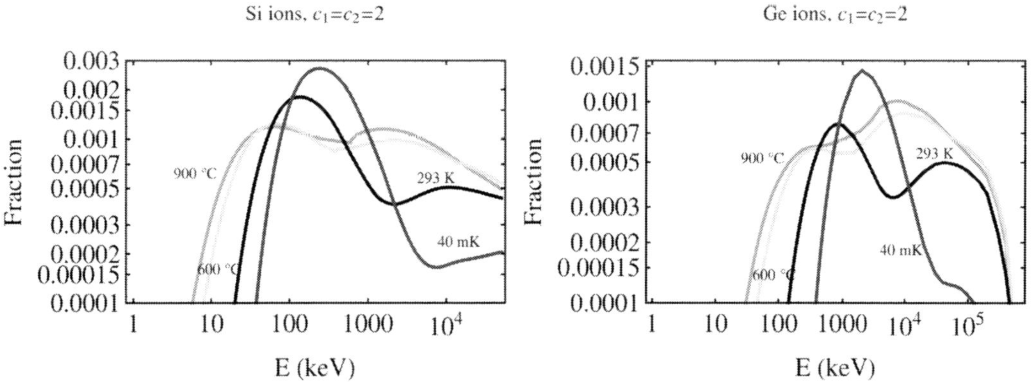

Figure 6. Same as Fig. 5 but for $c = 2$.

However, due to vibrations in the crystal, the atom that interacts with a WIMP may be displaced from its position in a perfect lattice, and there is a non-zero probability of channeling.

As seen in Fig. 4(a) for NaI, without including dechanneling, the channeling fraction is never larger than 5% and the maximum happens at 100's of keV. This maximum occurs because the critical distances decrease with the ion energy E, making channeling more probable, and the critical angles also decrease with E, making channeling less probable. The simple extreme model of dechanneling we used for NaI predicts much smaller fractions, at most in the 0.1% level, with the maximum shifted to small energies, less than 10 keV (see Fig. 4(b)). This reduction may eventually prove to be too extreme and at present we do not have a better formalism to model dechanneling. With the simple model of dechanneling we used for NaI, we could reproduce the channeling fractions computed by the DAMA collaboration (which, however, apply to ions which start their motion close to the middle of a channel and not to the case of WIMP direct detection).

If the values found by Hobler [13], and also by us above (see Fig. 1), to reproduce measured channeling angles in B and P propagating in Si apply also to the propagation of Si ions in Si, then the case of $c = 2$ should be chosen and the channeling fractions would never be larger than 0.3%. Moreover, increasing the temperature of a crystal usually increases the fraction of channeled recoiling ions (see Fig. 5), but not always. Sometimes the opposite happens (see

Fig. 6). The $c = 1$ choice leads to channeling fractions close to 1% for Si and Ge.

The analytical approach used here can successfully describe qualitative features of the channeling and blocking effects, but should be complemented by data fitting of parameters and by simulations to obtain a good quantitative description too. Thus our results should in the last instance be checked by using some of the many sophisticated Monte Carlo simulation programs.

Acknowledgments

N.B. and Graciela Gelmini were supported in part by the US Department of Energy Grant DE-FG03-91ER40662, Task C. Paolo Gondolo was supported in part by the NFS grant PHY-0756962 at the University of Utah.

References

[1] Altman R, Dietrich H B, Murray R B and Rock T J 1973 *Phys. Rev.* B **7** 1743
[2] Drobyshevski E M 2008 *Mod. Phys. Lett.* A **23** 3077 (*Preprint* 0706.3095)
[3] Bernabei R *et al* 2008 *Eur. Phys. J.* C **53** 205 (*Preprint* 0710.0288)
[4] Avignone F T, Creswick R J and Nussinov S 2008 *Preprint* 0807.3758
Creswick R J, Nussinov S and Avignone F T 2010 *Preprint* 1007.0214
[5] Bozorgnia N, Gelmini G B and Gondolo P 2010 *J. Cosmol. Astropart. Phys.* JCAP11(2010)019 (*Preprint* 1006.3110)
Bozorgnia N, Gelmini G B and Gondolo P 2010 *J. Cosmol. Astropart. Phys.* JCAP11(2010)028 (*Preprint* 1008.3676)
[6] Lindhard J 1965 *Kongel. Dan. Vidensk. Selsk., Mat.-Fys. Medd.* **34** No. 14
[7] Andersen J U 1967 *Kongel. Dan. Vidensk. Selsk., Mat.-Fys. Medd.* **36** No. 7
[8] Morgan D V and Van Vliet D 1963 *Can. J. Phys.* **46** 503
Morgan D V and Van Vliet D 1971 *Radiat. Effects and Defects in Solids* **8** 51
[9] Van Vliet D 1973 *Channeling*, ed D V Morgan (London: Wiley) p 37
[10] Andersen J U and Feldman L C 1970 *Phys. Rev.* B **1** 2063
[11] Komaki K and Fujimoto F 1970 *Phys. Stat. Sol.* (a) **2** 875
[12] Appleton B R and Foti G 1977 *Ion Beam Handbook for Material Analysis*, ed J W Mayer and E Rimini (New York: Academic) p 67
[13] Hobler G 1996 *Radiat. effects and defects in solids* **139** 21
Hobler G 1996 *Nucl. Instrum. Methods Phys. Research* (NIM) B **115** 323
[14] Górski K M *et al* 2005 *ApJ* **622** 759
[15] Gemmell D S 1974 *Rev. Mod. Phys.* **46** 129
[16] Komaki K *et al* 1971 *Phys. Stat. Sol.* (a) **4** 495

R^2 Dark Matter[1]

Jose A. R. Cembranos

William I. Fine Theoretical Physics Institute, University of Minnesota, Minneapolis, 55455, USA

,

School of Physics and Astronomy, University of Minnesota, Minneapolis, 55455, USA

Abstract. There is a non-trivial four-derivative extension of the gravitational spectrum that is free of ghosts and phenomenologically viable. It is the so called R^2-gravity since it is defined by the only addition of a term proportional to the square of the scalar curvature. Just the presence of this term does not improve the ultraviolet behaviour of Einstein gravity but introduces one additional scalar degree of freedom that can account for the dark matter of our Universe.

1. Introduction

The non-unitarity and non-renormalizability of the gravitational interaction described by the Einstein-Hilbert action (EHA) demands its modification at high energies. It has been pointed out that this correction cannot be accomplished without the introduction of new states [1]; these states typically interact with SM fields through Planck scale suppressed couplings and potentially work as dark matter (DM).

In spite of many and continuous efforts, the ultraviolet (UV) completion of the gravitational interaction is still an open question. In these conditions, it is difficult to make general statements about its phenomenology although different types of new scalar fields are commonly predicted [2, 3]. We can adopt a conservative and minimal approach in order to capture the fundamental physics of this fact [1]. The simplest correction to the EHA at high energies is provided by the inclusion of four-derivative terms in the metric that preserve general covariance. The most general four-derivative action supports, in addition to the usual massless spin-two graviton, a massive spin-two and a massive scalar mode, with a total of eight degrees of freedom (in the *physical* or *transverse* gauge [4, 5]). Indeed, four-derivative gravity is renormalizable, although the massive spin-two gravitons are ghost-like particles that generate new unitarity violations, breaking of causality, and inadmissible instabilities [6].

2. R^2 gravity

However, we can work with R^2-gravity, that is defined by the only addition of a term proportional to the square of the scalar curvature to the EHA. It illustrates the idea in a consistent and minimal way since it only introduces one additional scalar degree of freedom, whose mass m_0 is

[1] To appear in the Proceedings of the sixth International Workshop on the Dark Side of the Universe (DSU20010) Leon, Guanajuato, Mexico 1-6 June 2010.

given by the corresponding new constant in the action:

$$S_G = \int \sqrt{g}\left\{-\Lambda^4 - \frac{M_P^2}{2}R + \frac{M_P^2}{12\,m_0^2}R^2 + ...\right\} \tag{1}$$

where $M_P \equiv (8\pi G_N)^{-1/2} \simeq 2.4 \times 10^{18}$ GeV, $\Lambda \simeq 2.3 \times 10^{-3}$ eV, and the dots refer to higher energy corrections that must be present in the model to complete the UV limit. In [1], it has been shown that just the Action (1) can explain the late time cosmology since the first term can account for the dark energy (DE) content, while the third term is able to explain the DM one.

R^2-gravity modifies Einstein's Equations (EEs) as [7, 8] (following notation from [9]):

$$\left[1 - \frac{1}{3\,m_0^2}R\right]R_{\mu\nu} - \frac{1}{2}\left[R - \frac{1}{6\,m_0^2}R^2\right]g_{\mu\nu} - \mathcal{I}_{\alpha\beta\mu\nu}\nabla^\alpha\nabla^\beta\left[\frac{1}{3\,m_0^2}R\right] = \frac{T_{\mu\nu}}{M_P^2}, \tag{2}$$

where $\mathcal{I}_{\alpha\beta\mu\nu} \equiv (g_{\alpha\beta}g_{\mu\nu} - g_{\alpha\mu}g_{\beta\nu})$. The new terms do not modify the standard EEs at low energies except for the mentioned introduction of a new mode. It is straight forward to check that the metric $g_{\mu\nu} = [1 + c_1\sin(m_0 t)]\eta_{\mu\nu}$ is solution of the linearized Eq. (2), i.e. for $c_1 \ll 1$, without any kind of energy source. It has been argue that the energy stored in such oscillations behaves exactly as cold DM and can explain the missing matter problem of the Universe [1].

3. Scalar graviton couplings

The phenomenology of the new scalar can be computed inside the *Jordan* or the *Einstein frame* by expanding the metric perturbatively:

$$g_{\mu\nu} = \hat{g}_{\mu\nu} + \frac{2}{M_P}h_{\mu\nu} - \sqrt{\frac{2}{3}}\frac{1}{M_P}\,\phi\,\hat{g}_{\mu\nu}\,, \tag{3}$$

where $\hat{g}_{\mu\nu}$ is its classical background solution, $h_{\mu\nu}$ takes into account the standard two degrees of freedom associated with the spin-two (traceless) graviton, and ϕ corresponds to the new mode, which owns a standard kinetic term.

The couplings of this scalar graviton with the SM fields have been computed in [1] by supposing that gravity is minimally coupled to matter (in the Jordan frame). In such a case, there is a linear coupling to matter through the trace of the standard energy-momentum tensor:

$$\mathcal{L}_{\phi-T_{\mu\nu}} = \frac{1}{M_P\sqrt{6}}\,\phi\,T^\mu_\mu\,. \tag{4}$$

Therefore, the couplings with the massive SM particles -Higgs boson (Φ), (Dirac) fermions (ψ), and electroweak gauge bosons- are:

$$\begin{aligned}\mathcal{L}_{\phi-SM}^{\text{tree-level}} = {} & \frac{1}{M_P\sqrt{6}}\,\phi\,\left\{2\,m_\Phi^2\Phi^2 - \nabla_\mu\Phi\nabla^\mu\Phi\right. \\ & + \left.\sum_\psi m_\psi\,\bar{\psi}\psi - 2\,m_W^2\,W_\mu^+W^{-\mu} - m_Z^2\,Z_\mu Z^\mu\right\}\,.\end{aligned} \tag{5}$$

In addition, this scalar graviton couples to photons and gluons due to the *conformal anomaly* [1]:

$$\mathcal{L}_{\phi-SM}^{\text{one-loop}} = \frac{1}{M_P\sqrt{6}}\,\phi\,\left\{\frac{\alpha_{EM}c_{EM}}{8\pi}\,F_{\mu\nu}F^{\mu\nu} + \frac{\alpha_s c_G}{8\pi}\,G^a_{\mu\nu}G^{\mu\nu}_a\right\}\,. \tag{6}$$

The particular value of the couplings (c_{EM} and c_G) depends on the energy and possible heavy particles, charged with respect to these gauge interactions, that may extend the SM at high energies.

4. Scalar graviton abundance

The thermal abundance that this field can achieve depends on the UV completion of the theory. However, there is no reason to expect that the initial value of the scalar field (ϕ_1) should coincide with the minimum of its potential ($\phi = 0$) if $H(T) \gg m_0$. It implies that the scalar graviton may have associated big abundances through the so called *misalignment mechanism*. Below the temperature T_1 for which $3H(T_1) \simeq m_0$, ϕ behaves as a standard scalar. It oscillates around the minimum. These oscillations correspond to a zero-momentum condensate, whose initial number density: $n_\phi \sim m_0\phi_1^2/2$ (where $\phi_1 = \sqrt{\langle\phi(T_1)^2\rangle}$), will evolve as the typical one associated to standard non-relativistic matter. The abundance o this particle has been computed in [1]:

$$\Omega_\phi h^2 \simeq 0.86 \left[\frac{m_0}{1\,\mathrm{eV}}\right]^{\frac{1}{2}} \left[\frac{\phi_1}{10^{12}\,\mathrm{GeV}}\right]^2 \left[\frac{100\, g_{e\,1}^3}{(\gamma_{s1} g_{s1})^4}\right]^{\frac{1}{4}}, \qquad (7)$$

where $g_{e\,1}$ (g_{s1}) are the effective energy (entropy) number of relativistic degrees of freedom at T_1, $h \simeq 0.70$ is the Hubble parameter, and γ_{s1} is the factor that this entropy has increased in a comoving volume since the onset of scalar oscillations. We see that initial conditions of order of $\phi_1 \sim 10^{12}$ GeV can lead to the non-baryonic DM (NBDM) abundance depending on the rest of parameters and the early physics of the Universe (see Fig. 1).

5. Scalar graviton signatures

On the other hand, Eqs. (4,5) imply that the new scalar graviton mediates an attractive Yukawa force between two non-relativistic particles of masses M_a and M_b:

$$V_{ab} = -\frac{1}{24\pi M_{\rm P}^2}\frac{M_a M_b}{r} e^{-m_0\, r}. \qquad (8)$$

Since it has not been observed, torsion-balance measurements are able to constraint the scalar mass [1]:

$$m_0 \geq 2.7\times 10^{-3}\mathrm{eV} \quad \text{at } 95\% \text{ c.l.} \qquad (9)$$

This is the most important lower bound on this mass, and it is independent of its abundance. Depending on its abundance, m_0 is constrained from above. The decay in e^+e^- is the most constraining if ϕ constitutes the total NBDM. From (5), it is possible to calculate the ϕ decay rate into a generic pair fermion anti-fermion [1]. Restrictions are set by the observations of the SPI spectrometer on the INTEGRAL (International Gamma-ray Astrophysics Laboratory) satellite, which has measured a 511 keV line emission of $1.05\pm0.06\times10^{-3}$ photons $\mathrm{cm}^{-2}\,\mathrm{s}^{-1}$ from the Galactic center (GC) [10], confirming previous measurements. This 511 keV line flux is fully consistent with an e^+e^- annihilation spectrum although the source of positrons is unknown.

If $m_0 \geq 1.2$ MeV, the scalar mode cannot constitute the total local DM since we should observe a bigger excess of the 511 line coming from the GC. On the other hand, decaying DM (DDM) has been already proposed in different works [11, 12] as a possible source of the inferred positrons if its mass is lighter than $M_{\rm DDM} \lesssim 10$ MeV [13] and its decay rate in e^+e^- verifies [1]:

$$\frac{\Omega_{\rm DDM}h^2\,\Gamma_{\rm DDM}}{M_{\rm DDM}} \simeq \left[(0.2-4)\times 10^{27}\ \mathrm{s\,MeV}\right]^{-1}. \qquad (10)$$

The most important uncertainty for this interval comes from the dark halo profile, although a cuspy density is definitely needed (with a inner slope $\gamma \gtrsim 1.5$ [12]). If m_0 is tuned to $2\,m_e$ with an accuracy of 5-10%, the line could be explained by R^2-gravity. The same gravitational DM can explain the 511 line with a less tuned mass (up to $m_0 \sim 10$ MeV) if $\phi_1 \sim 10^9$ GeV, i.e. with a lower abundance (See Fig. 1).

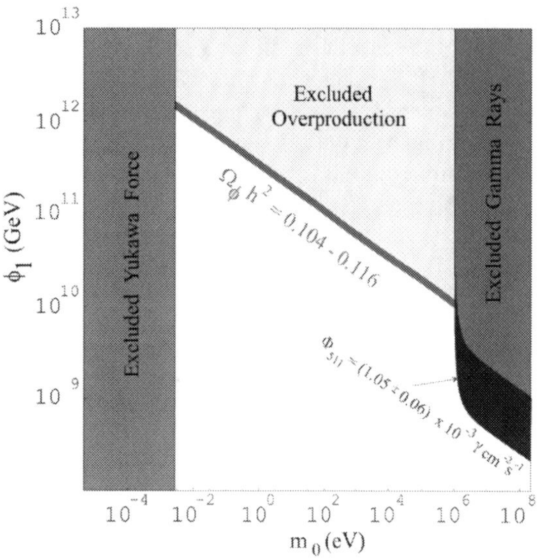

Figure 1. Different regions of the parameter space of R^2 gravity: m_0 is the mass of the scalar graviton and ϕ_1 is its misalignment when $3H \sim m_0$ (we assume $g_{e\,1} = g_{s1} \simeq 106.75$, and $\gamma_{s1} \simeq 1$). The left side is excluded by modifications of Newton's law. The right one is excluded by cosmic ray observations. In the limit of this region, R^2-gravity can account for the positron production in order to explain the 511 keV line coming from the GC confirmed by INTEGRAL [10] (up to $m_0 \sim 10$ MeV). The upper area is ruled out by DM overproduction. The diagonal line corresponds to the NBDM abundance fitted with WMAP data (Figure taken from [1]).

If $m_0 < 2m_e$, the only decay channel that may be observable is in two photons. If $m_0 \lesssim 1$ MeV, it is difficult to detect these gravitational decays in the isotropic diffuse photon background (iDPB) [14, 12]. However, a more promising analysis is associated with the search of photon lines at $E_\gamma = m_0/2$ from localized sources. The iDPB is continuum since it suffers the cosmological redshift, but the mono-energetic photons originated by local sources may give a clear signal of R^2-gravity in future experiments if the scalar graviton is inside the heavier allowed region of the model [14].

6. Conclusions

Although there are other possibilities [15], DM is usually assumed to be in the form of stable Weakly-interacting massive particles (WIMPs) that naturally freeze-out with the right thermal abundance. One of the most interesting features of WIMPs, is that they emerge in well-motivated particle physics scenarios as in R-parity conserving supersymmetry (SUSY) models [16, 17], universal extra dimensions (UED) [18, 19], or brane-worlds [20, 21, 22]. In this analysis, we have studied the possibility that the DM origin resides in UV modifications of gravity. We have focused on R^2-gravity, but the low energy phenomenology of the studied scalar mode is present in the same well-motivated frameworks such as string theory, supersymmetry or extra dimensional models (in form of dilatons, graviscalars of radions). Another interesting property of WIMPs, it is that they can be tested with high energy experiments as the new generation of colliders [23]. This possibility seems remote for the type of gravitational DM discussed in this work. However, indirect observations as modifications of Newton's law or cosmic rays can provide the first signatures of this type of DM.

Acknowledgments

I am grateful to K. Olive, M. Peloso, and M. Voloshin for useful discussions. DOE grant DOE/DE-FG02-94ER40823, FPA 2005-02327 project (DGICYT, Spain), CAM/UCM 910309 project, and MICINN Consolider-Ingenio MULTIDARK CSD2009-00064.

References

[1] J. A. R. Cembranos, Phys. Rev. Lett. **102**, 141301 (2009) [arXiv:0809.1653 [hep-ph]].
[2] P. Jordan, Nature (London) **164**, 637 (1949); M. Fierz, Helv. Phys. Acta **29**, 128 (1956); C. Brans and R. Dicke, Phys. Rev. **124**, 925 (1961); P. G. Bergmann, Int. J. Theor. Phys. **1**, 25 (1968); K. Nordtvedt, Astrophys. J. **161**, 1059 (1970); R. Wagoner, Phys. Rev. **D1**, 3209 (1970); T. Damour and K. Nordtvedt, Phys. Rev. Lett. **70**, 2217 (1993); *ibid.*, Phys. Rev. D **48**, 3436 (1993); T. Damour and B. Pichon, Phys. Rev. D **59**, 123502 (1999) [astro-ph/9807176]; C. Schimd, J. P. Uzan, and A. Riazuelo, Phys. Rev. D **71**, 083512 (2005); J. A. R. Cembranos, K. A. Olive, M. Peloso and J. P. Uzan, JCAP **0907**, 025 (2009) [0905.1989 [astro-ph.CO]]; J. A. R. Cembranos, AIP Conf. Proc. **1182**, 288 (2009) [arXiv:0906.5297 [astro-ph.CO]]; A. de la Cruz-Dombriz *et al.*, Phys. Rev. D **74**, 087501 (2006) [gr-qc/0607118]; Phys. Rev. D **77**, 123515 (2008) [0802.2999 [astro-ph]]; A. Coc, K. A. Olive, J. P. Uzan and E. Vangioni, Phys. Rev. D **73**, 083525 (2006) [astro-ph/0601299]; [0811.1845 [astro-ph]].
[3] E. Witten, Nucl. Phys. B **268**, 253 (1986); V. A. Kostelecky and S. Samuel, Phys. Lett. B **207**, 169 (1988); Nucl. Phys. B **336**, 263 (1990); I. Y. Aref'eva, A. S. Koshelev, D. M. Belov and P. B. Medvedev, Nucl. Phys. B **638**, 3 (2002) [arXiv:hep th/0011117]; I. Y. Aref'eva, L. V. Joukovskaya and A. S. Koshelev, JHEP **0309**, 012 (2003) [arXiv:hep-th/0301137]; M. Fujita and H. Hata, Phys. Rev. D **70**, 086010 (2004) [arXiv:hep-th/0403031]; T. Biswas, J. A. R. Cembranos and J. I. Kapusta, Phys. Rev. Lett. **104**, 021601 (2010) [arXiv:0910.2274 [hep-th]]; arXiv:1005.0430 [hep-th]; arXiv:1006.4098 [hep-th].
[4] K. S. Stelle, Phys. Rev. D **16**, 953 (1977).
[5] K. S. Stelle, Gen. Rel. Grav. **9**, 353 (1978).
[6] J. Z. Simon, Phys. Rev. D **45**, 1953 (1992).
[7] A. A. Starobinsky, Phys. Lett. B **91**, 99 (1980).
[8] S. Gottlober, V. Muller and A. A. Starobinsky, Phys. Rev. D **43**, 2510 (1991).
[9] J. A. R. Cembranos, Phys. Rev. D **73**, 064029 (2006) [arXiv:gr-qc/0507039].
[10] J. Knodlseder *et al.*, Astron. Astrophys. **441**, 513 (2005) [arXiv:astro-ph/0506026].
[11] C. Picciotto and M. Pospelov, Phys. Lett. B **605**, 15 (2005) [arXiv:hep-ph/0402178]; D. Hooper and L. T. Wang, Phys. Rev. D **70**, 063506 (2004) [arXiv:hep-ph/0402220]; S. Kasuya and M. Kawasaki, Phys. Rev. D **73**, 063007 (2006) [arXiv:astro-ph/0602296]; M. Pospelov and A. Ritz, Phys. Lett. B **651**, 208 (2007) [arXiv:hep-ph/0703128].
[12] J. A. R. Cembranos and L. E. Strigari, Phys. Rev. D **77**, 123519 (2008) [arXiv:0801.0630 [astro-ph]].
[13] J. F. Beacom and H. Yuksel, Phys. Rev. Lett. **97**, 071102 (2006) [arXiv:astro-ph/0512411].
[14] J. A. R. Cembranos, J. L. Feng and L. E. Strigari, Phys. Rev. Lett. **99**, 191301 (2007) [arXiv:0704.1658 [astro-ph]].
[15] L. Covi, J. E. Kim and L. Roszkowski, Phys. Rev. Lett. **82**, 4180 (1999) [hep-ph/9905212]; J. L. Feng, A. Rajaraman and F. Takayama, Phys. Rev. D **68**, 085018 (2003) [hep-ph/0307375]; J. L. Feng, A. Rajaraman and F. Takayama, Int. J. Mod. Phys. D **13**, 2355 (2004) [hep-th/0405248]; J. A. R. Cembranos, J. L. Feng, A. Rajaraman and F. Takayama, Phys. Rev. Lett. **95**, 181301 (2005) [hep-ph/0507150]; J. A. R. Cembranos, J. L. Feng, L E. Strigari, Phys. Rev. D **75**, 036004 (2007) [hep-ph/0612157]; 0708.0247 [hep-ph]; 0708.0239[hep-ph].
[16] H. Goldberg, Phys. Rev. Lett. **50**, 1419 (1983).
[17] J. R. Ellis *et al.*, Nucl. Phys. B **238**, 453 (1984).
[18] G. Servant and T. M. P. Tait, Nucl. Phys. B **650**, 391 (2003) [hep-ph/0206071].
[19] H. C. Cheng, J. L. Feng and K. T. Matchev, Phys. Rev. Lett. **89**, 211301 (2002) [hep-ph/0207125].
[20] J. A. R. Cembranos, A. Dobado and A. L. Maroto, Phys. Rev. Lett. **90**, 241301 (2003) [hep-ph/0302041]; and Phys. Rev. D **68**, 103505 (2003) [hep-ph/0307062].
[21] J. Alcaraz *et al.*, Phys. Rev.**D67**, 075010 (2003) [hep-ph/0212269]; J. A. R. Cembranos, A. Dobado and A. L. Maroto, AIP Conf.Proc. **670**, 235 (2003) [hep-ph/0301009]; hep-ph/0402142; hep-ph/0406076; Int. J. Mod. Phys. **D13**, 2275 (2004) [hep-ph/0405165]; hep-ph/0411076; astro-ph/0411262; Phys. Rev. D **73**, 035008 (2006) [hep-ph/0510399]; astro-ph/0503622; astro-ph/0512569; astro-ph/0611911; A. L. Maroto, Phys. Rev. D **69**, 043509 (2004) [hep-ph/0310272]; Phys. Rev. D **69**, 101304 (2004) [hep-ph/0402278]; J. A. R. Cembranos *et al.*, 0708.0235 [hep-ph]; JCAP **0810**, 039 (2008) [0803.0694 [astro-ph]];
[22] T. E. Clark *et al.*, Phys. Lett. B **671**, 383 (2009) [0709.4023 [hep-th]]; Nucl. Phys. B **810**, 97 (2009) [0809.1083 [hep-th]].

[23] P. Achard *et al.*, Phys. Lett. **B597**, 145 (2004) [hep-ex/0407017]; S. Heinemeyer *et al.*, hep-ph/0511332; J. A. R. Cembranos, A. Rajaraman and F. Takayama, hep-ph/0512020; Europhys. Lett. **82**, 21001 (2008) [hep-ph/0609244]; J. A. R. Cembranos, A. Dobado and A. L. Maroto, hep-ph/0107155; Phys. Rev. **D65** 026005 (2002) [hep-ph/0106322]; hep-ph/0307015; Phys. Rev. **D70**, 096001 (2004) [hep-ph/0405286]; hep-ph/0512302; Phys. Rev. D **73**, 057303 (2006) [hep-ph/0507066]; J. Phys. A **40**, 6631 (2007) [hep-ph/0611024]; A. Juste *et al.*, hep-ph/0601112; J. A. R. Cembranos *et al.*, hep-ph/0603067; AIP Conf. Proc. **903**, 591 (2007) [hep-ph/0701011]; ILC Collaboration, 0709.1893 [hep-ph]; 0712.1950 [physics.acc-ph]; 0712.2356 [physics.ins-det]; and T. E. Clark *et al.*, Phys. Rev. D **78**, 115004 (2008) [0809.3999 [hep-ph]].

Journal of Physics: Conference Series **315** (2011) 012005 doi:10.1088/1742-6596/315/1/012005

Dark Matter Universal Properties in Galaxies

Christiane Frigerio Martins

Universidade Federal do ABC, Rua Catequese 242, 09090-400 Santo André-SP, Brazil

E-mail: cfrigerio@ufabc.edu.br

Abstract. In the past years a wealth of observations has unraveled the structural properties of dark and luminous mass distribution in galaxies, a benchmark for understanding dark matter and the process of galaxy formation. The study of the kinematics of over thousand spirals has evidenced a dark-luminous matter coupling and the presence of a series of scaling laws, pictured by the Universal Rotation Curve paradigm, an intriguing observational scenario not easily explained by present theories of galaxy formation.

1. Introduction

It has been known for several decades that the kinematics of disk galaxies exhibit a mass discrepancy [1, 2, 3]: spirals show an inner baryon dominance region [4, 5, 6], whose size ranges between 1 and 3 disk exponential length-scales according to the galaxy luminosity [7], inside which the observed ordinary baryonic matter accounts for the rotation curve (RC), but outside which, the distribution of the baryonic components cannot justify the observed profiles and sometimes the amplitudes of the measured circular velocities [8, 9, 10]. This is usually solved by adding an extra mass component, the dark matter (DM) halo. RCs have been used to assess the existance, the amount and the distribution of this dark component.

In Spirals we have a unique opportunity to investigate their global mass distribution: the gravitational potentials of a spherical stellar bulge, a dark halo, a stellar disk and a gaseous disk give rise to an observed equilibrium circular velocity

$$V^2_{tot}(r) = r\frac{d}{dr}\phi_{tot} = V_b^2 + V^2_{DM} + V^2_\star + V^2_{HI}. \quad (1)$$

The Poisson equation relates the surface (spatial) densities of these components to the corresponding gravitational potentials. It is not difficult to estimate their contributions. For example the surface disk stellar density is proportional (by the mass-to-light ratio) to the observed Freeman surface brightness [11], $\Sigma_\star(r) \equiv \frac{M_D}{2\pi R_D^2} e^{-r/R_D}$, where M_D is the disk mass and R_D is the disk length-scale [1], the latter being measured directly from the observations. Then $V_\star^2 \equiv \frac{GM_D}{2R_D} x^2 B(\frac{x}{2})$, where $x \equiv r/R_D$, G is the gravitational constant and the quantity $B = I_0K_0 - I_1K_1$ is a combination of Bessel functions. The surface gaseous density is directly obtained by HI measurements.

The assumption that the RC of a spiral leads to a fair measure of its underlying gravitational potential is well justified: i) in their very inner regions the light well traces the gravitational

[1] It is useful to define the optical radius (the radius enclosing 83% of the total light [12]), $R_{opt} \equiv 3.2R_D$, as the "size" of the stellar disk.

mass [13] ii) there exists, at any galactocentric radii measured in terms of disk length-scale $R_n \equiv (n/5)R_{opt}$, a *radial* Tully-Fisher relation [14] linking with very low scatter the local rotational velocity $V_n \equiv V_{rot}(R_n)$ with the total galaxy luminosity (see Fig.1) $M_{band} = a_n log V_n + b_n$ (a_n and B_n are the slope and zero-point of the relations) iii) non-circular motions are only present at the level of a few km s^{-1} (and are not associated with a global elongation of the potential), as found from harmonic decomposition by [15] and by [16] of the velocity field of many galaxies from the HI Nearby Galaxy Survey [17] iv) disk mass estimates from spectro-photometric and kinematical methods are statistically equivalent (see e.g. [18]).

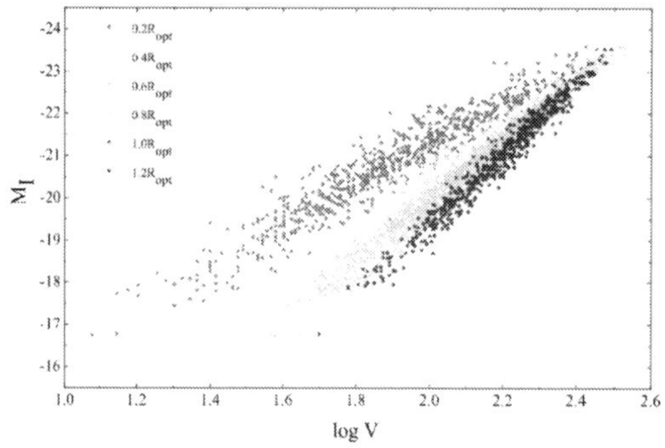

Figure 1. The Radial TF. The relations at different radii are indicated with different colours [14].

Contrary to what is often claimed, the observational evidence indicates that RCs are not asymptotically flat [19, 20] neither flat inside R_{opt} [21]. When in the late 1970s the phenomenon of DM was discovered a few truly flat RCs were highlighted in order to rule out the claim that non Keplerian velocity profiles originate from a faint baryonic component distributed at large radii. At that time a large part of the evidence for DM was provided by extended, low-resolution HI RCs of very luminous spirals whose velocity profile did show small radial variations.

The increase in the quality of the RCs though soon leads to the conclusion that baryonic (dark) matter was not a plausible candidate for the cosmological DM and that the RCs did show variation with radius, even at large radii. Later numerical simulations in the Cold Dark Matter (CDM) scenario also predicted asymptotically declining RCs [22]. The flat RC paradigm was hence dismissed in the 1990s (e.g. [5, 23, 24]). Today, the DM halo structure and its rotation speed has a central role in Cosmology and is thought to have a strong link to Elementary Particles via the Nature of their constituents (e.g. [25]). A careful interpretation of the spiral RCs is then crucial.

Note that the circular velocity due to a Freeman stellar disk has a flattish profile between 2 and 3 disk length-scales implying that a flat RC is not necessarily a proof for the existence of DM. Its most solid evidence instead originates from the fact that even in very faint galaxies the RCs are often steeply rising already in their optical regions. Fig. 2 shows, for a large sample of galaxies, the logarithmic slope ∇ of the circular velocity at R_{opt} as a function of M_B and V_{opt}: it takes almost all the values allowed by Newtonian gravity, from -0.5 (Keplerian regime) to 1 (solid body regime) and, furthermore, it strongly correlates with galaxy luminosity and V_{opt} [12, 19, 26, 27].

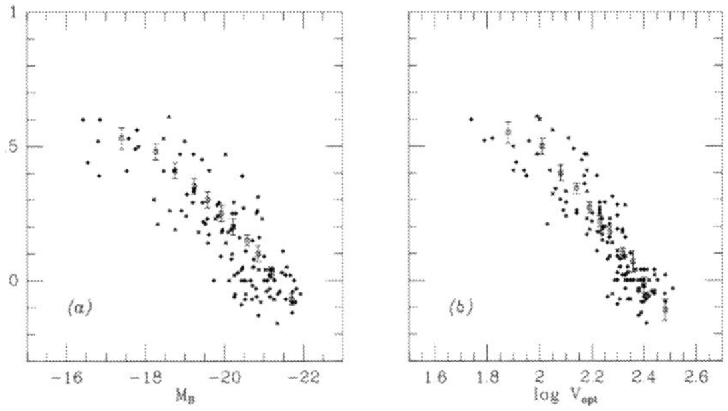

Figure 2. The RC log slope as a function of M_B and V_{opt} of 1100 coadded and 100 individual objects [12].

2. Dark Halos from Observations: the Universal Rotation Curve

The study of the systematics of spiral kinematics, pioneered by Persic & Salucci 1991, further developed by Persic, Salucci & Stel 1996 and by Salucci *et al.* 2007 has evidenced that these systems present universal features well correlating with global galactic properties. This has led to the construction of the "Universal Rotation Curve", $V_{URC}(r; P)$, i.e. an empirical function of galactocentric radius r, that, tuned by a global galaxy property (e.g. luminosity), is able to reproduce the RC of any object[2] (see Fig. 3). Additional kinematical data and virial velocities [28], $V_{vir} \equiv (GM_{vir}/R_{vir})^{1/2}$, have determined the URC out to the virial radii.

V_{URC} is then the observational counterpart of the velocity profile that emerges out of CDM large N-body numerical simulations of structure formation. As individual RCs, it implies a mass model including a Freeman disk and a DM halo with an empirical Burkert cored profile [29],

$$\rho(R) = \frac{\rho_0 \, r_0^3}{(R + r_0)\,(R^2 + r_0^2)}, \tag{2}$$

where r_0 is the core radius and ρ_0 its central density.

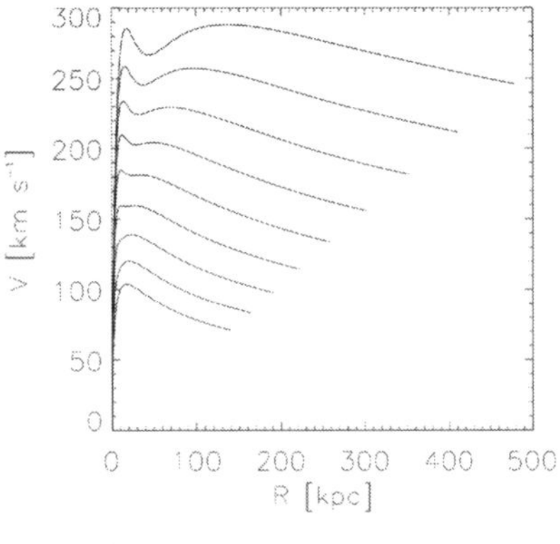

Figure 3. The URC [20]. Each curve corresponds to a galaxy mass $M_{vir} = 10^{11} 10^{n/5} M_\odot$, with $n = 1...9$ from the lowest to the highest curve.

[2] A 3D visualization can be found at http://www.youtube.com/watch?v=YcgafVb-WJI.

Recent debate in the literature has focused on the "cuspiness" of the DM density distribution in the centres of galaxy halos predicted from the simulations [22, 30, 31, 32], commonly represented by the Navarro, Frenk & White (NFW) profile, but not seen in observed data, as well as in the various systematics of the DM distribution (e.g. [9, 15, 20, 33, 34, 35, 36]).

It is worth illustrating one example of this disagreement: the nearby spiral dwarf galaxy DDO 47 [15]. The RC mass modeling (Fig. 4) finds the dark halo with a core radius of about 7 kpc and a central density of 1.4×10^{-24}g cm^{-3}: the underlying DM density profile is much shallower than the NFW predictions, totally unable to fit the RC. Presently there are about a hundred high quality, extended and free from deviations from axial symmetry RCs whose careful analysis has strongly disfavored the disk + NFW halo mass model (in favor of cored profiles) that: i) fits the RC poorly ii) implies an implausibly low stellar mass-to-light ratio and in some case iii) an unphysical high halo mass (e.g. [9, 15, 36, 37, 38, 39, 40, 41]). Furthermore, these evidences are accompanied by investigations that have ruled out that it may arise from (neglected) systematical effects [15, 16, 42] (for a recent review on the cusp versus core issue see [43]).

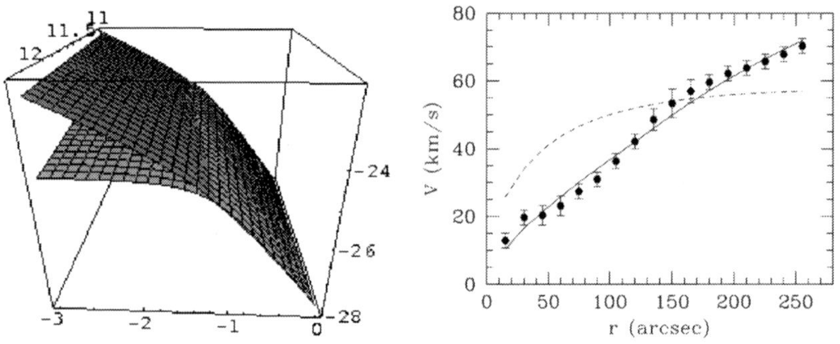

Figure 4. Left: The URC halo density *vs* the NFW halo density of objects of the same mass, as a function of normalized radius and virial mass [20]. The axes labels are R/R$_{vir}$, log M$_{vir}$/M$_\odot$ and $\log(\rho/(\mathrm{g\ cm^{-3}}))$. Right: RCs best-fits of DDO 47 [15]: Burkert halo + stellar disk (solid line), NFW halo + stellar disk (dashed line).

The structural parameters ρ_0, r_0 and M_D are obtained for the URC and for any individual RC by χ^2 fitting. As result, a clear scenario of the mass distribution emerges[3], with a cored DM distribution and a set of scaling laws among local and global galaxy quantities:

- Spirals have an inner baryon dominance region where the stellar disk dominates the total gravitational potential, while the DM halo emerges farther out
- At any radii, objects with lower luminosities have a larger dark-to-stellar mass ratio. The baryonic fraction in spirals is always much smaller than the cosmological value $\Omega_b/\Omega_{matter} \simeq 1/6$, and it ranges between $(7-50) \times 10^{-3}$, suggesting that processes such as SN explosions must have removed a large fraction of the original hydrogen
- Smaller spirals are denser, with their central density spanning 2 orders of magnitude over the mass sequence of spirals
- The stellar mass-to-light ratio (in the B band) lies between 0.5 and 4 and increases with galaxy luminosity as $L_B^{0.2}$, in agreement with the values obtained by fitting their SED with spectro-photometric models

[3] A worldwide initiative specially devised to increase awareness of the phenomenology of DM in galaxies can be found in [44] http://darkmatteringalaxies.selfip.org.

- Dark and luminous matter are remarkably linked together (see Fig. 5).

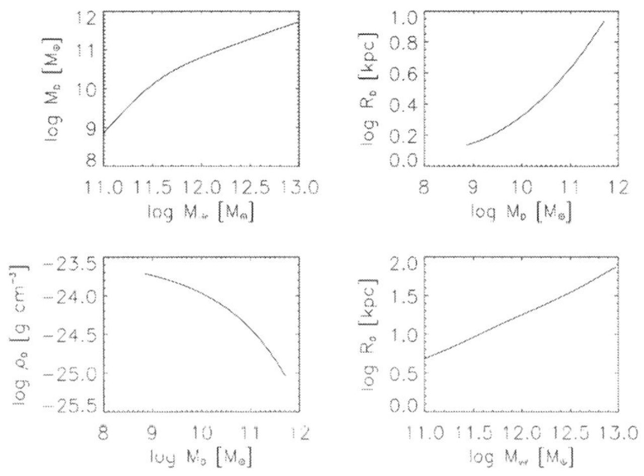

Figure 5. Scaling relations between the structural parameters of the dark and luminous mass distribution in spirals [20].

3. A constant Dark Matter halo surface density in galaxies

An intriguing finding among the DM structural properties is that the central surface density, $\mu_{0D} \equiv r_0\rho_0$, is nearly constant and independent of galaxy luminosity. Based on the co-added RCs of $\sim$ 1000 spirals, mass models of individual dwarf irregular and spiral galaxies of late and early types with high-quality RCs, and galaxy-galaxy weak lensing signals from a sample of spirals and ellipticals, we find that [45]

$$\mu_{0D} = 2.15 \pm 0.2, \tag{3}$$

in units of log($M_\odot$ pc^{-2}). We also show that the observed internal kinematics of Local Group dwarf spheroidal galaxies are consistent with this value. Our results are obtained for galactic systems spanning over 14 magnitudes, belonging to different Hubble Types, and whose mass profiles have been determined by several independent methods.

The approximate constancy of μ_{0D} is particularly relevant also because it is in sharp contrast to the observed systematical variations, by several orders of magnitude, of the central stellar surface density, i.e. of its luminous counterpart, as illustrated in Fig. 6. As a consequence, the central surface density is the only DM quantity which is not correlated with its stellar analagous, different from any other (core radius, central spatial density, mass, etc). Moreover it is remarkable that the constancy of μ_{0D} can be related to the above scaling laws of spirals. As an example, let us define M_{h0} and V_{h0} as the enclosed halo mass inside r_0 and the halo circular velocity at r_0. Since for a Burkert halo $M_{h0} \propto \rho_0 r_0^3$, then $M_{h0} \propto V_{h0}^4$, which immediately reminds a sort of Tully-Fisher relation.

It is surprising and presently difficult to understand how the DM halo central surface density can be maintained constant across galaxies with very different evolutionary histories (e.g. numbers of mergers, significance of baryon cooling, stellar feedback) and with range from

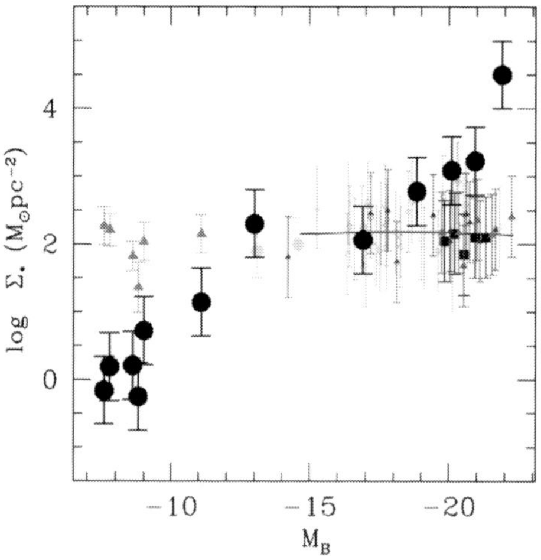

Figure 6. Dark matter central surface density in units of $M_\odot pc^{-2}$ as a function of galaxy magnitude, for different galaxies and Hubble Types [45]. As a comparison the values of the same quantity of the stellar component is also shown (big filled circles).

DM-dominated to baryon-dominated in the inner regions. This quantity may hide an important physical meaning in the nature itself of the DM, as it seems to indicate recent theoretical work [46, 47, 48].

4. Conclusions

More then eighty years ago E. Hubble established the expansion of the Universe with his pioneering observations of galaxies. Since then galaxies have been fundamental tools for understanding the structure and evolution of our Universe. Today they are crucial laboratories where microphysics phenomena, up to now not detected by particle physics experiments, emerge with unprecedented clarity.

The distribution of luminous and dark matter in galaxies shows amazing properties and a remarkable systematics that make it as one of the hottest cosmological issues. There is no doubt that this emerging observational scenario will be decisive in guiding how the CDM-based theory of galaxy formation must evolve to meet the challenge that the observational data are posing.

Acknowledgments

CFM is a postdoctoral researcher of the FAPESP (Brasil).

References

[1] A. Bosma, Ph.D. thesis, Groningen University (1978).
[2] A. Bosma, and P. C. van der Kruit, *Astron. Astrophys.* **79**, 281 (1979).
[3] V.C. Rubin, N. Thonnard, and W.K.Jr. Ford, *Astrophys. J.* **238**, 471 (1980).
[4] E. Athanassoula, A. Bosma, and S. Papaioannou, *Astron. Astrophys.* **179**, 23 (1987).
[5] M. Persic, and P. Salucci, *Mon. Not. R. Astron. Soc.* **234**, 131 (1988).
[6] P. Palunas, and T. B. Williams, *Astron. J.* **120**, 2884 (2000).
[7] P. Salucci, and M. Persic, *Astron. Astrophys.* **351**, 442 (1999).
[8] A. Bosma, *Astron. J.***86**, 1825 (1981).
[9] G. Gentile, P. Salucci, U. Klein, and G. L. Granato, *Mon. Not. R. Astron. Soc.* **375**, 199 (2007).

[10] G. Gentile, C. Tonini, and P. Salucci, *Astron. Astrophys.* **467**, 925 (2007).
[11] K. C. Freeman, *Astrophys. J.* **160**, 811 (1970).
[12] M. Persic, P. Salucci, and F. Stel, *Mon. Not. R. Astron. Soc.* **281**, 27 (1996).
[13] C. Ratnam, and P. Salucci, *New Astronomy* **5**, 427 (2000).
[14] I.A. Yegorova, and P. Salucci, *Mon. Not. R. Astron. Soc.* **377**, 507 (2007).
[15] G. Gentile, A. Burkert, P. Salucci, U. Klein, and F. Walter, *Astrophys. J.* **634**, L145 (2005).
[16] C. Trachternach, W.J.G. de Blok, F. Walter, E. Brinks, and R.C. Kennicutt, *Astron. J.* **136**, 2720 (2008).
[17] F. Walter, E. Brinks, W.J.G. de Blok, R.C. Kennicutt, M.D. Thornley, and A. Leroy, *Astron. J.* **136**, 2563 (2008).
[18] P. Salucci, I.A. Yegorova, and N. Drory, *Mon. Not. R. Astron. Soc.* **388**, 159 (2008).
[19] P. Salucci, and G. Gentile, *Phys. Rev.D* **73**, 128501 (2006).
[20] P. Salucci, A. Lapi, C. Tonini, G. Gentile, I.A. Yegorova, and U. Klein, *Mon. Not. R. Astron. Soc.* **378**, 41 (2007).
[21] M. Persic, and P. Salucci, *Astrophys. J.* **368**, 60 (1991).
[22] J. F. Navarro, C. S. Frenk and S. D. M. White, *Astrophys. J.* **462**, 563 (1996).
[23] K.M. Ashman, *Publ. Astron. Soc. Pac.* **104**, 1109 (1992).
[24] D. Weinberg, *ASP Conf. Series* **117**, 578 (1997).
[25] K.A. Olive, *preprint* arXiv: astro-ph/0503065.
[26] M.A.W. Verheijen, Ph.D. thesis, Groningen University (1997).
[27] J. Zavala, V. Avila-Reese, H. Hernández-Toledo, and C. Firmani, *Astron. Astrophys.* **412**, 633 (2003).
[28] F. Shankar, A. Lapi, P. Salucci, G. De Zotti, and L. Danese, *Astrophys. J.*, **643**, 14 (2006).
[29] P. Salucci, and A. Burkert, *Astrophys. J.* **537**, L9 (2000).
[30] B. Moore, T. Quinn, F. Governato, J. Stadel, and G. Lake, *Mon. Not. R. Astron. Soc.* **310**, 1147 (1999).
[31] J.F. Navarro, E. Hayashi, C. Power, A.R. Jenkins, C.S. Frenk, S. D. M. White, V. Springel, J. Stadel, and T.R. Quinn, *Mon. Not. R. Astron. Soc.* **349**, 1039 (2004).
[32] A. F. Neto, L. Gao, P. Bett, S. Cole, J.F. Navarro, C.S. Frenk, S.D.M. White, V. Springel, and A. Jenkins, *Mon. Not. R. Astron. Soc.* **381**, 1450 (2007).
[33] W. J. G. de Blok, S.S. McGaugh, and V.C. Rubin, *Astron. J.* **122**, 2396 (2001).
[34] W.J.G. de Blok, A. Bosma, *Astron. Astrophys.* **385**, 816 (2002).
[35] D. Marchesini, E. D'Onghia, G. Chincarini, C. Firmani, P. Conconi, E. Molinari, and A. Zacchei, *Astrophys. J.* **575**, 801 (2002).
[36] G. Gentile, P. Salucci, U. Klein, D. Vergani, and P. Kalberla, *Mon. Not. R. Astron. Soc.* **351**, 903 (2004).
[37] K. Spekkens, R. Giovanelli, and M.P. Haynes, *Astron. J.* **129**, 2119 (2005).
[38] J. D. Simon, A.D. Bolatto, A. Leroy, L. Blitz, and E.L. Gates, *Astrophys. J.* **621**, 757 (2005).
[39] M. Spano, M. Marcelin, and P. Amram, *Mon. Not. Roy. Astron. Soc.* **383**, 297 (2008).
[40] W. J. G. de Blok, F. Walter, E. Brinks, C. Trachternach, S.-H. Oh, and R.C. Kennicutt, *Astron. J.* **136** (6) 2648 (2008).
[41] R. Kuzio de Naray, S.S. McGaugh, and W.J.G. de Blok, *Astrophys. J.* **676**, 920 (2008).
[42] S-H. Oh, W.J.G. de Blok, F. Walter, E. Brinks, and R.C. Kennicutt, *Astron. J.* **136** (6) 2761 (2008).
[43] W. J. G. de Blok, *Advances in Astronomy* **2010**, 1 (2010).
[44] P. Salucci, C. Frigerio Martins, A. Lapi, arXiv: astro-ph/1102.1184 (2011).
[45] F. Donato, G. Gentile, P. Salucci, C. Frigerio Martins, M.I. Wilkinson, G. Gilmore, E. K. Grebel, A. Koch, and R. Wyse, *Mon. Not. Roy. Astron. Soc.* **397**, 1169 (2009).
[46] H. J. de Vega, and N. G. Sanchez, *Mon. Not. Roy. Astron. Soc.* **404**, 885 (2010).
[47] H. J. de Vega, P. Salucci, and N. G. Sanchez, arXiv:1004.1908 (2010).
[48] G. Gentile, B. Famaey, H. Zhao, and P. Salucci, *Nature* **461**, 627 (2009).

Journal of Physics: Conference Series **315** (2011) 012006 doi:10.1088/1742-6596/315/1/012006

Dark Left-Right Model: CDMS, LHC, etc.

Ernest Ma

Physics and Astronomy Department, University of California, Riverside, California 92521, USA

E-mail: ernest.ma@ucr.edu

Abstract. The Standard Model of particle interactions is extended to include fermion doublets $(n, e)_R$ transforming under the gauge group $SU(2)_R$ such that n is a Dirac scotino (dark-matter fermion), with odd R parity. Based on recent CDMS data, it is shown how this new dark left-right model (DLRM2) favors a Z' gauge boson at around 1 or 2 TeV and be observable at the LHC. The new $W_R^{\pm}$ gauge bosons may also contribute significantly to lepton-flavor-changing processes such as $\mu \to e\gamma$ and $\mu - e$ conversion in a nucleus or muonic atom.

1. Left-right Extension of Standard Model

If the Standard Model (SM) of particle interactions is extended to accommodate $SU(3)_C \times SU(2)_L \times SU(2)_R \times U(1)_X$, then the conventional assignment of

$$(\nu, l)_L \sim (1, 2, 1, -1/2), \qquad (\nu, l)_R \sim (1, 1, 2, -1/2), \tag{1}$$

$$(u, d)_L \sim (3, 2, 1, 1/6), \qquad (u, d)_R \sim (3, 1, 2, 1/6), \tag{2}$$

implies the well-known result that $X = (B - L)/2$ and $Y = T_{3R} + (B - L)/2$. There must then be Higgs bidoublets

$$\Phi = \begin{pmatrix} \phi_1^0 & \phi_2^+ \\ \phi_1^- & \phi_2^0 \end{pmatrix}, \quad \tilde{\Phi} = \begin{pmatrix} \bar{\phi}_2^0 & -\phi_1^+ \\ -\phi_2^- & \bar{\phi}_1^0 \end{pmatrix}, \tag{3}$$

both transforming as $(1, 2, 2, 0)$, yielding lepton Dirac mass terms

$$m_l = f_l \langle \phi_2^0 \rangle + f_l' \langle \bar{\phi}_1^0 \rangle, \quad m_\nu = f_l \langle \phi_1^0 \rangle + f_l' \langle \bar{\phi}_2^0 \rangle, \tag{4}$$

and similarly in the quark sector. This results in the appearance of phenomenologically undesirable tree-level flavor-changing neutral currents from Higgs exchange, as well as inevitable $W_L - W_R$ mixing. If supersymmetry is imposed, then $\tilde{\Phi}$ can be eliminated, but then $(M_\nu)_{ij} \propto (M_l)_{ij}$ as well as $(M_u)_{ij} \propto (M_d)_{ij}$, contrary to what is observed. Hence the prevalent thinking is that $SU(2)_R \times U(1)_{B-L}$ is actually broken down to $U(1)_Y$ at a very high scale from an $SU(2)_R$ Higgs triplet $(\Delta_R^{++}, \Delta_R^+, \Delta_R^0) \sim (1, 1, 3, 1)$ which provides ν_R at the same time with a large Majorana mass from $\langle \Delta_R^0 \rangle$.

The canonical seesaw mechanism for neutrino mass is thus implemented and everyone should be happy. But wait, no remnant of the $SU(2)_R$ gauge symmetry is detectable at the TeV scale and we will not know if ν_R really exists. Is there a natural way to lower the $SU(2)_R \times U(1)_{B-L}$ breaking scale?

The answer was already provided 23 years ago [1] in the context of the superstring-inspired supersymmetric E_6 model. The fundamental $\underline{27}$ fermion representation here is decomposed under $[(SO(10), SU(5)]$ as

$$\underline{27} = (16,10) + (16,5^*) + (16,1) + (10,5) + (10,5^*) + (1,1). \tag{5}$$

Under its maximum subgroup $SU(3)_C \times SU(3)_L \times SU(3)_R$, the $\underline{27}$ is organized instead as $(3,3^*,1) + (1,3,3^*) + (3^*,1,3)$, i.e.

$$\begin{pmatrix} d & u & h \\ d & u & h \\ d & u & h \end{pmatrix} + \begin{pmatrix} N & E^c & \nu \\ E & N^c & e \\ \nu^c & e^c & n^c \end{pmatrix} + \begin{pmatrix} d^c & d^c & d^c \\ u^c & u^c & u^c \\ h^c & h^c & h^c \end{pmatrix}. \tag{6}$$

It was realized [1] in 1987 that there are actually two left-right options: (A) Let E_6 break down to the fermion content of the conventional $SO(10)$, given by $(16,10) + (16,5^*) + (16,1)$, which is the usual left-right model which everybody knows. (B) Let E_6 break down to the fermion content given by $(16,10) + (10,5^*) + (1,1)$ instead, thereby switching the first and third rows of $(3^*,1,3)$ and the first and third columns of $(1,3,3^*)$. Thus $(\nu,e)_R$ becomes $(n,e)_R$ and n_R is not the mass partner of ν_L. This is referred to by the Particle Data Group as the Alternative Left-Right Model (ALRM). Here the usual left-handed lepton doublet is part of a bidoublet:

$$\begin{pmatrix} \nu & E^c \\ e & N^c \end{pmatrix}_L \sim (1,2,2,0). \tag{7}$$

In this supersymmetric model, ν_L is still the Dirac mass partner of ν_R and gets a seesaw mass, whereas [2] n_R (which couples to e_R through W_R) mixes with the usual neutralinos, the lightest of which is a dark-matter candidate.

2. Dark left-right model

Two simpler nonsupersymmetric versions of the ALRM with n_R as dark matter have recently been proposed [3, 4]. We call them Dark Left-Right Models (DLRM and DLRM2). We impose a global U(1) symmetry S, so that under $SU(3)_C \times SU(2)_L \times SU(2)_R \times U(1)_X \times S$, where $Q = T_{3L} + T_{3R} + X$, a generalized lepton number is conserved, such that $L = S - T_{3R}$ in DLRM and $L = S + T_{3R}$ in DLRM2. The resulting dark-matter fermion n_R has $L = 0$ (Majorana) in DLRM and $L = 2$ (Dirac) in DLRM2. This talk is on DLRM2, with particle content [4] under $SU(3)_C \times SU(2)_L \times SU(2)_R \times U(1)_X \times S$ given below:

$$\psi_L = (\nu, e)_L \sim (1,2,1,-1/2;1), \quad \nu_R \sim (1,1,1,0;1), \tag{8}$$
$$\psi_R = (n, e)_R \sim (1,1,2,-1/2;3/2), \quad n_L \sim (1,1,1,0;2), \tag{9}$$
$$Q_L = (u, d)_L \sim (3,2,1,1/6;0), \quad d_R \sim (3,1,1,-1/3;0), \tag{10}$$
$$Q_R = (u, h)_R \sim (3,1,2,1/6;-1/2), \quad h_L \sim (3,1,1,-1/3;-1), \tag{11}$$
$$\Phi \sim (1,2,2,0;-1/2), \quad \tilde{\Phi} \sim (1,2,2,0;1/2), \tag{12}$$
$$\Phi_L = (\phi_L^+, \phi_L^0) \sim (1,2,1,1/2;0), \quad \Phi_R = (\phi_R^+, \phi_R^0) \sim (1,1,2,1/2;1/2). \tag{13}$$

As a result, the Yukawa terms $\bar{\psi}_L \Phi \psi_R$, $\bar{\psi}_L \tilde{\Phi}_L \nu_R$, $\bar{\psi}_R \tilde{\Phi}_R n_L$, $\bar{Q}_L \tilde{\Phi} Q_R$, $\bar{Q}_L \Phi_L d_R$, $\bar{Q}_R \Phi_R h_L$ are allowed, whereas $\bar{\psi}_L \tilde{\Phi} \psi_R$, $\bar{Q}_L \Phi Q_R$ are forbidden together with the bilinear terms $\bar{n}_L \nu_R$, $\bar{h}_L d_R$. The breaking of $SU(2)_R \times U(1)_X \to U(1)_Y$ leaves $L = S + T_{3R}$ unbroken, so that $v_2 = \langle \phi_2^0 \rangle \neq 0$ [ϕ_2^0 has $L = 0$], but $\langle \phi_1^0 \rangle = 0$ [ϕ_1^0 has $L = -1$]. The former contributes to m_e and m_u, whereas the latter means that ν_L and n_R are not Dirac mass partners and can be completely different particles. In fact, m_ν, m_d come from $v_3 = \langle \phi_L^0 \rangle$, and m_n, m_h from $v_4 = \langle \phi_R^0 \rangle$. This structure

Journal of Physics: Conference Series **315** (2011) 012006 doi:10.1088/1742-6596/315/1/012006

guarantees the absence of tree-level flavor-changing neutral currents. As for the gauge bosons and their interactions, let $e/g_L = s_L = \sin\theta_W$, $e/g_R = s_R$, $e/g_X = \sqrt{1 - s_L^2 - s_R^2} = \sqrt{c_L^2 - s_R^2}$, then

$$A = s_L W_L^0 + s_R W_R^0 + \sqrt{c_L^2 - s_R^2} X, \tag{14}$$

$$Z = c_L W_L^0 - (s_L s_R/c_L) W_R^0 - (s_L \sqrt{c_L^2 - s_R^2}/c_L) X, \tag{15}$$

$$Z' = (\sqrt{c_L^2 - s_R^2}/c_L) W_R^0 - (s_R/c_L) X, \tag{16}$$

$$g_Z = e/s_L c_L, \quad J_Z = J_{3L} - s_L^2 J_{em}, \tag{17}$$

$$g_{Z'} = e/s_R c_L \sqrt{c_L^2 - s_R^2}, \quad J_{Z'} = s_R^2 J_{3L} + c_L^2 J_{3R} - s_R^2 J_{em}. \tag{18}$$

To avoid $Z - Z'$ mixing at tree level, the condition $v_2^2/(v_2^2 + v_3^2) = s_R^2/c_L^2$ must be imposed. In that case, $M_{W_R} \simeq (\sqrt{c_L^2 - s_R^2}/c_L) M_{Z'}$. Note that W_R does not mix with W_L because they have different R parity. In Fig. 1, the present Tevatron bound on $M_{Z'}$ is shown for various values of s_R^2, showing a typical bound of about 1 TeV.

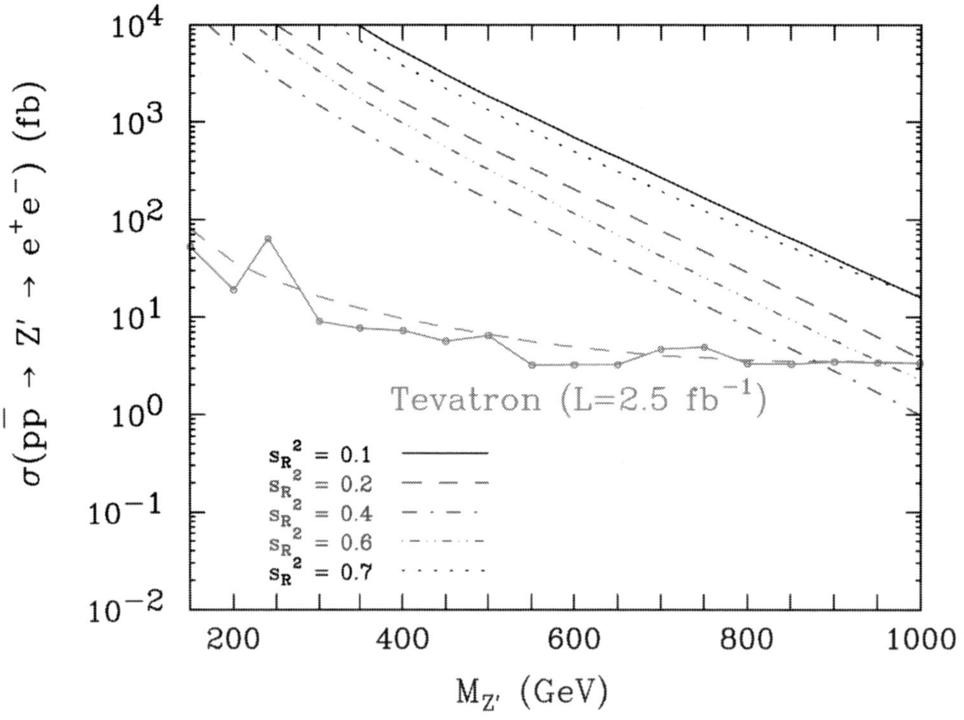

Figure 1. $\sigma(p\bar{p} \to Z' \to e^+e^-)$ vs $M_{Z'}$ compared against Tevatron data.

3. CDMS and More

The usual leptons have $L = 1$ as expected, but there are now new particles also with lepton number: $W_R^+, \phi_R^+, \phi_1^+$ have $L = 1$ and h has $L = -1$ as well as $B = 1/3$. Thus they all have odd R parity, i.e. $R = (-)^{3B+L+2j} = -1$, even though the model is nonsupersymmetric. The scotino n has $L = 2$ and thus also odd R. The lightest n is a dark-matter candidate, and will

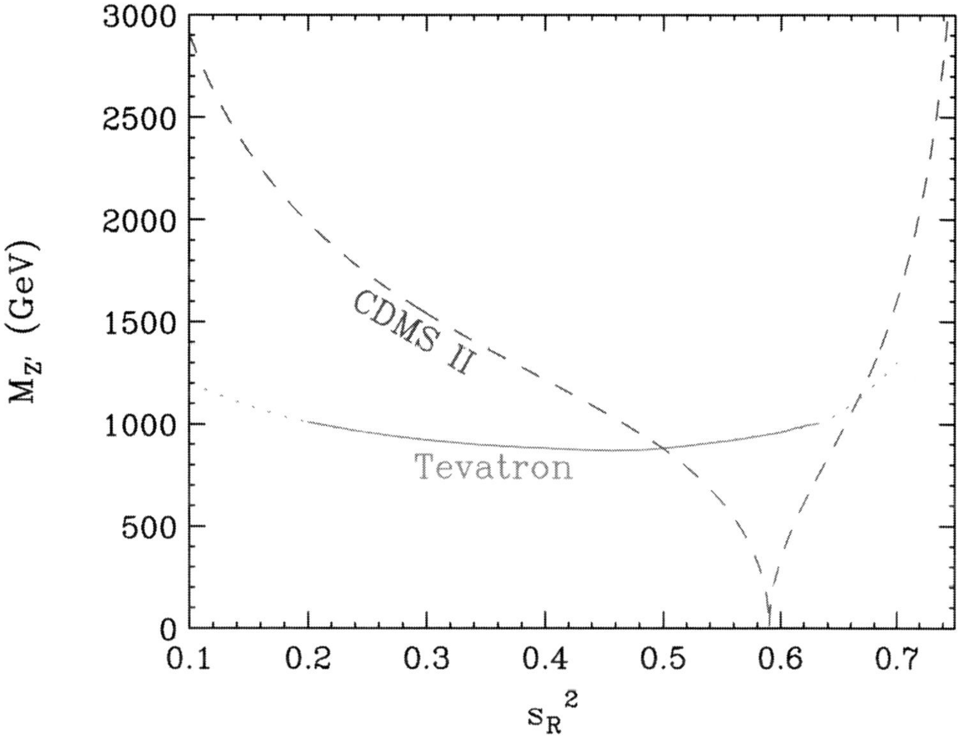

Figure 2. Lower bounds on $M_{Z'}$ vs s_R^2 from the Tevatron search (red solid line) and from the CDMS search at $m_n = 70$ GeV (blue dashed line). The dotted segments assume a simple extrapolation of the Tevatron data.

be considered below in the context of recent data from the CDMS-II collaboration [5]. Two possible dark-matter signal events were observed with an expected background of 0.9 ± 0.1. The most stringent bound on the elastic spin-independent scattering cross section of $nq \to nq$ occurs at $m_n = 70$ GeV, and it is 3.8×10^{-8} pb. In the DLRM2,

$$\mathcal{L} = \frac{g_{Z'}^2 n_V}{M_{Z'}^2}(\bar{n}\gamma_\mu n)(u_V \bar{u}\gamma^\mu u + d_V \bar{d}\gamma^\mu d), \tag{19}$$

where $n_V = c_L^2/4$, $u_V = c_L^2/4 - 5s_R^2/12$, $d_V = s_R^2/12$. Let $f_P = g_{Z'}^2 n_V(2u_V + d_V)/M_{Z'}^2$, $f_N = g_{Z'}^2 n_V(u_V + 2d_V)/M_{Z'}^2$, then

$$\sigma_0 \simeq \frac{4m_P^2}{\pi}\frac{[Zf_P + (A-Z)f_N]^2}{A^2}. \tag{20}$$

Using ^{73}Ge, i.e. $Z = 32$ and $A - Z = 41$, as an estimate, the CDMS bound (of 3.8×10^{-8} pb at $m_n = 70$ GeV) implies a bound on $M_{Z'}$. In Fig. 2, the resulting lower bounds on $M_{Z'}$ from the Tevatron search and from CDMS are plotted as functions of s_R^2.

To obtain the correct dark-matter relic abundance, the annihilation of $n\bar{n} \to Z' \to$ SM fermions is considered. The thermally averaged cross section multiplied by the relative velocity of the annihilating particles is given by

$$\langle \sigma v_{rel} \rangle_{Z'} = \frac{\pi\alpha^2(3 - 9r + 10r^2)m_n^2}{2c_L^4 r^2(1-r)^2(4m_n^2 - M_{Z'}^2)^2}, \tag{21}$$

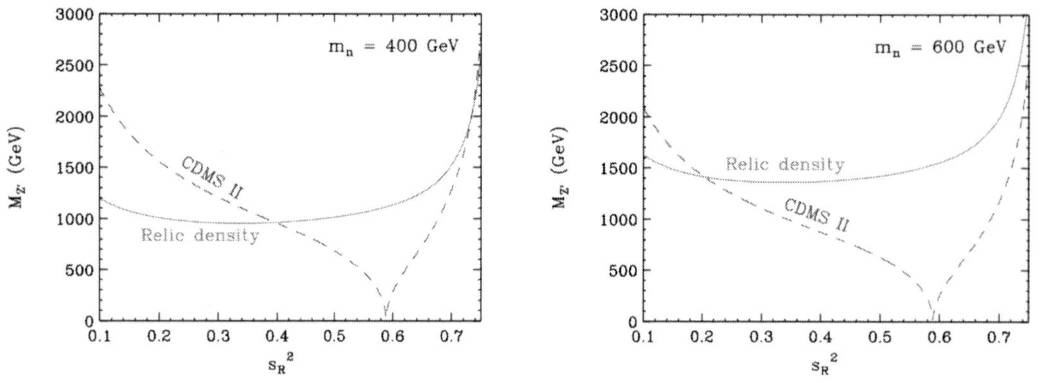

Figure 3. The CDMS bound on $M_{Z'}$ (blue dashed line) and its value (red solid line) from $\langle \sigma v_{rel} \rangle_{Z'} = 1$ pb vs s_R^2 for $m_n = 400$ and 600 GeV.

where $r = s_R^2/c_L^2$. Fixing the above at 1 pb, the values of m_n and $M_{Z'}$ are constrained as a function of s_R^2. For $m_n = 70$ GeV, there is no solution, but if m_n is greater than about 300 GeV, solutions exist which are consistent with the Tevatron bound as well as the CDMS bound. In the range $0.3 < m_n < 1.0$ TeV, the latter is well approximated by $\sigma_0 < 2.2 \times 10^{-7}$ pb $(m_n/1$ TeV$)^{0.86}$. In Fig. 3, the cases $m_n = 400$ and 600 GeV are shown.

The $n\bar{n}$ annihilation to leptons through $W_R^\pm$ exchange also contributes, i.e.

$$\langle \sigma v_{rel} \rangle_{W_R} = \frac{3 g_R^4 m_n^2}{64\pi (m_n^2 + M_{W_R}^2)^2}, \tag{22}$$

but it is subdominant and has been neglected.

4. LHC and More

At the LHC ($E_{cm} = 14$ GeV), Z' may be discovered with 10 dilepton events. Using the cuts

- $p_T > 20$ GeV for each lepton,
- $|\eta| < 2.4$ for each lepton,
- $|M_{l^- l^+} - M_{Z'}| < 3\Gamma_{Z'}$,

the SM background is negligible. With an integrated luminosity of 1 fb^{-1}, the discovery reach of the Z' of the DLRM2 is about 2 TeV, as shown in Fig. 4.

To distinguish this Z' from others, the following ratios [6] may be considered:

$$\frac{\Gamma(Z' \to t\bar{t})}{\Gamma(Z' \to \mu^- \mu^+)} = \frac{(9 - 24r + 17r^2)}{3(1 - 4r + 5r^2)} = 4.44 \ (g_L = g_R), \tag{23}$$

$$\frac{\Gamma(Z' \to b\bar{b})}{\Gamma(Z' \to \mu^- \mu^+)} = \frac{5r^2}{3(1 - 4r + 5r^2)} = 0.60 \ (g_L = g_R), \tag{24}$$

where $r = s_R^2/c_L^2$. In the conventional left-right model, the numerator for $b\bar{b}$ is changed to $(9 - 12r + 8r^2)$, i.e. 13.6 larger $(g_L = g_R)$. In the ALRM, the denominator for both is changed to $3(2 - 6r + 5r^2)$, i.e. 2.6 larger $(g_L = g_R)$.

There are also important loop effects [2] on rare processes from the new interactions $\gamma W_R^+ W_R^-$, $Z W_R^+ W_R^-$, $W_R^+ \bar{n}_R e_R$, $W_R^+ \bar{u}_R h_R$, etc. The anomalous magnetic moment of the muon receives a

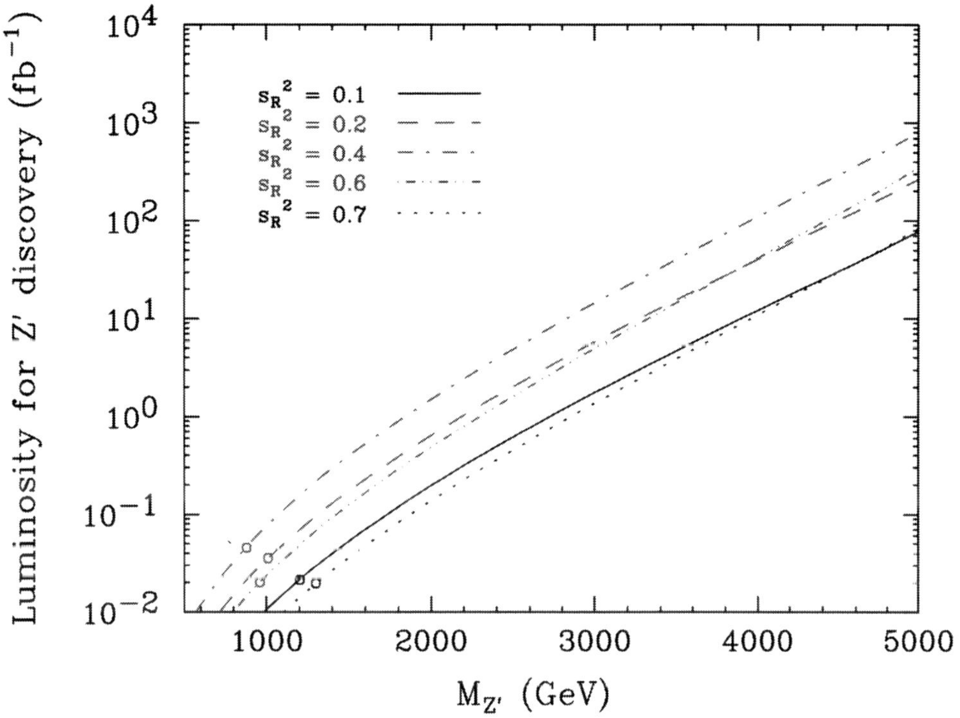

Figure 4. Luminosity for Z' discovery by 10 dielectron events at LHC. Small circles are Tevatron limits.

contribution of order 10^{-10}, below the present experimental sensitivity of 10^{-9}. The flavor-changing radiative decay $\mu \to e\gamma$ has the branching fraction

$$B(\mu \to e\gamma) = \frac{3\alpha}{32\pi}\left(\frac{s_L M_{W_L}}{s_R M_{W_R}}\right)^4 |\sum_i U_{\mu i} U_{ei} F_\gamma(r_i)|^2, \tag{25}$$

where $r_i = m^2_{n_i}/M^2_{W_R}$ and

$$F_\gamma(r_i) = \frac{r_i(-1+5r_i+2r_i^2)}{(1-r_i)^3} + \frac{6r_i^3 \ln r_i}{(1-r_i)^4}. \tag{26}$$

It is suppressed by M_{W_R} and for $M_{W_R} = 1.17$ TeV (corresponding to $M_{Z'} = 1.4$ TeV), the current experimental bound of 1.2×10^{-11} implies $|\sum_i U_{\mu i} U_{ei} F_\gamma(r_i)| < 0.05$ for $g_L = g_R$. Another probe of the existence of these new interactions is $\mu \to eee$ or $\mu - e$ conversion in a nucleus or muonic atom [7]. There is now also an effective $\mu \to eZ$ vertex from $ZW_R^+W_R^-$ and $W_R^+ n_R e_R$, given by

$$g_{\mu e Z} = \frac{e^3 s_L}{16\pi^2 s_R^2 c_L} \sum_i U_{\mu i} U_{ei} F_Z(r_i), \tag{27}$$

where $F_Z(r_i)$ is yet to be calculated. The current experimental bound of 1.0×10^{-12} on $B(\mu \to eee)$ implies $|\sum_i U_{\mu i} U_{ei} F_Z(r_i)| < 1.44 \times 10^{-3}$ for $g_L = g_R$.

5. Conclusion

The presence of ν_R is unavoidable in a left-right gauge extension of the Standard Model. However, it does not have to be the Dirac mass partner of ν_L. In that case, it should be renamed n_R and could function as a scotino, i.e. a dark-matter fermion. In the context of the recently proposed new dark left-right model (DLRM2), latest CDMS observations are shown to be consistent with the lightest n at about a few hundred GeV in mass with the new Z' gauge boson at less than 2 TeV. The latter should then be accessible directly at the LHC, while the $W_R^\pm$ gauge bosn may contribute indirectly to enhancing rare lepton-flavor-changing processes such as $\mu \to eee$ and $\mu - e$ conversion in a nucleus or muonic atom.

Acknowledgments

This work was supported in part by the U. S. Department of Energy under Grant No. DE-FG03-94ER40837. I thank D. Delepine and the other organizers of the VI International Workshop on the Dark Side of the Universe for their great hospitality and a stimulating meeting in Leon.

References

[1] E. Ma, Phys. Rev. **D36**, 274 (1987).
[2] E. Ma, Phys. Rev. **D62**, 093022 (2000).
[3] S. Khalil, H.-S. Lee, and E. Ma, Phys. Rev. **D79**, 041701(R) (2009).
[4] S. Khalil, H.-S. Lee, and E. Ma, Phys. Rev. **D81**, 051702(R) (2010).
[5] Z. Ahmed *et al.*, Science **327**, 1619 (2010).
[6] S. Godfrey and T. A. W. Martin, Phys. Rev. Lett. **101**, 151803 (2008).
[7] M. Yamanaka, These Proceedings.

Journal of Physics: Conference Series **315** (2011) 012007 doi:10.1088/1742-6596/315/1/012007

Toroidal Dipole Moment of the Lightest Neutralino in the MSSM

L.G. Cabral-Rosetti

Depto. de Posgrado, Centro Interdisciplinario de Investigación y Docencia en Educación Tecnológica (CIIDET), Av. Universidad 282 Pte., Col. Centro, A. Postal 752, C.P. 76000, Santiago de Querétaro, Qro., México

M. Mondragón

Instituto de Física, Universidad Nacional Autónoma de México (IF-UNAM), A. Postal 20-364, 04510 México D.F., México

E. Reyes Pérez

Instituto de Física, Universidad Nacional Autónoma de México (IF-UNAM), A. Postal 20-364, 04510 México D.F., México

Abstract. In order to characterize one of the most favored candidates for dark matter, we calculate the anapole form factor of the lightest neutralino in the Minimal Supersymmetric Standard Model (MSSM) at the one-loop level. As a Majorana fermion, this particle only shows one electromagnetic property, the toroidal dipole moment, which is directly related to the anapole form factor. We obtain the result analitically in terms of two- and three-points Passarino-Veltman scalar functions and evaluate it for a given spectrum of supersymmetric masses and matrix elements. This work is part of a broader project still in progress.

1. Introduction

The most favored candidates for cold dark matter (non relativistic) are the so called WIMPs (Weakly Interactive Massive Particle), with masses between 10 GeV and a few TeV, and with interaction cross sections around the electroweak scale ($\sim$ 100 GeV). Currently, the best motivated candidate for WIMP is the lightest supersymmetric particle (LSP) of the supersymmetric models. However, candidates like axions, sneutrinos or gravitinos have not been discarded yet.

Goldberg [1] and Ellis et al. [2] were the first to suggest that the neutral gauginos and higgsinos, with the same quantum numbers, can mix. This superposition is called neutralino, $\tilde{\chi}_l^0$; there are four of them and one is the LSP in some models. Any WIMP candidate to dark matter must be an electrically neutral particle without strong interactions. The LSP is stable in supersymmetric models with R parity, which is a condition of the Minimal Supersymmetric Standard Model (MSSM), where the lightest neutralino is the LSP.

If the Universe is vastly populated by neutralinos, their detection must be possible, directly (via inelastic collisions with nucleons) or indirectly (observing the fermion-antifermion pairs

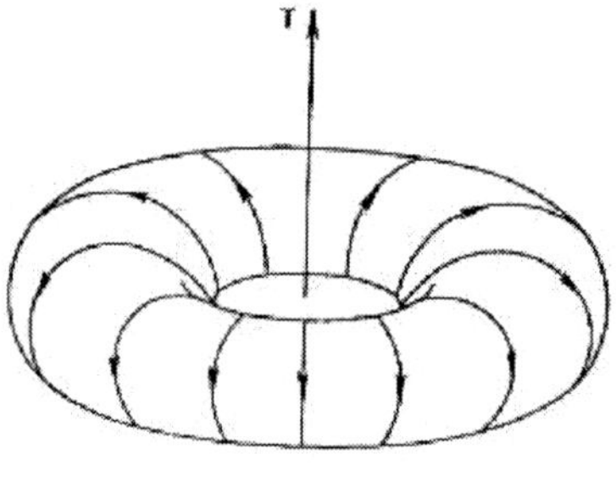

Figure 1. Configuration of a current having a toroidal dipole moment. The arrows on the torus indicate the current direction, and the TDM is directed along the symmetry axis of the torus.

when neutralinos are annihilated). The expected cross sections for such interactions are determined by the allowed regions in the paramether space of the SUSY model used.

2. Toroidal Dipole Moment

It is well known that the electromagnetic properties of a j-spin particle are described by $2(2j+1)$ form factors. For spin 1/2-particles, several electromagnetic current parametrizations have therefore four independent form factors. Thus, the most general expression for an electromagnetic vertex function is:

$$\Gamma_\mu(q) = f_Q(q^2)\gamma_\mu + f_\mu(q^2) i\sigma_{\mu\nu}q^\nu - f_E(q^2)\sigma_{\mu\nu}q^\nu\gamma_5 + f_A(q^2)(q^2\gamma_\mu - \not{q}q_\mu)\gamma_5, \qquad (1)$$

where $f_Q(q^2)$, $f_M(q^2)$, $f_E(q^2)$ y $f_A(q^2)$ are called the charge, magnetic dipole, electric dipole and anapole dipole form factors, respectively; $q_\mu = p'_\mu - p_\mu$ is the transferred 4-momentum; and $\sigma_{\mu\nu} = (i/2)\,[\gamma_\mu, \gamma_\nu]$. These form factors are physically observable quantities as $q^2 \to 0$, and their combinations define the well-known magnetic(μ), electric (d) and anapole (a) dipole moments [3].

The anapole moment was introduced by Zel'dovich [4] for a T-invariant interaction which does not conserve P-parity and C-parity individually. Subsequently, a more convenient characteristic was pointed out to describe this kind of interaction, the toroidal dipole moment (TDM). The TDM (τ) is a general case of the anapole moment, it coincides with an anapole on the mass-shell of the particle under consideration and has a simple classical analogue. The TDM, as the electric dipole and magnetic dipole moments, is the first term of the third multipole family [3], the toroidal moments. This type of static multipole momenta does not produce any external electromagnetic fields in vacuum but generates a free-field (gauge-invariant) potential which is responsible for topological effects such as the Aharonov-Bohm effect.

The simplest model of TDM (anapole) was given by Zel'dovich as a conventional solenoid folded into a torus and having only a poloidal current (see figure 1). For such a stationary solenoid, having neither an azimuthal (toroidal) component of the current nor electric fields around the torus, there is only a nonzero azimuthal magnetic field inside the torus [5].

Besides not having electrical charge, the neutralino does not have electric and magnetic dipole moments because that would violate CPT conservation (remember it is a Majorana fermion). Thus, the neutralino electromagnetic vertex function is

$$\Gamma_\mu(q) = f_A(q^2)(q^2\gamma_\mu - \not{q}q_\mu)\gamma_5. \qquad (2)$$

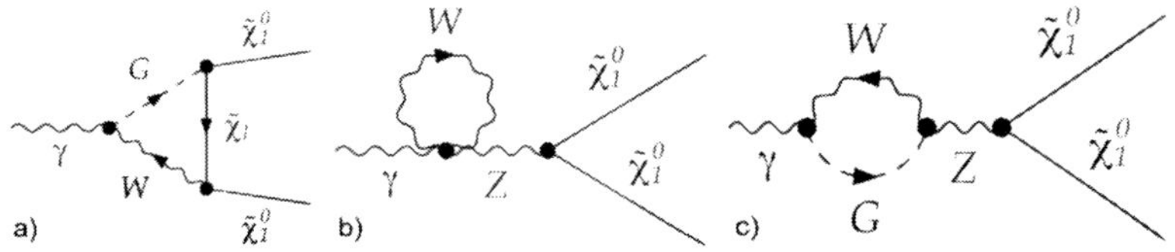

Figure 2. Examples of diagrams that contribute to the process $\gamma \longrightarrow \tilde{\chi}_1^0\tilde{\chi}_1^0$ at the one-loop level. a) is a vertex correction, b) is a sea-gull type diagram, and c) is a self-energy.

In this work, we calculate the anapolar form factor and, therefore, the TDM of the neutralino in the framework of the MSSM at the one-loop level aproximation. Using the *FeynArts* [6] software we obtained 94 Feynman diagrams which contribute to this calculation, 28 vertex corrections, 6 sea-gulls and 60 self-energies.

3. Results

Using *FeynCalc* we calculate the amplitude of each of the 94 diagrams for the process $\gamma \longrightarrow \tilde{\chi}_1^0\tilde{\chi}_1^0$ at the one-loop level. In figure 2 we show from left to right an example of one of the 28 vertex correction diagrams, one of the 6 sea-gulls and one of the 60 self-energies. From these amplitudes we isolate the terms that go as $\gamma_\mu\gamma_5$. The first result we get is that the $\gamma - H^0$, $\gamma - h^0$, $\gamma - A^0$ and $\gamma - G^0$ self-energies do not contribute to the calculation.

For diagrams in figure 2, for example, we get

$$\Xi_{24} = \sum_l \frac{e^3}{32\pi^2 c_W s_W^2 \left(q^2 - 4M_{\tilde{\chi}_1^0}^2\right)} \left[\alpha B_0\left(q^2, M_W^2, M_W^2\right)\right. \tag{3}$$

$$\left. +\beta B_0\left(M_{\tilde{\chi}_1^0}^2, M_{\tilde{\chi}_l}^2, M_W^2\right) + \delta C_0\left(q^2, M_{\tilde{\chi}_1^0}^2, M_{\tilde{\chi}_1^0}^2, M_W^2, M_W^2, M_{\tilde{\chi}_l}^2\right)\right]$$

$$\Xi_{34} = \frac{-M_W^2 e^3 \epsilon}{16\pi^2 s_W^2 \left(q^2 - M_Z^2\right)} \left[3B_0\left(0, M_W^2, M_W^2\right) + 1\right] \tag{4}$$

$$\Xi_{94} = \frac{e^3\zeta}{\pi^2 s_W^2 \left(q^2 - M_Z^2\right)} \left[-\frac{M_W^2}{192} B_0\left(0, M_W^2, M_W^2\right) + \frac{q^2 - 4M_W^2}{384} B_0\left(q^2, M_W^2, M_W^2\right) + \frac{q^2 - 9M_W^2}{576}\right]. \tag{5}$$

respectively, where α, β, δ, ϵ and ζ are coefficients that depend on the neutralino mixing matrix elements; B_0 and C_0 are two- and three-points Passarino-Veltman scalar functions; and

$$\sum_i \Xi_i = f_A(q^2)q^2 \qquad i = 1, 2, ..., 94. \tag{6}$$

To get $\tau = f_A(0)$ we use l'Hopital's rule. Derivatives of the C_0 functions appear in the calculation. That is why we developed an expansion in power series of q^2, around $q^2 = 0$, for the generic C_0 that appears in every vertex correction, in terms of logarithms of the masses involved. This will be useful for the evaluation of the expressions above. In figure 3, we show the plot of this expansion for $C_0\left(q^2, M_{\tilde{\chi}_1^0}^2, M_{\tilde{\chi}_1^0}^2, M_W^2, M_W^2, M_{\tilde{\chi}_l}^2\right)$.

Using the supersymmetric spectrum given by *SuSpect* [7] for the mSUGRA model and the parameter values $m_0 = 100\text{GeV}$, $m_{1/2} = 250\text{GeV}$, $\tan\beta = 10$, sign$\mu = +$ and $A_0 = -100$ we get the following value for the toroidal dipole moment of the lightest neutralino:

$$\tau = f_A(0) = -1.03899 \times 10^{-5}\text{GeV}^{-2} = -4.05798 \times 10^{-33}\text{cm}^2. \tag{7}$$

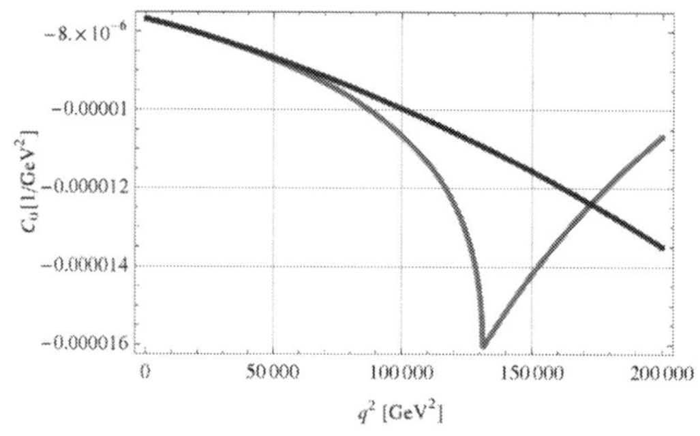

Figure 3. Comparison between exact and approximate solutions of $C_0\left(q^2, M^2_{\tilde{\chi}^0_1}, M^2_{\tilde{\chi}^0_1}, M^2_W, M^2_W, M^2_{\tilde{\chi}_l}\right)$, using the masses we wet using *SuSpect* for the mSUGRA model and the parameter values given below. The monotonic decreasing line represents the approximate solution valid only when $q^2 \to 0$, while the other represents the exact solution valid for all q^2.

4. Connection to Cosmology

In here we assume the neutralino is the main component of dark matter. This work is part of a broader analysis which intends to connect our calculation to cosmology and astrophysics.

For a given cosmological model, the WIMP relic density is determined uniquely by its annihilation cross section into SM particles, if all WIMPs were produced thermally. Under standard cosmology conditions the uncertainty of the current WIMP relic density is essentially given by the uncertainty of the WIMP annihilation cross section [8]. In order to calculate this cross section to percent level accuracy, at least leading radiative corrections will have to be included.

Thus, we can write the 1-loop corrected WIMP annihilation cross section as

$$\sigma = \sigma_0 + \delta\sigma, \tag{8}$$

where σ_0 is the cross section at the tree level. Part of the one-loop level correction to the neutralino cross section, and, therefore, to its relic density, would be obtained using our results.

We also have the hamiltonian of interaction due to the anapole moment in the non-relativistic limit:

$$\mathcal{H}_{int} \propto -a\,(\sigma \cdot \nabla \times B) \tag{9}$$

where a stands for the anapole dipole moment, σ is the spin and B is the strength of the magnetic field [5]. Although the value of this correction is very small, it could lead to an observable effect in regions with high density of dark matter.

Finally, there are experimental bounds we can compare our result with. By using a phenomenological approach to WIMPs, Pospelov and ter Veldhuis[9] got an upper bound for the anapole form factor of $\left|\frac{a}{\mu_n}\right|\sqrt{\frac{(S+1)}{3S}} \leq 4\times 10^{-2}$ fm using DAMA experiment data and lower than 3×10^{-2} fm using CDMS experiment data. New data from CDMS experiment can provide more restrictive bounds. In particular for mSUGRA (minimal Supergravity) model we can get the TDM of the neutralino in terms of five parameters: m_0, the unified mass for scalars, $m_{1/2}$, the unified mass for gauginos, A_0, the unified trilinear coupling, $\tan\beta$, the ratio of the vacuum expectation values for the neutral supersymmetric Higgs and the sign of μ, where μ is the higgsino mass parameter. By scanning over this set of parameters and comparing this theoretical result with the experimental bounds, we can constrain the allowed regions in the parameter space.

References

[1] H. Goldberg, 1983, *Phys. Rev. Lett.*, 50:1419.
[2] J.R. Ellis et al., 1984, *Nucl. Phys. B*, 238:453.
[3] V.M. Dubovik and V.E. Kuznetsov, 1998, *Int. J. Mod. Phys. A* 13: 5257-5278.
[4] Ya. B. Zel'dovich, 1958, *Sov. Phys. JETP*, 6, 118.
[5] V.M. Dubovik and A.A. Cheshkov, 1974, *Phys. Part. Nucl.*, 5, 791.
[6] T. Hahn, 2001, *Comput.Phys.Commun.* 140, 418-431.
[7] A. Djouadi et al., 2007, *Comput.Phys.Commun.* 176, 426-455.
[8] M. Dress et al., *arXiv:0911.3795v1*, 2009.
[9] M. Pospelov and T. ter Veldhuis, 2000, *Phys. Lett.* B480, 181-186.

Journal of Physics: Conference Series **315** (2011) 012008 doi:10.1088/1742-6596/315/1/012008

Electroweak scale neutrinos and decaying Majorons

Alfredo Aranda[1], Francisco J. de Anda[2] and Mónica F. Ramírez[2]

[1] Facultad de Ciencias - CUICBAS, Universidad de Colima, México,
Dual C-P Institute of High Energy Physics, México

[2] Departamento de Física - CUCEI, Universidad de Guadalajara, México

E-mail: `fefo@ucol.mx`

E-mail: `panchjosedea@hotmail.com`

Abstract. We study the Higgs phenomenology of a model of electroweak scale neutrinos that incorporates the presence of a lepton number violating singlet scalar. In order to verify its viability, we perform an analysis of the pseudoscalar-Majoron field. In particular we study the Majoron decay $J \to \nu\nu$ and use the bounds on the Majoron mass and width obtained in a modified Majoron Decaying Dark Matter scenario.

We would like to be able to explore neutrino masses at Colliders. One way this could be achieved is if right-handed neutrinos exist that have a mass in the electroweak scale range. Then it is possible to generate light masses with the use of the seesaw mechanism [1, 2, 3, 4, 5, 6, 7, 8, 9, 10, 11]. In this presentation we review the results in [12] where a model has been presented that incorporates a minimum set of additions to the Standard Model (SM), namely the addition of three right-handed neutrinos and a lepton number violating singlet scalar. The model contains a pseudoscalar-Majoron field that can be a candidate for decaying dark matter. An analysis regarding this possibility is presented in order to verify the viability of the model.

1. The Model

The model adds to the SM particle content the following: three right-handed neutrinos (N_{Ri}, $i = 1, 2, 3$) with a mass of electroweak (EW) scale size, and a SM singlet complex scalar field (η) that breaks global $U(1)_L$ symmetry. The field η has lepton number -2. The relevant terms in the Lagrangian are given by

$$\begin{aligned} \mathcal{L}_{\nu H} &= \mathcal{L}_{\nu y} - V \,, & (1)\\ \mathcal{L}_{\nu y} &= -y_{\alpha i}\overline{L}_\alpha \tilde{\Phi} N_{Ri} - \frac{1}{2} Z_{ij}\eta \overline{N}^c_{Ri} N_{Rj} + h.c. \,, & (2)\\ V &= \mu_D^2 \Phi^\dagger\Phi + \frac{\lambda}{2}(\Phi^\dagger\Phi)^2 + \mu_S^2 \eta^*\eta + \lambda'(\eta^*\eta)^2 \\ &+ \kappa(\eta\Phi^\dagger\Phi + \eta^*\Phi^\dagger\Phi) + \lambda_m(\Phi^\dagger\Phi)(\eta^*\eta) \,, & (3) \end{aligned}$$

where $\tilde{\Phi} = i\sigma^2\Phi^*$, $(\Psi_R)^c = P_L\Psi^c$, $\Psi^c = C\gamma^0\Psi^*$, and where

$$\Phi = \begin{pmatrix} 0 \\ \frac{\phi^0 + v}{\sqrt{2}} \end{pmatrix}, \quad \eta = \frac{\rho + u + iJ}{\sqrt{2}} \,, \qquad (4)$$

where $v/\sqrt{2}$ and $u/\sqrt{2}$ are the vacuum expectation values (vevs) of Φ and η respectively ($v(u)$ denoting vev of $\Phi(\eta)$)

And we define the mass eigenstates:

$$\begin{pmatrix} \phi^0 \\ \rho \end{pmatrix} = \begin{pmatrix} \cos\alpha & -\sin\alpha \\ \sin\alpha & \cos\alpha \end{pmatrix} \begin{pmatrix} h \\ H \end{pmatrix} \tag{5}$$

$$\begin{pmatrix} L_i \\ N_i \end{pmatrix} = \begin{pmatrix} \cos\theta & \sin\theta \\ -\sin\theta & \cos\theta \end{pmatrix} \begin{pmatrix} \nu_{Li} \\ \nu_{Ri} \end{pmatrix} \tag{6}$$

The model assumes that there is only one high energy scale, the EW scale and both u and v are of this order. We assume too that the explicit $\mathrm{U}(1)_L$ breaking is very small so $|\kappa| \ll$ EW scale.

The model has eight free parameters: $y, Z, u, v, \kappa, \lambda, \lambda', \lambda_m$, ($\mu_D$ and μ_S are fixed by minimization conditions). First we simplify the neutrino terms and neglect inter-family interactions. We make the assumption that the coupling constants of each family are of the same order, that is: $Z_{ij} = y_{ij} = 0$ for $i \neq j$ and $Z = Z_{11} \approx Z_{22} \approx Z_{33}$, $y = y_{11} \approx y_{22} \approx y_{33}$. We take $m_L \sim O(eV)$ and we obtain $\theta \approx 10^{-6}$ which gives us EW scale masses for right-handed neutrinos [5].

2. Analysis

Now we assume that dark matter is composed of Majorons J. With the modified ΛCDM which has the decaying dark matter (DDM) idea [13], we fix the Majoron mass and its decay rate [14]:

$$\Gamma_\sigma < 1.3 \times 10^{-19} s^{-1} \tag{7}$$

$$0.12keV < m_\sigma < 0.17keV \tag{8}$$

From the model we obtain the decay rate and mass of the Majoron

$$\Gamma_J = \frac{Z^2 \sin^4\theta}{4\pi M_J} \left(1 - \frac{4m_\nu^2}{M_J^2}\right)^{1/2} m_\nu^2, \tag{9}$$

$$M_J^2 = \frac{-\kappa v^2}{\sqrt{2}u}, \tag{10}$$

Saturating the bounds on the width and mass of the Majoron requires Z to be of $\mathrm{O}(10^{-2})$. Together with assuming $u \sim \mathrm{O(EW)}$, which gives us nicely the masses of right-handed neutrinos in the EW scale, and $|\kappa|/(GeV) \sim 10^{-9} \ll 1$. Now we explore the phenomenology of the Higgs in this model. All SM processes get suppressed by a factor $\cos(\alpha)^2$ [5], thus we look for situations where $\cos(\alpha) \sim 1$. The parameters used in this case are $\lambda = 1.5$, $\lambda' = 0.09$, and $\lambda_m = 0.002$ which represent a typical set under these requirements.

The Higgs mass is then given as a function of u, and for the required values of α we obtain plausible values (see figure 1).

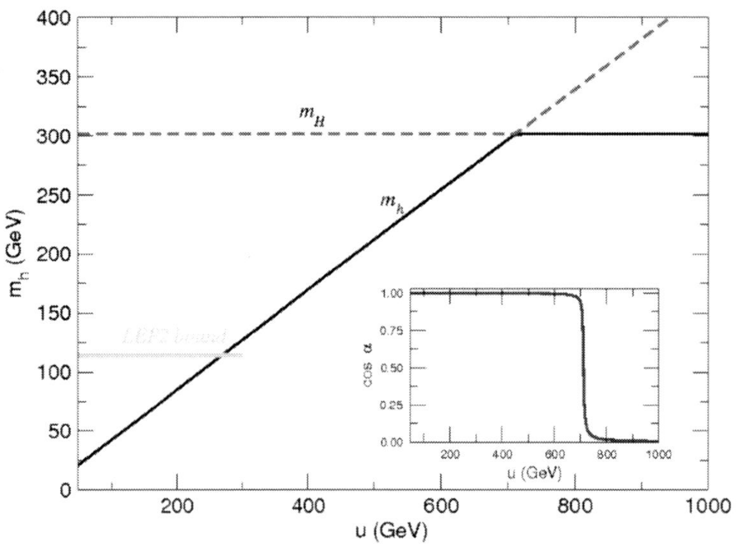

Figure 1. Scalar spectrum for the case described in the text. The horizontal line in the lower left corresponds to the 114 GeV lower bound from LEP2 [15].

The Higgs branching ratios are shown in figure 2 for a characteristic set of parameters (see [12]). We generally observe that the invisible decay to Majorons dominates over most of the parameter space.

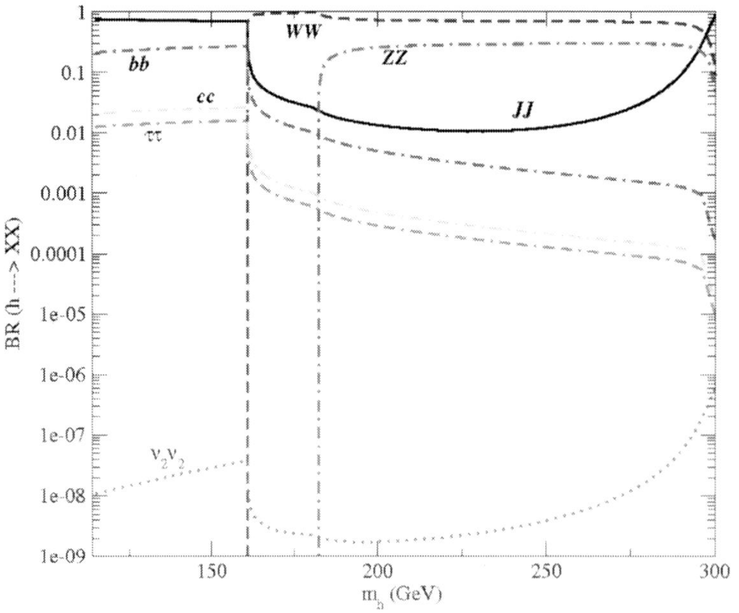

Figure 2. Branching ratios for the decay $h \to XX$ for the set of parameters shown in Figure 1.

3. Conclusions

We have shown that the model meets all the desired requirements for it to be plausible and well motivated. The parameter space was restricted by imposing Higgs mass constraints as well as information from the cosmological microwave observations of WMAP. Furthermore we have shown that its phenomenology is dominated by the invisible Higgs decay to Majorons and that it is within the reach of present colliders

Acknowledgements

This work was supported by CONACYT and SNI.

References

[1] M. L. Graesser, arXiv:0705.2190 [hep-ph].
[2] P. Q. Hung, Phys. Lett. B **649**, 275 (2007) [arXiv:hep-ph/0612004].
[3] J. Kersten and A. Y. Smirnov, Phys. Rev. D **76**, 073005 (2007) [arXiv:0705.3221 [hep-ph]].
[4] A. de Gouvea, arXiv:0706.1732 [hep-ph].
[5] A. Aranda, O. Blanno and J. Lorenzo Diaz-Cruz, Phys. Lett. B **660**, 62 (2008) [arXiv:0707.3662 [hep-ph]].
[6] J. P. Bu, Y. Liao and J. Y. Liu, Phys. Lett. B **665**, 39 (2008) [arXiv:0802.3241 [hep-ph]].
[7] P. Q. Hung, Nucl. Phys. B **805**, 326 (2008) [arXiv:0805.3486 [hep-ph]].
[8] F. del Aguila and J. A. Aguilar-Saavedra, Phys. Lett. B **672**, 158 (2009) [arXiv:0809.2096 [hep-ph]].
[9] A. Aranda, J. Hernandez Sanchez and P. Q. Hung, JHEP **0811**, 092 (2008) [arXiv:0809.2791 [hep-ph]].
[10] A. Aranda, AIP Conf. Proc. **1116**, 227 (2009) [arXiv:0812.2958 [hep-ph]].
[11] P. F. Perez, T. Han and T. Li, arXiv:0907.4186 [hep-ph].
[12] Alfredo Aranda and Francisco J. de Anda. *Phys. Lett.*, B683:183–185, 2010. arXiv:0909.2667 [hep-ph].
[13] V. Berezinsky and J. W. F. Valle, Phys. Lett. B **318**, 360 (1993) [arXiv:hep-ph/9309214].
[14] M. Lattanzi and J. W. F. Valle, Phys. Rev. Lett. **99**, 121301 (2007) [arXiv:0705.2406 [astro-ph]].
[15] R. Barate *et al.* [LEP Working Group for Higgs boson searches and ALEPH Collaboration and and], Phys. Lett. B **565**, 61 (2003) [arXiv:hep-ex/0306033].

Unification of Dark fields and Inflation.

J. D. García-Aguilar[1], **A. Pérez-Lorenzana**[2]

1, 2 Departamento de Física, Centro de Investigación y de Estudios Avanzados del I.P.N., Apdo. Post. 14-740 07000 México D.F., México.

E-mail: [1]jdgarcia@fis.cinvestav.mx, [2]aplorenz@fis.cinvestav.mx

Abstract. We present a partial unified model for inflation, dark matter and dark energy, based on a global $SU(2) \times U(1)$ symmetry, where the corresponding cosmological scalar fields that source such components of the Universe appear as different degrees of freedom of fundamental representations. The protecting symmetry allows to build successful potentials to describe both the early and late eras of accelerated expansion of the Universe, and admit for the interpretation of Dark energy as quintessence or phantom matter. The model also provides a limit where the fields have a dual behavior.

1. Introduction

Measurements of the microwave cosmic background spectrum by the Cosmic Background Explorer (COBE) and most recently by the Wilkinson Microwave Anisotropy Probe (WMAP), and the observations of distant type Ia supernovae and galaxy cluster measurements, point towards the unavoidable necessity of dark components of the Universe. Both in the form of dark matter (DM) and dark energy (DE).

Apart from the cosmological constant (Λ), one of the favorite candidates for DE are scalar fields [1], for which acceleration from the Friedmann equation is easy to achieve by choosing an appropriate potential energy and tuning model parameters. For all such models the equation of state $P = \omega\rho$ leads to $\omega > -1$.

A simple way to realize this scenario is to introduce a scalar field φ minimally coupled to gravity, called quintessence, which gives pressure and energy density

$$P = \frac{1}{2}\dot{\varphi}^2 - V(\varphi), \tag{1}$$

$$\rho = \frac{1}{2}\dot{\varphi}^2 + V(\varphi), \tag{2}$$

such that ω becomes a dynamical parameter which in general varies through time and space. In contrast, a cosmological constant is static, with a fixed energy density, $\rho = -\Lambda$, and pressure, $P = \Lambda$, such that $\omega = -1$. Scalar fields had also been proposed as being dark matter. We postulate that all cosmology scalar fields could be interrelated by symmetries [2], and here we present a model that exemplifies the realization of such an idea.

In this context we start by considering a model with four real scalar fields, which latter on we will identify as sources for hybrid inflation, dark matter and dark energy. Thus, in order to clearly identify the natural symmetries of the system, we look upon the Lagrangian density without potential term

$$\mathcal{L} = \frac{1}{2}\partial^\mu \varphi_i \partial_\mu \varphi_i \qquad i = 1, \ldots, 4.$$

We notice that this Lagrangian density can be expressed as

$$\mathcal{L} = \frac{1}{2}\partial_\mu \Phi^T \cdot \partial^\mu \Phi = \frac{1}{2}\delta_{ij}\partial_\mu \Phi^i \partial^\mu \Phi^j,$$

where we have written the field components as

$$\Phi = \begin{pmatrix} \varphi_1 \\ \varphi_2 \\ \varphi_3 \\ \varphi_4 \end{pmatrix}, \tag{3}$$

and the background metric contraction is implicit.

As it is easy to see from this expression, the symmetry of the Lagrangian corresponds to the group of symmetry transformations on field space that preserve the Euclidean metric. That is the symmetry group $O(4)$, consisting of all four by four orthogonal real matrices.

Unfortunately given any two arbitrary quadruplet representations, Φ and Ψ, there is only one invariant bilinear form under $O(4)$ transformations, $\Phi^T \mathrm{I} \Psi = \delta^{ij}\Phi_i \Psi_j$. This indicates, that if one chooses $O\,(4)$ as a fundamental symmetry, the only possible potential one can admit for the system is a general function of $\Phi^T\Phi = \delta_{ij}\phi_i\phi_j$. That hardly builds any cosmological scenario. However the reduced subgroups $SO(4)$ and $SU(4)$ have extra invariants. Additional invariants are found for smaller groups. We shall only use the other associated invariants of a particular subgroup to build the potential of our unified theory. This way we assume that only part of the total symmetry of the kinetic term is actually fundamental for the theory.

2. Evolution equations of Universe

The dynamics of the universe whit the Robertson-Walker metric is described by the Einstein's equations. Let us consider an ideal perfect fluid as the source of the energy momentum tensor $T^\mu_\nu = Diag\,(-\rho, p, p, p)$. The Einstein's equations are [1]

$$\begin{aligned} H^2 &= \left(\frac{\dot{a}}{a}\right)^2 = \frac{1}{3M_{Pl}^2}\rho - \frac{K}{a^2}, \\ \dot{H} &= -\frac{1}{2M_{Pl}^2}(\rho + P) + \frac{K}{a^2}, \end{aligned}$$

where H is the Hubble parameter, ρ and P denote the total energy density and pressure of all the species present in the universe at a given epoch.

3. A simple model

As a working example, let is consider a complex scalar field system, where the kinetic term becomes

$$T = \frac{1}{2}\partial_\mu \Phi^* \partial^\mu \Phi,$$

and thus the largest symmetry becomes $O\,(4, \mathbb{C})$.

Next, we consider the subgroup $G = U(1) \times SU(2) \subset SU(4)$, whose elements can be represented as

$$G = \begin{pmatrix} e^{i\theta/2} & 0 & 0 & 0 \\ 0 & e^{-i\theta/2} & 0 & 0 \\ 0 & 0 & \cos(\theta'/2) & i\sin(\theta'/2) \\ 0 & 0 & i\sin(\theta'/2) & \cos(\theta'/2) \end{pmatrix},$$

for which the four scalars correspond to the representations $\phi_1\,(1,1)$; $\phi_2\,(1,-1)$ and $(\phi_3,\phi_4)\,(2,0)$.

So the G invariants given in terms of component fields, are

$$\phi_1^2, \qquad \phi_2^2, \qquad \delta_{ij}\phi_i^*\phi_j = |\phi_3|^2 + |\phi_4|^2, \qquad (\sigma_1)_{ij}\,\phi_i^*\phi_j = \phi_3^*\phi_4 + \phi_4^*\phi_3,$$

out of which we may build the simple potential

$$\begin{aligned} U &= \frac{1}{4\lambda}\left(M_{\phi_2}^2 - \lambda\phi_2^2\right) + \frac{m^2}{2}\phi_1^2 + \frac{g^2}{2}\phi_1^2\phi_2^2 \\ &\quad + M^2\left(|\phi_3|^2 + |\phi_4|^2\right) + \mu^2\left(\phi_3^*\phi_4 + \phi_4^*\phi_3\right). \end{aligned} \tag{4}$$

Thus we get two decoupled subsystems ($V_1(\phi_1,\phi_2)$ and $V_2(\phi_3,\phi_4)$).

4. Cosmology of the two subsystems

The potential $V_1(\phi_1,\phi_2)$ corresponds to the simple model due to Linde [3] for hybrid inflation. When inflation ends, the dynamics of potential $V_2(\phi_3,\phi_4)$ begins.

Now, the potential $V_2(\phi_3,\phi_4)$ can be rewritten in the form

$$V_2 = \begin{pmatrix} \phi_3^*, & \phi_4^* \end{pmatrix} \begin{pmatrix} M^2 & \mu^2 \\ \mu^2 & M^2 \end{pmatrix} \begin{pmatrix} \phi_3 \\ \phi_4 \end{pmatrix} \tag{5}$$

$$= \Phi^\dagger \mathbf{M^2} \Phi, \tag{6}$$

where $\mathbf{M^2}$ is the mass matrix of the potential and $\Phi = \begin{pmatrix} \phi_3, & \phi_4 \end{pmatrix}$.

Diagonalizing $\mathbf{M^2}$ we find the squared masses of field φ_+ and φ_- mass eigenstates, which are given by $M^2 \pm \mu^2$. If $\mu^2 = M^2 - \delta M^2$, then the potential in the eigenmode system is expressed as

$$V_2(\varphi_+,\varphi_-) = 2M^2|\varphi_+|^2 + \delta M^2|\varphi_-|^2.$$

This potential is steeper along φ_+, with an almost flat direction along φ_-, when $\delta M^2 \ll M^2$ (see Fig. 1).

If initial conditions are such that φ_3 and φ_4 are about M, then φ_+ shall roll down the potential towards the local minimum at $V_2(0,\varphi_-)$ then, dynamics of φ_+ ends and the dynamics of φ_- begins, which is a quintessential inflation stage, for which we first need to take δM^2 of the order of the observed cosmological constant. We can take φ_- as the quintessence field and φ_+ as the field of Dark Matter due to it is high mass which makes it a superheavy particle (a candidate of dark matter [4]). Due to small value of φ_+ mass (in comparison of M_{Pl}), Dark Matter and Dark Energy fields don't contribute to the value of Hubble's constant at the early Universe.

5. Summary

In this work we explore the unification of cosmological fields in the same way as for particle physics. We show, with an specific example, that it is possible to build adequate cosmological potentials that are invariant under a global symmetry, which may explain the currently dynamics of the Universe.

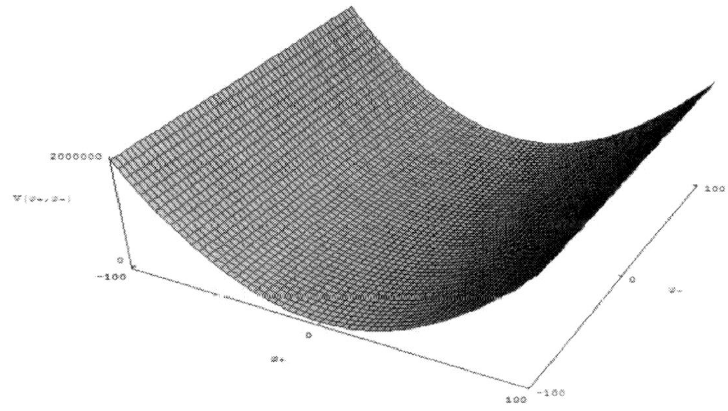

Figure 1. Potential $V_2(\varphi_+, \varphi_-)$ with $\delta M^2 \ll M^2$.

Acknowledgments

Work supported in part by Conacyt, México, under grant number 54576.

References

[1] E. J. Copeland, M. Sami, S. Tsujikawa, arXiv:hep-th/0603057v3 (2006)
[2] A. Peréz-Lorenzana, M. Montesinos and T. Matos, *Phys. Rev. D* **77**, 063507 (2008); J. D. García Aguilar, *MSc.Thesis*, CINVESTAV (2009).
[3] Andrei Linde, *Phys. Lett B* **259** 38 (1991)
[4] G. Bertone, D. Hooper and J. Silk arXiv:hep-ph/0404175v2 (2004)

VI International Workshop on the Dark side of the Universe (DSU 2010) IOP Publishing
Journal of Physics: Conference Series **315** (2011) 012010 doi:10.1088/1742-6596/315/1/012010

Direct Detection of Dark Matter with MiniCLEAN

R. Hennings-Yeomans, for the DEAP/CLEAN collaboration
Physics Division, Los Alamos National Laboratory, Los Alamos, New Mexico 87545, USA
E-mail: hennings@lanl.gov

Abstract. Overwhelming astrophysical evidence indicates that non-baryonic Dark Matter constitutes most of the mass of the Universe. Nevertheless, the particle nature of Dark Matter remains a long standing mystery. The use of noble liquids as scintillators in single and dual-phase detectors are some of the most promising scalable WIMP detectors currently planned and under construction. The MiniCLEAN experiment will have 92 photomultiplier tubes (PMTs) looking at a liquid Argon detector mass of over 500 kg in a single-phase configuration. It will use Pulse Shape Discrimination (PSD) techniques to search for low-energy WIMP nuclear recoils inside a fiducial volume. Liquid Argon would be interchangeable with liquid Neon to study A^2 dependence of a potential signal and examine backgrounds external to the cryogenic liquid. For the Argon run, MiniCLEAN projects a sensitivity in terms of spin-independent WIMP-nucleon cross-section of $2{\times}10^{-45}\text{cm}^2$ for a mass of 100 GeV/c^2. A status report of MiniCLEAN will be presented as well as plans to deploy the experiment at SNOLAB.

1. Introduction

It has been well established by cosmological observations [1] that the Universe is made of ~85% non-baryonic and non-luminous matter, called dark matter that had non-relativistic speeds at the time of structure formation. The nature of dark matter remains one of the biggest challenges in Astroparticle physics today. Extensions of the Standard Model of particle physics propose a variety of dark matter particle candidates. Among the well motivated candidates are the Weakly Interacting Massive particles (WIMPs) [2, 3, 4, 5]. Using standard assumptions of the WIMP halo that is gravitationally bound to the Milky Way we expect a mean nuclear recoil energy due to WIMP coherent scattering at a scale of tens of keV [6, 7]. The rate is currently limited by observation of less than 0.1 event/kg/day [8, 9] above ~40 GeV/c^2 WIMP mass. Recently, the DAMA/LIBRA collaboration has reported 1.17 ton-years of data from which an annual modulation with the correct phase was observed with a significance of 8.9σ [10]. When the DAMA/LIBRA modulation is confronted with the null observation of other direct detection experiments, light WIMPs have been invoked, if this modulation is indeed a signal due to dark matter [11].

The use of liquid noble elements as scintillators for direct detection of dark matter in single and dual phase configurations are one of the most promising techniques to reach sensitivities beyond 10^{-44}cm^2 in WIMP-nucleon spin-independent cross section. Particularly bright are the efforts by the WARP [12], LUX [13], XENON10/100 [9, 14] and ZEPLIN III [16] collaborations. The status of the MiniCLEAN experiment, a single phase liquid Argon and Neon detector, will be presented.

2. Experimental technique

The use of liquid noble gases as targets in direct detection experiments has bloomed in the last ten years due to their bright scintillation when exposed to radiation (40 photons per keV in Argon [15]), the ion mobilities, their high-Z for self-shielding purposes and the ability to purify the liquid with well established methods such as charcoal traps and getter technologies. The nuclear recoil discrimination is obtained either with timing of the scintillation light alone (single-phase experiments) or from the ratio of the produced scintillation to ionization (dual-phase experiments).

Natural argon being cheap and readily available ($\sim$ \$5 per kg) has ^{39}Ar, a β-emitter produced by cosmic-rays and with an activity of 1 Bq/kg and a half-life of 269 years. Events from the β-decays of ^{39}Ar produce $\sim 10^7$ electronic recoils/keV/year/100 kg of liquid Argon, therefore the need for high-efficiency discrimination method. Neon on the other hand, does not have any long-lived radioactive isotopes and would not require such high-efficiency discrimination methods to search for dark matter. Nevertheless, due to the A^2 enhancement of the spin-independent WIMP-nucleon cross section, Argon has the advantage over Neon on event rate although with a higher threshold.

Dual-phase Argon detectors using natural Argon are limited in size due to pile-up events from ^{39}Ar (500 kg to 1 ton). Recently, a sample of underground Argon has been shown to have less than 0.4 Bq/kg ^{39}Ar [17]. Large quantities of depleted Argon make ton-scale Argon TPCs viable for dark matter searches and extend the sensitivity of single-phase experiments by allowing to lower the threshold.

The MiniCLEAN experiment aims at using a single-phase configuration with over 500 kg of liquid Argon and use pulse shape discrimination methods as well as position reconstruction to look for WIMP scatters inside a fiducial volume. The liquid Argon target would be interchangeable with liquid Neon to examine any potential signal and external backgrounds taking advantage of the A^2 dependance.

When ionizing radiation interacts in any liquid noble gas, two types of dimer molecules are formed: singlet and triplet states. The singlet states quickly decay into two ground state atoms but, due to a forbidden spin flip, the triplet state survives a few microseconds more compared to the singlet in liquid Argon and Neon. The peak wavelength of the emitted photons as the singlet and triplet states decay is 129 nm and 80 nm for Ar and Ne respectively, and therefore the need to use scintillation-grade tetraphenyl-butadiene (TPB) wavelength shifter as means to convert the VUV light to the visible where it can be readily transported through light-guides and detected using PMTs. The fraction of light arriving in the first 90 ns as compared to the total amount of light is the parameter used for Pulse Shape Discrimination (PSD) in both liquid Argon and Neon [18, 19].

Eventually, a large multi-ton Neon detector would be sensitive to *pp* neutrinos as well as dark matter, in which case the main background to dark matter is the *pp* solar spectrum itself [19].

3. Calibration across the detector target

For the same reason that liquid noble gas based detectors use self shielding, calibrations with sources placed outside the detector or point sources inside of it make it difficult to have a uniform energy calibration through out the detector target. Furthermore given the energy scale of nuclear recoils expected from WIMPs and that of the *pp* neutrino flux, an energy line at this scales, preferably at or close to the threshold is needed. A low-energy radioactive source (^{83}Kr^m) that can be diffused through the target volume and at the same time have decay energies at tens of keV was developed and tested in a 4 kg liquid Argon chamber. ^{83}Kr^m was produced in the decay of ^{83}Rb, with a half-life of 86.2 days (see Figure 3 upper left schematic for the decay chain of ^{83}Rb). The half life of ^{83}Kr^m is 1.83$\pm$0.02 hrs through two electromagnetic decays (emission of a conversion electron accompanied by either an x-ray or another electron) with

Journal of Physics: Conference Series **315** (2011) 012010 doi:10.1088/1742-6596/315/1/012010

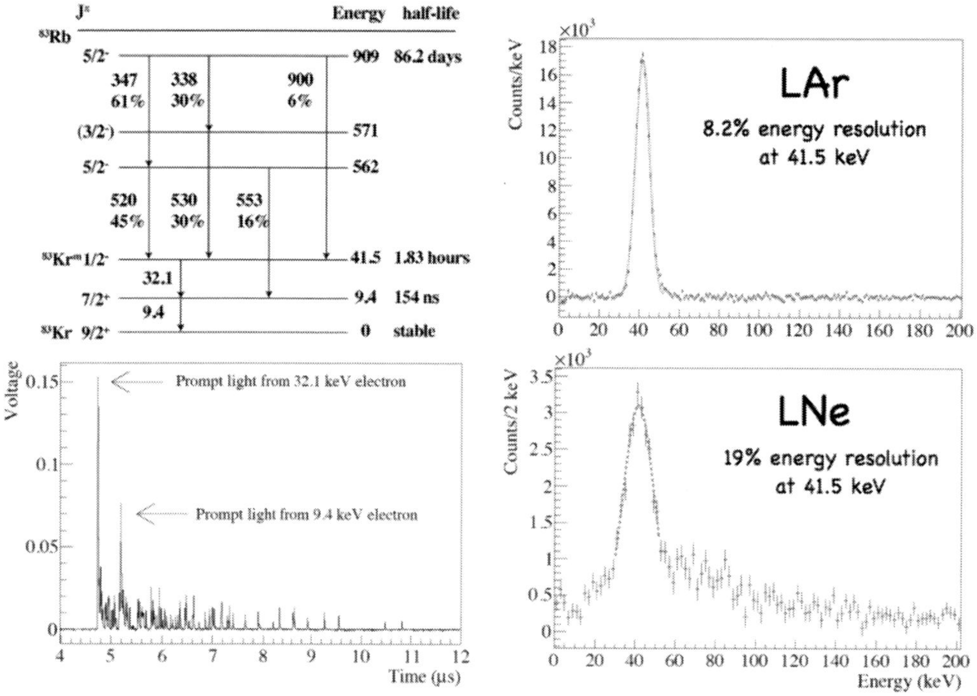

Figure 1. *Upper left:* Decay diagram of ^{83}Rb showing the energy levels in keV. There is a as 75% probability that ^{83}Rb decays into ^{83}Kr^{m} which then decays to the 32.1 and 9.4 keV electromagnetic transitions. *Lower left:* An example of a ^{83}Kr^{m} event that shows the 32.1 and 9.4 keV electromagnetic transitions. *Upper right:* The energy spectrum as observed in Argon for the ^{83}Kr^{m} run in which the sum of the electromagnetic transitions gives precisely a 41.5 keV peak with 8.2 % (σ/E) and 6.0 pe/keV light yield. *Lower right:* The ^{83}Kr^{m} run in Neon. A light yield of 3.0±0.3 pe/keV was observed and a 19% resolution at the 41.5 keV peak [21].

energy 32.1 keV and 9.4 keV. ^{83}Kr^{m} has been used before as a diagnostic tool in the study of beta decay of tritium [20]. On liquid Argon, after performing a background subtraction and fitting the resulting peak to a gaussian, an energy resolution of 8% (σ/E) is measured at 41.5 keV and 19% for the liquid Argon and liquid Neon run respectively. Due to the timing characteristics of Argon, it is not possible to separate the light generated by the 9.4 keV decay to that of the late component of the 32.1 keV transition (see Figure 3) [21].

4. The nuclear recoil scintillation efficiency

The energy threshold of the MiniCLEAN experiment is determined by the PSD in combination with the nuclear recoil scintillation efficiency $\mathcal{L}_{eff}$. A recent measurement of $\mathcal{L}_{eff}$ was made using a D-D generator that produces neutrons that in turn scattered off a 4 kg liquid Argon target inside a cryostat and from here into an organic liquid scintillator that was used as a coincidence trigger. By changing the orientation of the organic scintillator with respect to the trajectory from the neutron source to the liquid Argon target, a sampling of the scattering angles was made and from this an extraction of the nuclear recoil energy in the liquid Argon target nucleus. At every given angle, the electron-equivalent energy as well as the nuclear recoil energy (keV$_r$) was measured. The ratio of nuclear recoil scintillation response to the electronic recoil response was measured to be 0.25±0.02+0.01(correlated) for nuclear recoils between 10 to 250 keV_r [22]. A previous measurement of the nuclear recoil scintillation efficiency in liquid Argon was made by the WARP collaboration of 0.28±10% at 65 keV_r [12].

5. Measurement of LAr pulse shape discrimination

The DEAP-1 experiment is a 7 kg Argon stainless steel chamber with 6-inch diameter glass windows mounted on the sides and viewed by two 5-inch PMTs each coupled to a light guide 8-inch long and operated at room temperature. The purpose of DEAP-1 is to study the PSD in liquid Argon and to understand backgrounds such as Radon. The inner surfaces of the windows as well as an inserted acrylic sleeve coated with TiO_2 are coated with TPB. The outer vacuum chamber is made of PMMA acrylic and has a 12-inch diameter. Further details on the experimental setup can be found in Boulay et al. [23].

The electron equivalent energy calibration of DEAP-1 was made using a 10 μCi encapsulated ^{22}Na source located outside the Argon chamber. The ^{22}Na source emits a 1.2 MeV γ-ray as well as a positron that in turn annihilates in the source container into back-to-back 511 keV γ-rays. In order to reduce accidental backgrounds, a NaI detector in annular shape as well as another NaI detector as one of the caps of the annular detector is used. Therefore, events with a 511 keV γ ray scatters in Argon, with the two additional detectors tagging the other 511 keV and the 1.2 MeV γ-rays with the source being placed on the axis of the annular NaI detector. The measured light yield was 2.8±0.1 photoelectrons/keV_{ee}.

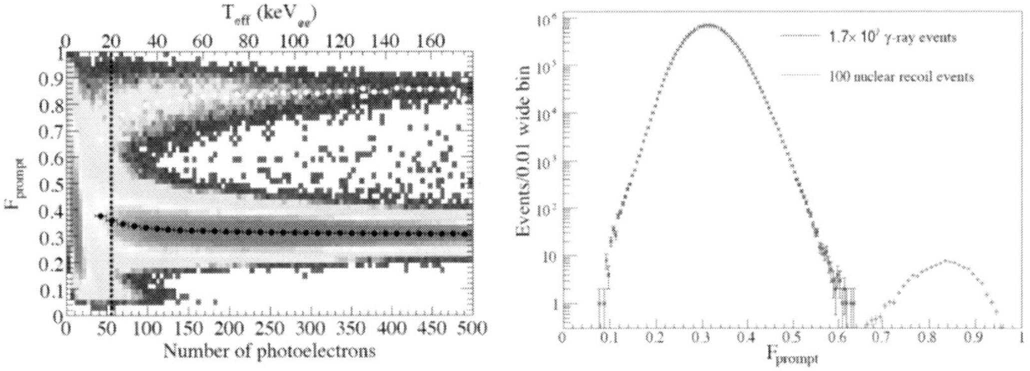

Figure 2. *Left:* F_{prompt} as a function of photoelectros as measured by the DEAP-1 prototype when shinned with an Am-Be source. Events in the upper band (0.7< F_{prompt} <1.0) are mostly nuclear recoils. *Right:* F_{prompt} histogram of 16.7 million electronic recoil events with 120-240 photoelectrons from the ^{22}Na calibration source (blue) and 100 nuclear recoils from the Am-Be neutron source (red) as measured by DEAP-1 [23].

Data taken with an Americium-Beryllium (Am-Be) calibration source is shown on the left plot of Figure 5. Nuclear recoil events show high-F_{prompt} (0.7-1.0) while γ and β-rays appear in the F_{prompt} band below it. Where F_{prompt} is defined as

$$F_{prompt} \equiv \frac{Q^{A}_{prompt} + Q^{B}_{prompt}}{\mathrm{Total\,PE}} \quad (1)$$

where $Q^{A,B}_{prompt}$ is the charge measured 50 ns before the leading edge of the pulse and until 150 ns after the leading edge of PMT A or B. Total PE refers to the total number of photoelectrons.

The right plot on Figure 5 shows the F_{prompt} distribution for γ-rays and nuclear recoils between 120-240 photoelectrons. From this data, Boulay et al. found no misidentification of the 16.7 million γ-rays. Furthermore, with a neutron efficiency of 50% between 70-240 photoelectrons ($\sim$ 25-86 keV_{ee}) the measured misidentification ratio between γ-rays and nuclear recoils was measured to be 1 in 4.7×10^{-8}.

Figure 3. *Upper-left:* Schematic drawing of the MiniCLEAN water tank as well as the outer and inner vessel. *Upper-right:* Cross section drawing of the outer vessel and inner vessel showing the "optical cassettes". *Lower-left:* SNOLAB Cube Hall, house of the DEAP-3600 and MiniCLEAN experiments showing the finished deck. *Lower-right:* Picture of the finished outer vessel.

6. MiniCLEAN engineering

The MiniCLEAN detector will have 92 PMTs (Hamamatsu R5912-02MOD tubes [24]) immersed in liquid Argon and will have over $\sim$500 kg of active target. Each of the PMTs will be assembled in "optical cassettes". An optical cassette is made of a PMT, the light guide and a thin acrylic window. In order to detect the VUV light form Argon and Neon, the light will be converted to the blue spectrum by using a layer of TPB wavelength shifter. The 92 optical cassettes will be cleaned and assembled inside a glove box under radon-free (or mitigated) environment to minimize the contact of the surfaces with radon daughters. After full assembly of the cassettes under vacuum, the PMTs will be submerged in liquid Argon (or Neon) and the liquid will be constantly recirculated in order to be purified. The purification system will be a conversion to the gas phase in order to feed it into a getter and charcoal trap. The optical cassettes are being designed and prototyped at Los Alamos National Laboratory using a custom made cryostat that can hold pressurized liquid Argon to 3 atmospheres. These tests will study the light collection by the light guide, the reflectivity of the aluminum coating and the efficiency of the PMT. This data will be used to benchmark the Monte Carlo simulation as well.

External to the outer vessel, MiniCLEAN will have a water shield ~5.6 m in diameter and ~7.8 m in height. The water tank will be instrumented with PMTs and its primary purpose will be to serve as a muon veto. In Figure 5 it is shown, in the upper-left a drawing of the MiniCLEAN water tank as well as outer and inner vessels. In the upper-right a cross-section of the outer vessel and inner vessels with the optical cassettes is shown. The lower-left and lower-right pictures show the location at SNOLAB of the MiniCLEAN experiment and the finished outer vessel respectively.

7. Projected sensitivity

In order to achieve a sensitivity of the order 10^{-45}cm^2 in terms of spin-independent WIMP-nucleon cross section, MiniCLEAN plans to run for two years in liquid Argon mode. In order to achieve zero-background at this level, MiniCLEAN will mainly have to discriminate the ^{39}Ar β-rays, shield or tag fast neutrons from (α, n) reactions in the PMT glass and reduce backgrounds associated with radon daughters at the level of 1 α/m^2/day on the acrylic.

In Figure 7 it is shown the projected sensitivity reach of MiniCLEAN assuming a zero-background experiment and that of various other competitive experiments using other targets and techniques [21]. The sensitivity of MiniCLEAN assumes a light yield of 7.0 pe/keV and a threshold of 44 keV$_r$. For DEAP-3600 and MiniCLEAN, a time period of 2 years of exposure is assumed. The commissioning of MiniCLEAN is scheduled for spring 2011. The outer vessel has been completed and the inner vessel is being fabricated, MiniCLEAN is expected to be taking data in Fall 2011.

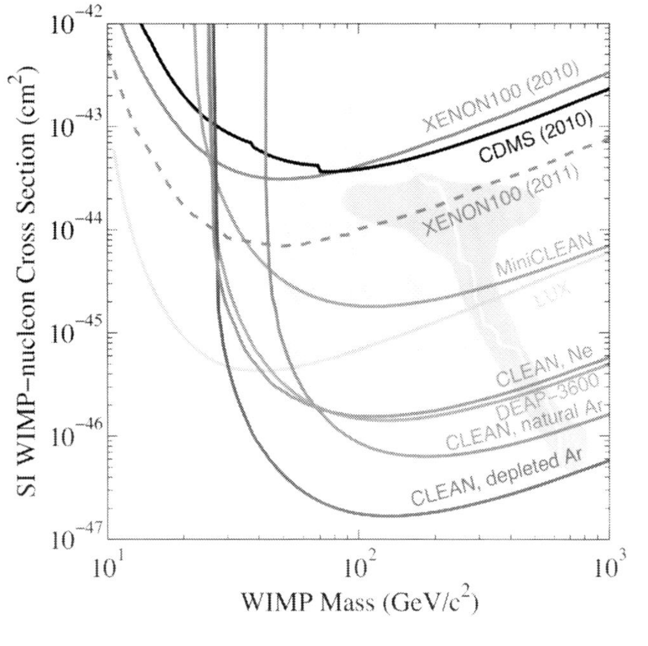

Figure 4. The projected sensitivities with respect to spin-independent WIMP-nucleon cross section in MiniCLEAN and DEAP-3600 [6, 21, 28] together with that of competitive experiments, SuperCDMS, XENON 100 and LUX. Model predictions are shown in the colored patches. The shaded pink and yellow patches correspond to allowed parameter space in model predictions for a SUSY WIMP [25, 26]. The light blue patch is the allowed parameter space for a KK WIMP dark matter [27].

References

[1] E. Komatsu et al. (WMAP), Astrophys. J. Suppl. **180**, 330 (2009).
[2] G. Steigman and M. S. Turner, Nucl. Phys. **B253**, 375 (1985).
[3] B.W. Lee and S. Weinberg, Phys. Rev. Lett. **39**, 165 (1977).
[4] S. Weinberg, Phys. Rev. Lett. **48**, 1976 (1982).
[5] G. Jungman, M. Kamionkowski, and K. Griest, Phys. Rept. **267**, 195 (1996).
[6] J. D. Lewin and P. F. Smith, Astropart. Phys. **6**, 87 (1996).
[7] R. J. Gaitskell, Ann. Rev. Nucl. Part. Sci. **54**, 315 (2004).
[8] Z. Ahmed et al. [The CDMS-II Collaboration], Science **327**, 1619-1621 (2010) [arXiv:0912.3592 [astro-ph.CO]].
[9] E. Aprile et al. [XENON100 Collaboration], [arXiv:1005.0380 [astro-ph.CO]].
[10] R. Bernabei, P. Belli, F. Cappella et al., Eur. Phys. J. **C67**, 39-49 (2010). [arXiv:1002.1028 [astro-ph.GA]].
[11] F. Petriello and K. M. Zurek, JHEP **0809**, 047 (2008) [arXiv:0806.3989 [hep-ph]]; C. Savage, K. Freese, P. Gondolo and D. Spolyar, JCAP **0909**, 036 (2009) [arXiv:0901.2713 [astro-ph]]; C. Savage, G. Gelmini, P. Gondolo and K. Freese, JCAP **0904**, 010 (2009) [arXiv:0808.3607 [astro-ph]].
[12] R. Brunetti, E. Calligarich, M. Cambiaghi, F. Carbonara, A. Cocco, C. De Vecchi, R. Dolfini, A. Ereditato, G. Fio- rillo, L. Grandi, et al., New Ast. Rev. **49**, 265 (2005).
[13] S. Fiorucci, et al. Proceedings of the SUSY09 conference, arXiv:0912.0482v1 [astro-ph.CO].
[14] J. Angle et al. (XENON10 Collaboration), Phys. Rev. Lett. **100**, 021303 (2008), 0706.0039.
[15] M. Miyajima et al., Phys. Rev. **A9**, 1438 (1974).
[16] V. N. Lebedenko et al. (UK Dark Matter Collaboration), Phys. Rev. **D80**, 052010 (2009), 0812.1150.
[17] C. Galbiati R. Purtschert et al., J. Phys. Conf. Ser. **120**, 042015 (2008).
[18] M. Boulay and A. Hime, Astroparticle Physics **25**, 179 (2006).
[19] D. N. McKinsey and J. M. Doyle, Journal of Low Temperature Physics **118**, 153 (2000); D. N. McKinsey and K. J. Coakley, Astroparticle Physics **22**, 355 (2005).
[20] J. F. Wilkerson et al., Phys.Rev. Lett. **58**, 2023 (1987); S. T. Staggs et al., Phys. Rev. **C39**, 1503 (1989); R. G. H. Robertson et al., Phys.Rev. Lett. **67**, 957 (1991); W. Stoeffl and D. J. Decman, Phys. Rev. Lett. **75**, 3237 (1995).
[21] W. Hugh Lippincott, *Direct detection of dark matter with liquid argon and neon*, Ph.D. thesis, Yale University (2010)
[22] D. Gastler et al., arXiv:1004.0373v1 [physics.ins-det].
[23] M. G. Boulay et al., arxiv:0904.2930v1[astro-ph.IM] (2009), submitted to PRC.
[24] Hamamatsu, http://www.hamamatsu.com/.
[25] J. Ellis, A. Ferstl, and K. A. Olive, Phys. Rev. D **63**, 065016 (2001).
[26] R. Trotta, F. Feroz, M. Hobson, L. Roszkowski, and R. R. de Austri, Journal of High Energy Physics **024** (2008).
[27] S. Arrenberg, L. Baudis, K. Kong, K. T. Matchev, and J. Yoo, Phys. Rev. D **78**, 056002 (2008).
[28] Dmtools, http://dmtools.brown.edu/.

Cosmic Ray Signatures of Decaying Dark Matter

Alejandro Ibarra
Physik-Department T30d, Technische Universität München ,
James-Franck-Straße, 85748 Garching, Germany
E-mail: alejandro.ibarra@ph.tum.de

Abstract. Astrophysical and cosmological observations do not require the dark matter particles to be absolutely stable. If they are indeed unstable, their decay into Standard Model particles might occur at a sufficiently large rate to allow the indirect detection of dark matter through an anomalous contribution to the high energy cosmic ray fluxes. We analyze the implications of the excess in the total electron plus positron flux and the positron fraction reported by the Fermi and PAMELA collaborations, respectively, for the scenario of decaying dark matter. We also discuss the constraints on this scenario from measurements of other cosmic ray species and the predictions for the diffuse gamma ray flux and the neutrino flux. In particular, we expect a sizable dipole-like anisotropy which may be observed in the near future by the Fermi-LAT.

1. Introduction

A series of astrophysical and cosmological observations have revealed the existence of a new particle which is not contained the Standard Model of Particle Physics: the dark matter particle. Despite the overwhelming evidences for its existence, very little is known about the fundamental properties of the dark matter particle, such as the mass, the spin, the lifetime or the interaction cross section with ordinary matter. There are different methods to constrain, or hopefully determine, these properties: direct detection, indirect detection and collider searches. The possibility of the indirect detection of dark matter has received a lot of attention in the last two years, after the PAMELA collaboration reported the existence of an excess in the positron fraction at energies larger than 7 GeV [1]. Furthermore, the measurements of the total electron plus positron flux by the Fermi collaboration hint towards an excess at high energies [2]. In this talk we will discuss the possibility that the decay of dark matter particles could be the origin of these anomalies.

Many models have been proposed where the dark matter particle is indeed predicted to decay with a cosmologically long lifetime, yielding potentially observable signatures in high-energy cosmic rays. Some examples are gravitinos in general supersymmetric models (namely, models without imposing R-parity conservation) [3, 4, 5], where the decay rate is doubly suppressed by the supersymmetry breaking scale and by the small R-parity violation; hidden sector $U(1)$ gauge bosons or gauginos [6, 7], where the decay rate is suppressed by a small kinetic mixing between a hidden $U(1)$ gauge group and the $U(1)$ of hypercharge; hidden sector $SU(2)$ gauge bosons, where the decay rate is suppressed by a possibly large breaking scale of the hidden $SU(2)$ "custodial symmetry" [8]; right-handed sneutrinos in scenarios with Dirac neutrino masses [9], where the decay rate is suppressed by the tiny neutrino Yukawa couplings; or hidden sector fermions [10] and bound states of strongly interacting particles [11], where the decay rate is suppressed by the scale of grand unification.

Instead of analyzing the cosmic-ray signatures for each of these scenarios separately, we will present here the results of a model-independent approach which encompasses the main features of all the scenarios listed above [12, 13]. We will first discuss the parameters which can reproduce the PAMELA and Fermi electron/positron excesses and later on we will discuss the predictions for the antiproton flux, the gamma-ray flux and the neutrino flux.

2. Decaying Dark Matter as possible explanation of the electron/positron anomalies

We will assume that the Milky way halo is populated with dark matter particles with mass $m_{\rm DM}$ and lifetime $\tau_{\rm DM}$ which are distributed following a Navarro-Frenk-White density profile [14]. In order to keep the analysis as model-independent as possible, we will consider the following decay channels for a fermionic dark matter particle:

$$\begin{aligned} \psi &\rightarrow Z^0\nu\,, \\ \psi &\rightarrow W^{\pm}\ell^{\mp}\,, \\ \psi &\rightarrow \ell^+\ell^-\nu\,; \end{aligned} \qquad (1)$$

the results for a scalar dark matter particle can be found in [12, 13]. The fragmentation of the weak gauge bosons produces a continuous spectrum of electrons and positrons (mainly from π^+ decay) that we have obtained using the event generator PYTHIA 6.4 [15]. Then, once a particular decay mode has been assumed, the only free parameters from the particle physics point of view in this analysis are the dark matter mass and lifetime.

Charged particles propagate through the halo in a complicated way before reaching the Earth. We will describe the propagation by means of a stationary two zone diffusion model with cylindrical boundary conditions [16]. For the positron propagation we will adopt the MED propagation model defined in [17], although our conclusions are rather insensitive to the choice of propagation parameters or to the choice of dark matter halo profile.

The predicted positron fraction in the case where the dark matter particles decay via $\psi_{\rm DM} \rightarrow Z^0\nu$ is shown in the left panel of Fig. 1, compared to the PAMELA, HEAT, CAPRICE and AMS-01 data, for the exemplary dark matter masses $m_{\rm DM} = 5$ and 100 TeV. In the right panel we show the corresponding total electron-plus-positron flux compared to the results from Fermi, H.E.S.S., PPB-BETS, BETS, ATIC, HEAT, CAPRICE and AMS-01. The dark matter lifetimes and the normalization of the primary electron flux have been chosen in each case to provide a reasonable fit to the PAMELA and Fermi data points. In this decay channel, the dominant source of electrons and positrons is the fragmentation of the Z^0 boson (with a rather small branching ratio of Z^0 decays into a pair of charged leptons), which produces relatively soft particles. As a result, even though this decay mode can produce a visible excess in the positron fraction, the energy spectrum is in general too flat to explain the steep rise observed by PAMELA. An exception occurs if the dark matter mass is very large, $m_{\rm DM} > 50$ TeV. In this case, the electrons and positrons from dark matter decay are boosted to high enough energies to produce the steep rise in the positron fraction. However, these large dark matter masses seem to be in conflict with the H.E.S.S. observations, which require a falloff in the total electron-plus-positron spectrum at ~ 1 TeV [18].

On the other hand, we show in Fig. 2 the results in the case that a fermionic dark matter particle decays into three leptons $\psi \rightarrow \ell^+\ell^-\nu$, via a virtual scalar particle. The decay $\psi \rightarrow e^+e^-\nu$ can accommodate the PAMELA anomaly when the dark matter mass is larger than ~ 300 GeV. However, this decay mode cannot be the origin of the excess observed by Fermi since it predicts a rather prominent bump in the spectrum which is not observed. More promising is the decay

Journal of Physics: Conference Series **315** (2011) 012011 doi:10.1088/1742-6596/315/1/012011

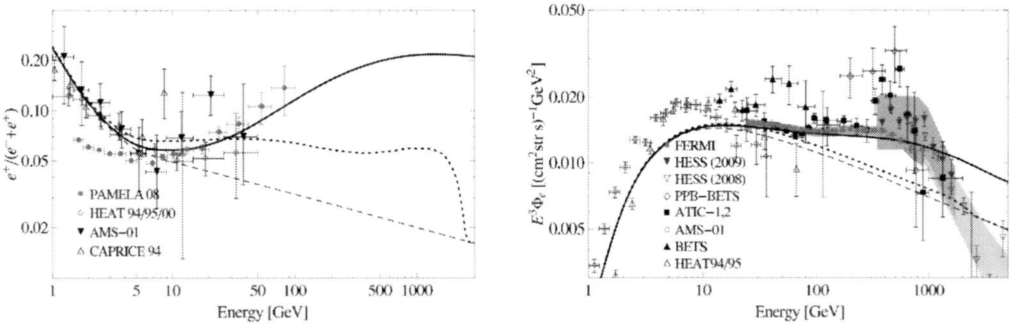

Figure 1. Positron fraction (*left panel*) and total electron-plus-positron flux (*right panel*) for the decay channel $\psi_{\rm DM} \to Z^0\nu$ with $m_{\rm DM} = 100\,{\rm TeV}$ (solid) and $5\,{\rm TeV}$ (dotted). The dashed line shows the background fluxes as discussed in the text. Solar modulation is taken into account using the force field approximation with $\phi_F = 550\,{\rm MV}$.

channel $\psi_{\rm DM} \to \mu^-\mu^+\nu$, which can reproduce quite nicely the Fermi electron-plus-positron spectrum and the steep rise in the positron fraction observed by PAMELA when the dark matter mass is $m_{\rm DM} \simeq 3500$ GeV and the lifetime is $\tau_{\rm DM} \simeq 1.1 \times 10^{26}$ s. Lastly, decays into tau flavor can qualitatively reproduce the steep rise in the positron fraction for dark matter masses above ~ 2.5 TeV, although as apparent from Fig. 2, lower right panel, the resulting electron-plus-positron spectrum has an energy dependence much steeper than $E^{-3.0}$ at high energies, in tension with the Fermi measurements. In this case an additional source of high-energy positrons, coming *e.g.* from pulsars, must be invoked in order to reproduce the Fermi energy spectrum.

Decay Channel	$m_{\rm DM}$ [GeV]	$\tau_{\rm DM}$ [10^{26}s]
$\psi_{\rm DM} \to \mu^+\mu^-\nu$	3500	1.1
$\psi_{\rm DM} \to \ell^+\ell^-\nu$	2500	1.5
$\phi_{\rm DM} \to \mu^+\mu^-$	2500	1.8
$\phi_{\rm DM} \to \tau^+\tau^-$	5000	0.9
$\psi_{\rm DM} \to W^\pm\mu^\mp$	3000	2.1

Table 1. Decay channels for fermionic and scalar dark matter, $\psi_{\rm DM}$ and $\phi_{\rm DM}$, respectively, that best fit the Fermi and PAMELA data for the MED propagation model and the NFW halo profile.

The decaying dark matter channels that can qualitatively reproduce the PAMELA and Fermi measurements, assuming the conventional diffusive model for the background electrons and positrons [19], are listed in Table 1. In this table there are two striking facts: *i)* the dark matter particle must be fairly heavy and *ii)* it must have a lifetime which is of the order of 10^{26} s.

More importantly, for a given decay channel, the requirement that the decay of dark matter particles is the origin of the PAMELA and Fermi excesses fixes the dark matter mass and lifetime, which are the only free parameters from the particle physics point of view. Therefore, it is possible to make *predictions* for the fluxes of antiprotons, gamma-rays and neutrinos in order to test the decaying dark matter interpretation of the electron/positron excesses.

3. Predictions for the antiproton flux

Decay modes into weak gauge bosons produce an associated antiproton flux, which is severely constrained by present experiments. Indeed, the measurements of the antiproton flux and

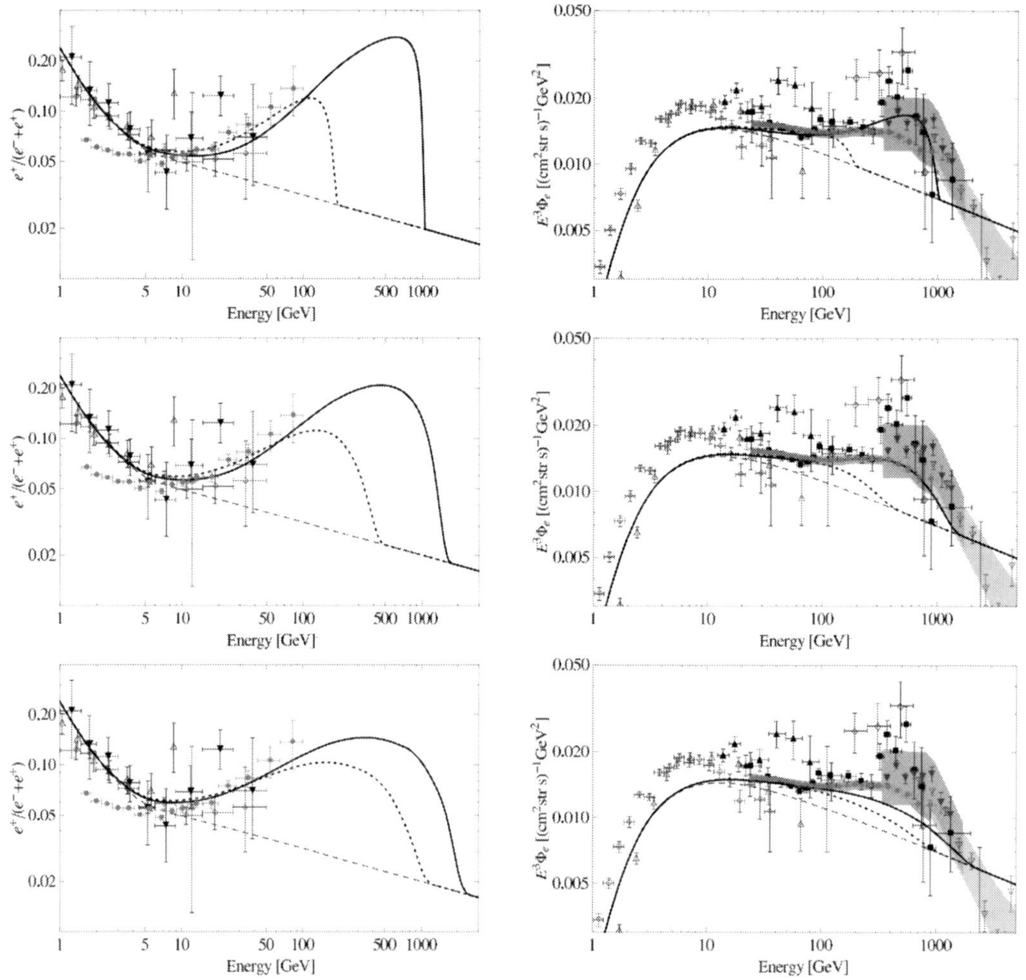

Figure 2. Same as Fig. 1, but for the decay channels $\psi_{\rm DM} \to \ell^\pm \ell^\mp \nu$. *Upper panels:* $\psi_{\rm DM} \to e^- e^+ \nu$ with $m_{\rm DM} = 2000\,{\rm GeV}$ (solid) and $400\,{\rm GeV}$ (dotted). *Middle panels:* $\psi_{\rm DM} \to \mu^- \mu^+ \nu$ with $m_{\rm DM} = 3500\,{\rm GeV}$ (solid) and $1000\,{\rm GeV}$ (dotted). *Lower panels:* $\psi_{\rm DM} \to \tau^- \tau^+ \nu$ with $m_{\rm DM} = 5000\,{\rm GeV}$ (solid) and $2500\,{\rm GeV}$ (dotted).

the antiproton-to-proton fraction do not show any deviation with respect to the theoretical expectations from spallation of cosmic rays on the interstellar medium [20], thus constraining the size of any exotic contribution. The decay mode $\psi \to W^\pm e^\mp$ predicts an antiproton-to-proton fraction which is in tension with the experimental measurements [13]. Therefore, purely leptonic dark matter decays are favored over decays into weak gauge bosons.

4. Predictions for the gamma-ray flux

The gamma ray flux from dark matter decay consists of two components. First, the prompt radiation of gamma-rays produced in the decay of the dark matter particle itself, coming from final state radiation, pion decay, etc. The prompt decay can also include in certain models monoenergetic lines which, if observed, would constitute an inequivocal signature of dark matter [5, 7, 8]. The prompt radiation is in turn constituted by two components: the halo

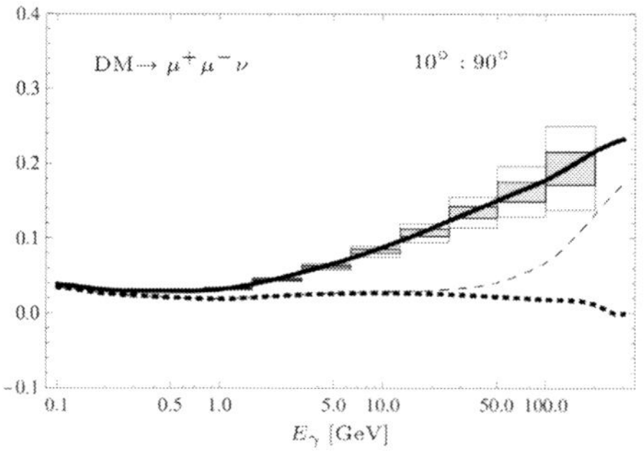

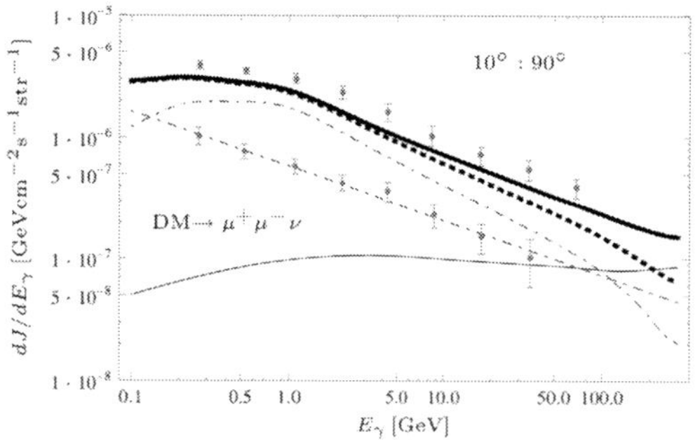

Figure 3. *Left panel:* Anisotropy of the gamma-ray flux from dark matter decay $\psi \to \mu^+\mu^-\nu$ as a function of energy. The *dotted line* shows the background anisotropy as expected from the Galactic foreground, while the *solid line* shows the anisotropy of signal + background. We also show the signal + background anisotropy neglecting gamma rays from inverse Compton scattering (*dashed*, overlapping with solid line). The boxes show estimates of the statistical errors for one-year and five-year Fermi LAT observations. *Right panel:* Gamma-ray fluxes averaged over all Galactic longitudes, excluding the region $|b| < 10°$, as a function of energy. The *thin solid line* shows the gamma rays from dark matter decay. The two *dash-dotted lines* show the astrophysical extragalactic background and the Galactic foreground separately. The *thick solid line* shows the sum of all contributions, whereas the *dotted line* shows the sum without contributions from dark matter. The data points show preliminary Fermi LAT data [24] for the total diffuse flux (*upper points*, in the $10° : 90°$ region we averaged the presented preliminary results appropriately) and the extragalactic background (*lower points*).

component and the extragalactic component. The former depends on the dark matter profile, displaying a strong dependence in the direction of the galactic center and being much milder at high galactic latitudes ($|b| > 10°$). Being the Earth 8.5 kpc away from the Galactic Center, this contribution is anisotropic, even if the halo profile is spherically symmetric, which may be detected in experiments as will be discussed below. On the other hand, the extragalactic contribution is assumed to be perfectly isotropic and is in most regions of the sky numerically smaller than the halo contribution. Moreover, the flux of gamma-rays of cosmological origin is attenuated by the electron-positron pair production on the extragalactic background light emitted by galaxies in the ultraviolet, optical and infrared frequencies [21]. However, the flux of gamma-rays originating from the decay of dark matter particles in the halo is barely attenuated by pair production on the Galactic interstellar radiation field at energies below 10 TeV [22]. Thus, the total flux is dominated by the halo component, yielding a slightly anisotropic gamma-ray flux.

In addition to the prompt radiation, the total gamma ray flux also receives a contribution from the inverse Compton scattering radiation of electrons and positrons produced in the decay on the interstellar radiation field (which includes the cosmic microwave background, thermal radiation of dust and starlight), which generates gamma rays with energies between 100 MeV and 1 TeV [23].

If decaying dark matter is the origin of the electron/positron excesses, the predicted gamma ray flux should not exceed the total measured gamma-ray flux in any direction of the sky. Furthermore, since the flux from dark matter decay is roughly (but not completely) isotropic, the flux from dark matter decay should not be much larger than the diffuse extragalactic gamma ray background determined assuming as only background the diffuse emission from the galaxy. We show in Fig.3, right panel, the prediction for the averaged diffuse gamma ray flux in the region $10° \leq |b| \leq 90°$, $0 \leq \ell < 360°$, for the decay mode $\psi \to \mu^+\mu^-\nu$ and for the mass and lifetime which reproduce the PAMELA and Fermi measurements. Indeed, the prediction for the total gamma-ray flux *including* the contribution from dark matter decay does not exceed the present measurements on the total diffuse gamma ray flux in that region of the sky. Furthermore, the contribution to the gamma-ray flux from dark matter decay (which resembles a diffuse extragalactic gamma-ray flux) is also below the extracted data points by the Fermi collaboration [24], thus allowing this decay mode as a possible explanation of the electron/positron excesses [25]. Notice that this decay channel predicts a rise in the energy spectrum of the diffuse extragalactic gamma ray flux at energies larger than 100 GeV.

The Earth is not located in the center of the Milky Way dark matter halo, therefore the flux from the decay of dark matter particles from the hemisphere in the direction of the galactic center should be larger than the flux from the hemisphere in the direction of the galactic anticenter. We show in Fig.3, left panel, the predicted anisotropy for the channel $\psi \to \mu^+\mu^-\nu$ calculated as $A = (J_{GC} - J_{GA})/(J_{GC} + J_{GA})$, where J_{GC} is the flux in the region $10° \leq |b| \leq 90°$, $-90° \leq \ell < 90°$ and J_{GA} is the flux in the region $10° \leq |b| \leq 90°$, $90° \leq \ell < 270°$. At high energies the anisotropy predicted by the conventional GALPROP model is at most a few percent (dotted line), whereas the anisotropy predicted by the scenario with dark matter decay can be as large as 20% (solid line). Remarkably, this conclusion holds for all decaying dark matter scenarios which explain the Fermi and PAMELA results, thus providing a very powerful test of the decaying dark matter interpretation of the electron/positron excesses [25].

5. Predictions for the neutrino flux

The decay of dark matter particles also produces neutrinos, which travel in straight lines through the Galaxy, essentially without any interactions. The only modifications to the fluxes during the propagation are due to flavour oscillations, which yield similar neutrino fluxes in all flavors regardless of the initial flavour composition.

The backgrounds for the neutrino signal from dark matter decay is produced by cosmic-ray interactions with the Earth's atmosphere; In our analysis we will use the atmospheric neutrino fluxes calculated by Honda *et al.* [26].

We show in Fig.4, left panel, the 90% exclusion region in the lifetime vs. mass plane for different decaying dark matter scenarios from the non-observation of an excess in the Super-Kamiokande data [27]. It is apparent from the figure that the masses and lifetimes required to explain the electron/positron excesses are allowed by present experiments. On the other hand, we show in the right panel the projected 90% exclusion region at IceCube in one year from the non-observation of a significant excess in the total rate of neutrino induced upward through-going muons. It is interesting to note that IceCube will probe part of the region of the parameter space which is of relevance for the decaying dark matter interpretation of the PAMELA and Fermi anomalies.

6. Conclusions

Many well-motivated particle physics scenarios predict that dark matter particles decay with a lifetime much longer than the age of the Universe. We have shown that the decay of dark matter particles could be the explanation of the cosmic ray anomalies reported by the PAMELA and Fermi collaborations, if the dark matter mass is in the TeV range and the lifetime is $\mathcal{O}(10^{26}$ s). We have also discussed the predictions of this interpretation for the antiproton, gamma-ray and neutrino fluxes.

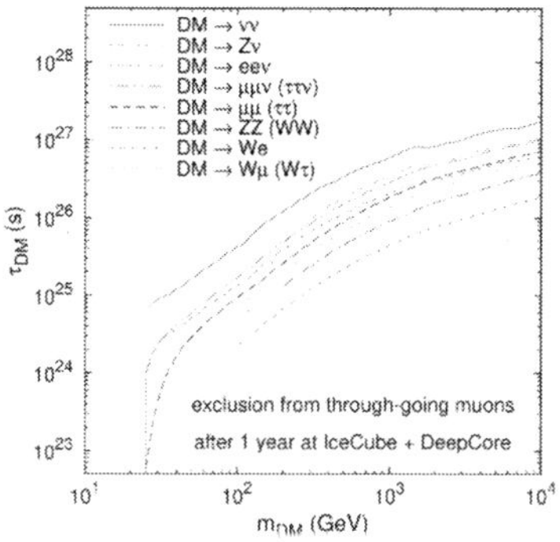

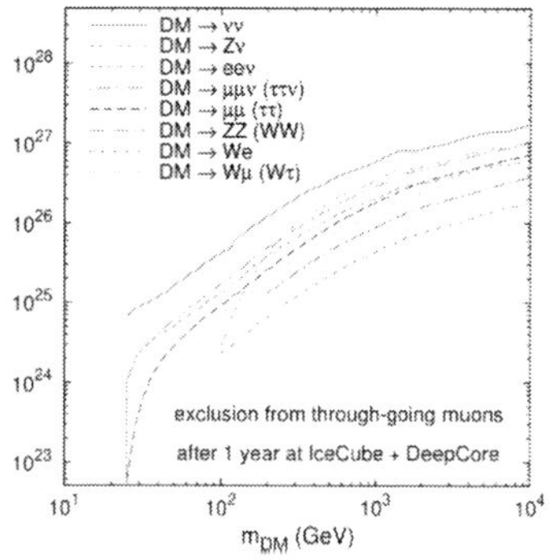

Figure 4. *Left panel:* 90 % C.L. exclusion region in the lifetime vs. mass plane for a decaying dark matter candidate from the non-observation of a significant excess in the total rate of neutrinos observed at Super-Kamiokande. *Right panel:* 90 % C.L. exclusion prospects after one year at IceCube.

Acknowledgments

The author is grateful to the DSU 2010 organizers for their kind invitation to participate in this workshop. The author is also grateful to David Tran and Christoph Weniger for very pleasant collaborations on which most of this talk is based. This work was partially supported by the DFG cluster of excellence "Origin and Structure of the Universe."

References

[1] O. Adriani *et al.* [PAMELA Collaboration], Nature **458** (2009) 607.
[2] A. A. Abdo *et al.* [The Fermi LAT Collaboration], Phys. Rev. Lett. **102**, 181101 (2009).
[3] F. Takayama and M. Yamaguchi, Phys. Lett. B **485**, 388 (2000); W. Buchmuller, L. Covi, K. Hamaguchi, A. Ibarra and T. Yanagida, JHEP **0703**, 037 (2007);
[4] A. Ibarra and D. Tran, JCAP **0807** (2008) 002; S. L. Chen, R. N. Mohapatra, S. Nussinov and Y. Zhang, Phys. Lett. B **677** (2009) 311; L. Covi, M. Grefe, A. Ibarra and D. Tran, JCAP **0901**, 029 (2009).
[5] A. Ibarra and D. Tran, Phys. Rev. Lett. **100**, 061301 (2008), K. Ishiwata, S. Matsumoto and T. Moroi, Phys. Rev. D **78** (2008) 063505. W. Buchmuller, *et al.* JCAP **0909**, 021 (2009). S. Lola, P. Osland and A. R. Raklev, Phys. Lett. B **656** (2007) 83;
[6] C. R. Chen, F. Takahashi and T. T. Yanagida, Phys. Lett. B **671** (2009) 71, Phys. Lett. B **673** (2009) 255.
[7] A. Ibarra, A. Ringwald and C. Weniger, JCAP **0901** (2009) 003; A. Ibarra, A. Ringwald, D. Tran and C. Weniger, JCAP **0908**, 017 (2009).
[8] C. Arina, T. Hambye, A. Ibarra and C. Weniger, JCAP **1003**, 024 (2010).
[9] M. Pospelov and M. Trott, JHEP **0904** (2009) 044.
[10] K. Hamaguchi, S. Shirai and T. T. Yanagida, Phys. Lett. B **673**, 247 (2009); A. Arvanitaki *et al.* Phys. Rev. D **79**, 105022 (2009).
[11] K. Hamaguchi, E. Nakamura, S. Shirai and T. T. Yanagida, Phys. Lett. B **674**, 299 (2009).
[12] A. Ibarra and D. Tran, JCAP **0902** (2009) 021.
[13] A. Ibarra, D. Tran and C. Weniger, JCAP **1001**, 009 (2010).
[14] J. F. Navarro, C. S. Frenk and S. D. M. White, Astrophys. J. **462** (1996) 563.
[15] T. Sjöstrand, S. Mrenna and P. Skands, JHEP **0605** (2006) 026.
[16] See for example V. S. Berezinskii, S. V. Buolanov, V. A. Dogiel, V. L. Ginzburg, V. S. Ptuskin, Astrophysics of Cosmic Rays (Amsterdam: North–Holland, 1990).
[17] D. Maurin, F. Donato, R. Taillet and P. Salati, Astrophys. J. **555** (2001) 585.
[18] F. Aharonian *et al.* [H.E.S.S. Collaboration], Astron. Astrophys. **508** (2009) 561.
[19] I. V. Moskalenko and A. W. Strong, Astrophys. J. **493** (1998) 694.
[20] O. Adriani *et al.* [PAMELA Collaboration], arXiv:1007.0821 [astro-ph.HE].
[21] R. J. Gould and G. P. Schreder, Phys. Rev. **155**, 1404 (1967).
[22] I. V. Moskalenko, T. A. Porter and A. W. Strong, Astrophys. J. **640**, L155 (2006).
[23] For a review see G. R. Blumenthal and R. J. Gould, Rev. Mod. Phys. **42**, 237 (1970).

[24] We used for our analysis the preliminary results shown by M. Ackermann at TeV Particle Astrophysics 2009. The final results can be found in A. A. Abdo *et al.* [The Fermi-LAT collaboration], Phys. Rev. Lett. **104** (2010) 101101.
[25] A. Ibarra, D. Tran and C. Weniger, Phys. Rev. D **81**, 023529 (2010).
[26] M. Honda *et al.*, Phys. Rev. D **75** (2007) 043006.
[27] L. Covi, M. Grefe, A. Ibarra and D. Tran, JCAP **1004**, 017 (2010).

Journal of Physics: Conference Series **315** (2011) 012012 doi:10.1088/1742-6596/315/1/012012

Dark Matter and Higgs Sector [1]

Jose A. R. Cembranos

William I. Fine Theoretical Physics Institute, University of Minnesota, Minneapolis, 55455, USA

School of Physics and Astronomy, University of Minnesota, Minneapolis, 55455, USA

Jose H. Montes de Oca Y.

Facultad de Ciencias Físico-Matemáticas, Benemérita Universidad Autónoma de Puebla, Puebla, Pue., C.P. 72570, Mexico

Lilian Prado

Facultad de Ciencias Físico-Matemáticas, Benemérita Universidad Autónoma de Puebla, Puebla, Pue., C.P. 72570, Mexico

Abstract.
The inert doublet model is an extension of the Standard Model of Elementary Particles that is defined by the only addition of a second Higgs doublet without couplings to quarks or leptons. This minimal framework has been studied for many reasons. In particular, it has been suggested that the new degrees of freedom contained in this doublet can account for the Dark Matter of the Universe.

1. Introduction

In spite of many and continuous efforts, the ultraviolet (UV) completion of the gravitational interaction is still an open question. In these conditions, it is difficult to make general statements about very early cosmology (although, in general, different types of new scalar fields are commonly predicted [1, 2]). However, these details are not needed to compute the relic density of many Dark Matter (DM) candidates [3, 4, 5, 6]. Although there are other possibilities [7], DM is usually assumed to be in the form of stable Weakly-interacting massive particles (WIMPs) that naturally freeze-out with the right thermal abundance. One of the most interesting features of WIMPs, is that they emerge in well-motivated particle physics scenarios as in R-parity conserving supersymmetry (SUSY) models [8, 9], universal extra dimensions (UED) [10, 11], or brane-worlds [12, 13]. Another interesting property of WIMPs, it is that they can be tested with high energy experiments as the new generation of colliders [14]. All these interesting features are also shared by the DM provided inside the Inert Higgs Doublet Model.

[1] To appear in the Proceedings of the sixth International Workshop on the Dark Side of the Universe (DSU2010) Leon, Guanajuato, Mexico 1-6 June 2010.

2. Inert Higgs Doublet Model (IHDM)

The two-Higgs-doublet model (2HDM) is one of the simplest extensions of the Higgs mechanism of the electroweak symmetry breaking beyond the standard model (SM) [15, 16, 17, 18, 19]. In this model, an additional Higgs doublet field is introduced in the Higgs sector of the SM with the same quantum numbers. We use the notation

$$\Phi_a = \begin{pmatrix} \phi_a^+ \\ \phi_a + i\chi_a \end{pmatrix}, \tag{1}$$

where $a = 1, 2$ for the Higgs doublets before the spontaneous symmetry breaking (SSB); that is, in the electroweak basis. Here ϕ_a^+, ϕ_a and χ_a represent Higgs fields. After SSB, these Higgs fields interact with the matter fields and also self-interact via an appropriate Higgs potential. This model presents new physics such as Flavor Changing Neutral Currents (FCNC) and new types of CP violation [15, 16, 17, 18, 19]. Nevertheless, we are interested in a DM candidate; so this candidate must be a neutral scalar field which weakly interacts with other particles.

In the literature, the classification of the 2HDM is done considering the Higgs-fermion couplings given by the Yukawa terms in the Lagrangian [20]. For the IHDM just one Higgs doublet interacts with the fermions in the Yukawa terms as follows

$$\mathcal{L}_Y = Y_{ij}^{(u)} \bar{q}_{Li} \tilde{\Phi}_1 u_{Rj} + Y_{ij}^{(d)} \bar{q}_{Li} \Phi_1 d_{Rj} + h.c. + \text{leptons}, \tag{2}$$

where $\widetilde{\Phi}_1 = i\sigma_2 \Phi_1^*$, $Y_{ij}^{(u,d)}$ are the Yukawa couplings for $i, j = 1, 2, 3$; u_{Ri}, d_{Ri} are the right singlets quarks and q_{Li} are left doublets under the electroweak symmetry group [20].

Now, for the potential we shall assume an additional Z_2 discrete symmetry[2] in order to avoid the mixing term $\Phi_1^\dagger \Phi_2$, which could introduce CP violation in the potential. Then, the potential under this last condition is

$$\begin{aligned} \mathcal{V} &= \mu_1^2 \Phi_1^\dagger \Phi_1 + \mu_2^2 \Phi_2^\dagger \Phi_2 + \lambda_1 (\Phi_1^\dagger \Phi_1)^2 + \lambda_2 (\Phi_2^\dagger \Phi_2)^2 \\ &+ \lambda_3 (\Phi_1^\dagger \Phi_1)(\Phi_2^\dagger \Phi_2) + \lambda_4 (\Phi_1^\dagger \Phi_2)(\Phi_2^\dagger \Phi_1) \\ &+ \frac{1}{2} \lambda_5 [(\Phi_1^\dagger \Phi_2)^2 + (\Phi_2^\dagger \Phi_1)^2], \end{aligned} \tag{3}$$

where $\mu_{1,2}^2$, $\lambda_{1,2,3,4}$ are real parameters, while λ_5 could be complex. The Φ_1 obtains a vacuum expectation value (VEV): $v/\sqrt{2} = 174$ GeV, as in the SM, while Φ_2 does not obtain a VEV [21, 22, 23]. Since this Z_2 symmetry is unbroken, the inert particles will be stable [26]. After SSB and by changing to the physical basis, the Yukawa terms (2) have the form [20]:

$$\begin{aligned} \mathcal{L}_{Hf\bar{f}} &= -\frac{g}{2M_W \sin\beta} \bar{d} M_d d (H^0 \sin\alpha + h^0 \cos\alpha) - \frac{ig\cot\beta}{2M_W} \bar{d} M_d \gamma_5 d A^0 \\ &- \frac{g}{2M_W \sin\beta} \bar{u} M_u u (H^0 \sin\alpha + h^0 \cos\alpha) + \frac{ig\cot\beta}{2M_W} \bar{u} M_u \gamma_5 u A^0 \\ &+ \frac{g\cot\beta}{2\sqrt{2} M_W} (H^+ \bar{u} [M_u V_{CKM} (1 - \gamma_5) - V_{CKM} M_d (1 - \gamma_5)] d + h.c. \end{aligned} \tag{4}$$

where α and β angles are the mixing angles for the Higgs bosons, V_{CKM} is the CKM matrix and H^0, h^0 are neutral CP even scalar Higgs bosons while A^0 is a neutral CP odd pseudo scalar Higgs boson.

Journal of Physics: Conference Series **315** (2011) 012012 doi:10.1088/1742-6596/315/1/012012

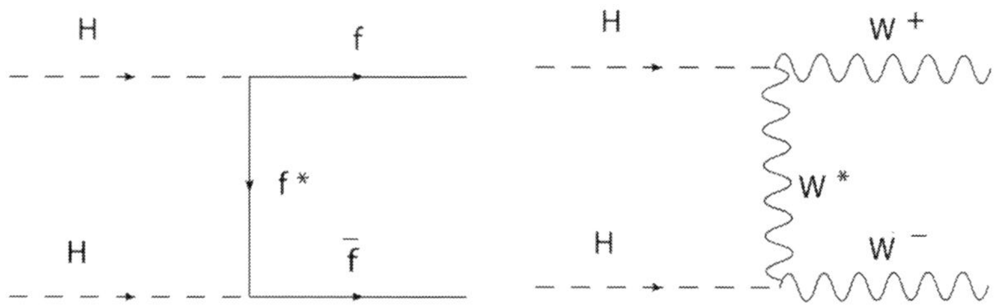

Figure 1. Feynman diagrams for the considered processes. H represents the scalar or pseudoscalar neutral Higgs bosons. On left, the products of the annihilation are fermion and antifermion pairs. On the right, the products are W charged bosons.

3. Neutral Higgs as Dark Matter candidate

The physical fields for the Higgs sector are given by two charged scalar Higgs bosons ($H^{\pm}$), two neutral CP even scalar Higgs bosons (H^0 and h^0) and one neutral CP odd pseudo scalar Higgs boson (A^0) [20]. Due to its characteristics of being inert and stable, the lightest neutral scalar Higgs boson or the pseudoscalar Higgs boson of the IHDM have been proposed as DM candidates [24, 25, 26, 27, 23, 28]. Only the scalar part h^0 has been considered for numerical results performed using the programs MICROMEGAS and CALCHEP; the pseudo-scalar part A^0 is only mentioned to behave in a similar way [23, 26, 27, 28].

In order to explore the possibilities and differences of the pseudo-scalar part, our aim is to calculate the relic abundance for both, h^0 and A^0, using an analytical method [5, 6]. The relic density computation requires the neutral Higgs bosons (of the IHDM) annihilations in standard model pairs. We shall consider the relevant processes for the case: scalar and pseudoscalar annihilations. In this approximation the relevant processes are $HH \to f\bar{f}$ and $HH \to W^+W^-$, where H could be h^0 or A^0 and f stands for fermion (we just take the top quark). For example, in the limit $E_{CM} > M_{top}$, the amplitude square for the scalar case is given by:

$$|\mathcal{M}_{\mathrm{scalar}}|^2 = \left(\frac{gM_t \sin\alpha}{2M_W \sin\beta}\right)^4 \frac{(6E_{CM}^2 M_t^2 - 8M_t^4)}{(E_{CM}^2 - M_t^2)^2} + \frac{(gM_W \sin(\beta-\alpha))^4}{(E_{CM}^2 - M_W^2)^2}\left(2 + \left(\frac{\frac{1}{2}E_{CM}^2 - M_W^2}{M_W^2}\right)^2\right) \quad (5)$$

and for the pseudoscalar:

$$|\mathcal{M}_{\mathrm{pseudoscalar}}|^2 = \left(\frac{gM_t \cot\beta}{2M_W}\right)^4 \frac{(2E_{CM}^2 M_t^2 - 8M_t^4)}{(E_{CM}^2 - M_t^2)^2}, \quad (6)$$

where E_{CM}, M_t and M_W are the energy in the center of mass frame, the top quark mass and W boson mass, respectively. The behavior of the amplitude for scalar case is shown in Figure 2. We have taken some characteristic values for the free parameters α and β based on [29].

4. Summary

In this work we have obtained the amplitude for the neutral Higgs bosons annihilation under a tree level approximation. The expressions for scalar and pseudoscalar cases have a pole when

[2] One doublet changes its sign under this Z_2 symmetry.

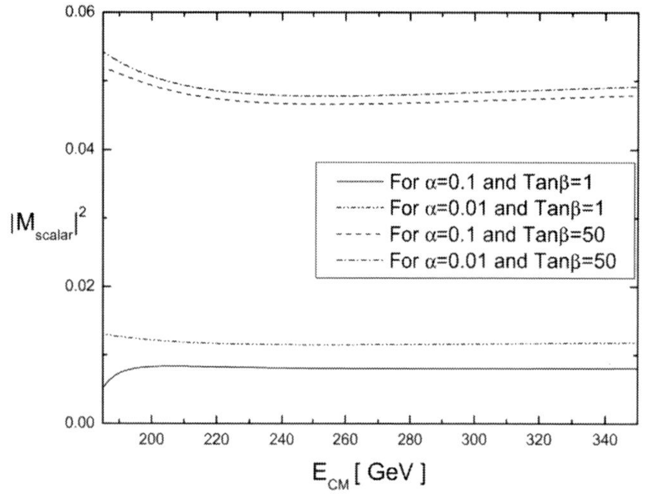

Figure 2. The graphic shows the numerical value for $|\mathcal{M}_{\text{scalar}}|^2$ as function of the energy in the center of mass frame. We are considering that $E_{CM} > M_{top}$.

the E_{CM} is equal to the top quark mass. The mixing angles are stable for the amplitude. For the pseudoscalar case we did not consider the second diagram contribution in order to maintain CP symmetry in this part. However, it is necessary to consider it to obtain a more precise value for the amplitude. Currently, we are working in a more complete expression for amplitudes in order to compute the discussed relic abundances.

Acknowledgments
This work is supported in part by CONACYT, DOE grant DOE/DE-FG02-94ER40823, FPA 2005-02327 project (DGICYT, Spain), CAM/UCM 910309 project, and MICINN Consolider-Ingenio MULTIDARK CSD2009-00064.

References
[1] P. Jordan, Nature (London) **164**, 637 (1949); M. Fierz, Helv. Phys. Acta **29**, 128 (1956); C. Brans and R. Dicke, Phys. Rev. **124**, 925 (1961); P. G. Bergmann, Int. J. Theor. Phys. **1**, 25 (1968); K. Nordtvedt, Astrophys. J. **161**, 1059 (1970); R. Wagoner, Phys. Rev. **D1**, 3209 (1970); T. Damour and K. Nordtvedt, Phys. Rev. Lett. **70**, 2217 (1993); *ibid.*, Phys. Rev. D **48**, 3436 (1993); T. Damour and B. Pichon, Phys. Rev. D **59**, 123502 (1999) [astro-ph/9807176]; C. Schimd, J. P. Uzan, and A. Riazuelo, Phys. Rev. D **71**, 083512 (2005); J. A. R. Cembranos, Phys. Rev. D **73**, 064029 (2006) [arXiv:gr-qc/0507039]; AIP Conf. Proc. **1182**, 288 (2009) [arXiv:0906.5297 [astro-ph.CO]]; J. A. R. Cembranos, K. A. Olive, M. Peloso and J. P. Uzan, JCAP **0907**, 025 (2009) [0905.1989 [astro-ph.CO]]; A. de la Cruz-Dombriz *et al.*, Phys. Rev. D **74**, 087501 (2006) [gr-qc/0607118]; Phys. Rev. D **77**, 123515 (2008) [0802.2999 [astro-ph]]; A. Coc, K. A. Olive, J. P. Uzan and E. Vangioni, Phys. Rev. D **73**, 083525 (2006) [astro-ph/0601299]; [0811.1845 [astro-ph]].
[2] E. Witten, Nucl. Phys. B **268**, 253 (1986); V. A. Kostelecky and S. Samuel, Phys. Lett. B **207**, 169 (1988); Nucl. Phys. B **336**, 263 (1990); I. Y. Aref'eva, A. S. Koshelev, D. M. Belov and P. B. Medvedev, Nucl. Phys. B **638**, 3 (2002) [arXiv:hep-th/0011117]; I. Y. Aref'eva, L. V. Joukovskaya and A. S. Koshelev, JHEP **0309**, 012 (2003) [arXiv:hep-th/0301137]; M. Fujita and H. Hata, Phys. Rev. D **70**, 086010 (2004)

[arXiv:hep-th/0403031]; T. Biswas, J. A. R. Cembranos and J. I. Kapusta, Phys. Rev. Lett. **104**, 021601 (2010) [arXiv:0910.2274 [hep-th]]; arXiv:1005.0430 [hep-th]; arXiv:1006.4098 [hep-th].
[3] L. Bergstrm and A.Goobar, *Cosmology and Particle Astrophysics* (Springer, 2006).
[4] E.W. Kolb and M.S. Turner, *The Early Universe* (Addison-Wesley, 1990)
[5] M. Srednicki, R. Watkins and K.A. Olive, *Nucl. Phys.* **B310**, 693 (1988).
[6] P. Gondolo and G. Gelmini, *Nucl. Phys.* **B360**, 145 (1991).
[7] L. Covi, J. E. Kim and L. Roszkowski, Phys. Rev. Lett. **82**, 4180 (1999) [hep-ph/9905212]; J. L. Feng, A. Rajaraman and F. Takayama, Phys. Rev. D **68**, 085018 (2003) [hep-ph/0307375]; J. L. Feng, A. Rajaraman and F. Takayama, Int. J. Mod. Phys. D **13**, 2355 (2004) [hep-th/0405248]; J. A. R. Cembranos, J. L. Feng, A. Rajaraman and F. Takayama, Phys. Rev. Lett. **95**, 181301 (2005) [hep-ph/0507150]; J. A. R. Cembranos, J. L. Feng, L E. Strigari, Phys. Rev. D **75**, 036004 (2007) [hep-ph/0612157]; 0708.0247 [hep-ph]; Phys. Rev. Lett. **99**, 191301 (2007) [0704.1658 [astro-ph]]; 0708.0239[hep-ph]; J. A. R. Cembranos, Phys. Rev. Lett. **102**, 141301 (2009) [arXiv:0809.1653 [hep-ph]].
[8] H. Goldberg, Phys. Rev. Lett. **50**, 1419 (1983).
[9] J. R. Ellis *et al.*, Nucl. Phys. B **238**, 453 (1984).
[10] G. Servant and T. M. P. Tait, Nucl. Phys. B **650**, 391 (2003) [hep-ph/0206071].
[11] H. C. Cheng, J. L. Feng and K. T. Matchev, Phys. Rev. Lett. **89**, 211301 (2002) [hep-ph/0207125].
[12] J. A. R. Cembranos, A. Dobado and A. L. Maroto, Phys. Rev. Lett. **90**, 241301 (2003) [hep-ph/0302041]; and Phys. Rev. D **68**, 103505 (2003) [hep-ph/0307062].
[13] J. Alcaraz *et al.*, Phys. Rev.**D67**, 075010 (2003) [hep-ph/0212269]; J. A. R. Cembranos, A. Dobado and A. L. Maroto, AIP Conf.Proc. **670**, 235 (2003) [hep-ph/0301009]; hep-ph/0402142; hep-ph/0406076; Int. J. Mod. Phys. **D13**, 2275 (2004) [hep-ph/0405165]; hep-ph/0411076; astro-ph/0411262; Phys. Rev. D **73**, 035008 (2006) [hep-ph/0510399]; astro-ph/0503622; astro-ph/0512569; astro-ph/0611911; A. L. Maroto, Phys. Rev. D **69**, 043509 (2004) [hep-ph/0310272]; Phys. Rev. D **69**, 101304 (2004) [hep-ph/0402278]; J. A. R. Cembranos *et al.*, 0708.0235 [hep-ph]; JCAP **0810**, 039 (2008) [0803.0694 [astro-ph]]; J. A. R. Cembranos and L. E. Strigari, Phys. Rev. D **77**, 123519 (2008) [0801.0630 [astro-ph]].
[14] P. Achard *et al.*, Phys. Lett. **B597**, 145 (2004) [hep-ex/0407017]; S. Heinemeyer *et al.*, hep-ph/0511332; J. A. R. Cembranos, A. Rajaraman and F. Takayama, hep-ph/0512020; Europhys. Lett. **82**, 21001 (2008) [hep-ph/0609244]; J. A. R. Cembranos, A. Dobado and A. L. Maroto, hep-ph/0107155; Phys. Rev. **D65** 026005 (2002) [hep-ph/0106322]; hep-ph/0307015; Phys. Rev. **D70**, 096001 (2004) [hep-ph/0405286]; hep-ph/0512302; Phys. Rev. D **73**, 057303 (2006) [hep-ph/0507066]; J. Phys. A **40**, 6631 (2007) [hep-ph/0611024]; A. Juste *et al.*, hep-ph/0601112; J. A. R. Cembranos *et al.*, hep-ph/0603067; AIP Conf. Proc. **903**, 591 (2007) [hep-ph/0701011]; ILC Collaboration, 0709.1893 [hep-ph]; 0712.1950 [physics.acc-ph]; 0712.2356 [physics.ins-det];
[15] T.D. Lee, *Phys. Rev.* **D8**, 1226 (1973).
[16] S.L. Glashow and S. Weinberg, *Phys. Rev.* **D15**, 1958 (1977).
[17] J.F. Donoghue and L. F. Li, *Phys. Rev.* **D19**, 945 (1979).
[18] H.E. Haber, G.L. Kane and T. Sterling, *Nucl. Phys.* **B161**, 493 (1979).
[19] L.J. Hall and M.B. Wise, *Nucl. Phys.* **B187**, 397 (1981).
[20] J. F. Gunion, H. E. Haber, G. Kane and S. Dawson, *The Higgs Hunters Guide* (Perseus Publishing, 1990).
[21] A. Dege and I.P. Ivanov, *Phys. Rev. D.* **81**, 015012 (2010) [arXiv:0910.4492 [hep-ph]].
[22] D. Kominis and R. Sekhar Chivukula, hep-ph/9301222
[23] E.M. Dolle and S. Su, *Phys. Rev. D* **80**, 055012 (2009) [arXiv:0906.1609v1 [hep-ph]].
[24] E.Ma, *Phys. Rev. D* **73**, 077301 (2006) [arXiv hep-ph/0601225].
[25] M. Cirelli, N. Fornengo and A. Sturmia, *Nucl. Phys.* **B753**, 178-194 (2006) [arXiv hep-ph/0512090].
[26] R.Barbieri, L. J. Hall and V. S. Rychkov, *Phys Rev D* **74** 015007 (2006) [arXiv hep-ph/0603188].
[27] L. Lopez-Honorez, E. Nezri, J. F. Oliver and M. H. G. Tytgat, *JCAP* **02** (2007) 028 .
[28] E. Dolle, X. Miao, S. Su and B. Thomas, arXiv:hep-ph 0909.3094(2009).
[29] M. Krawczyk, arXiv:hep-ph 0512371 (2005); M. Krawczyk and D. Sokolowska, arXiv:hep-ph 0911.2457 (2009).

Stability of Φ^4 oscillatons

S Valdez-Alvarado[1], L A Ureña-López[1] and R Becerril-Bárcenas[2]

[1] Departamento de Física, DCI, Campus León, CP 37150, Universidad de Guanajuato,León, Guanajuato, México

[2] Instituto de Física y Matemáticas, Universidad Michoacana de San Nicolás de Hidalgo. Edificio C-3, Cd. Universitaria, C. P. 58040 Morelia, Michoacán, México

E-mail: svaldez@fisica.ugto.mx,lurena@fisica.ugto.mx,becerril@ifm.umich.mx

Abstract. We investigate the stability of oscillatons with a quartic self-interaction. Oscillatons are spherically symmetric solutions of the coupled Einstein-Klein-Gordon for the case of a real scalar field. It is shown that there exist equilibrium configurations which are stables under small and strong (radial) perturbations. Some basic numerical tools are implemented that shall help in the study of arbitrary equilibrium configurations.

1. Introduction

Seidel and Suen were the first that found non-singular and asymptotically flat solutions of the coupled Einstein-Klein-Gordon (EKG) equations, in which both the (real) scalar field and the metric are time-dependent [1, 2]. These solutions are so-called *oscillatons* that is the short name for oscillating soliton stars. Time-dependent of oscillatons arise because we working with real scalar field, and this case is different from its complex counterpart, boson stars, for which the spacetime geometry is static[3].

In an exhaustive analysis presented in[4], it was shown that oscillatons are classified into stable (S-branch) and unstable (U-branch) configurations. S-oscillatons are stable configurations under small radial perturbations, and they typically migrate to other S-profiles if strongly perturbed. On the other hand, U-oscillatons are intrinsically unstable: they migrate to the S-branch if their mass is moderate, but they may collapse into black holes if their mass is large enough.

As we mention in [5], oscillatons may have some importance in both Astrophysics and Cosmology, because scalar fields have been proposed as candidates to be dark matter in the Universe [6, 7, 8, 9]. Scalar field dark matter would form cosmological structure with properties similar to those of oscillatons.More recently, it has been suggested that multistate bosonic objects could also provide an alternative explanation for the flat rotation curves in galaxies [10, 11, 12, 13].

In this work, we study the evolution of oscillatons with a quartic self-interaction in the scalar field potential, and analyze the stability in some selected cases following closely the numerical tools presented in Ref.[4].

Journal of Physics: Conference Series **315** (2011) 012013 doi:10.1088/1742-6596/315/1/012013

2. Mathematical background

We solve the system of the Einstein-Klein-Gordon (EKG) equations for a spherically-symmetric metric of the form

$$ds^2 = -\alpha^2(t,r)dt^2 + a^2(t,r)dr^2 + r^2\left(d\theta^2 + \sin^2\theta d\varphi^2\right), \tag{1}$$

where $a^2(t,r)$ is the radial metric function, and $\alpha^2(t,x)$ is the lapse function. The action that describes our self-gravitating system is

$$\int d^4x\sqrt{-g}\left(\frac{R}{16\pi G} - \frac{1}{2}g^{\mu\nu}\Phi_{;\mu}\Phi_{;\nu} - V(\Phi)\right), \tag{2}$$

where Φ is a real scalar field endowed with a scalar filed potential

$$V(\Phi) = \frac{1}{2}m_\Phi^2\Phi^2 + \frac{1}{4}\lambda\Phi^4. \tag{3}$$

The parameter m_Φ is the mass of the scalar field, and λ is the quartic interaction parameter.

The evolution of metric functions $a(t,r)$ and $\alpha(t,r)$ is governed by Einstein equations: $G_{\mu\nu} = 8\pi G T_{\mu\nu}$, where the energy-momentum tensor for the scalar field is

$$T_{\mu\nu} = \Phi_{,\mu}\Phi_{,\nu} - \frac{1}{2}g_{\mu\nu}[\Phi^{,\gamma}\Phi_{,\gamma} + m_\Phi^2\Phi^2 + \lambda\frac{\Phi^4}{2}] \tag{4}$$

The scalar field dynamics is described by the Klein-Gordon (KG) equation, which is obtained of the conservation equations for the scalar field energy-momentum tensor

$$T^{\mu\nu}_{;\nu} = \Phi^{;\nu}\left(\Box - \frac{dV}{d\Phi}\right)\Phi = 0 \tag{5}$$

where $\Box$ is the covariant d'Alembertian operator. It is convenient, for numerical purposes, to define the following set of dimensionless variables: $x = m_\Phi r$, $t \to m_\Phi t$ and $\Phi \to \sqrt{8\pi G}\Phi$. In terms of the new variables the Hamiltonian constraint is written as

$$a_{,x} = a\left[\frac{1-a^2}{2x} + \frac{x}{4}\left(\Pi^2 + \Psi^2 + a^2\left[\Phi^2 + \Lambda\frac{\Phi^4}{2}\right]\right)\right] \tag{6}$$

and the polar-areal slicing condition becomes

$$\alpha_{,x} = \alpha\left[\frac{a^2-1}{x} + \frac{a_{,x}}{a} - a^2x\left(\frac{\Phi^2}{2} + \Lambda\frac{\Phi^4}{4}\right)\right] \tag{7}$$

where we have also defined $\Lambda \equiv \lambda/(8\pi G m_\Phi^2)$. With the help of the first order variables

$$\begin{aligned} \Psi &= \Phi_{,r} \\ \Pi &= a\Phi_{,t}/\alpha. \end{aligned} \tag{8}$$

The KG equation is equivalent to the following set of first order differential equations

$$\Phi_{,t} = \frac{\alpha}{a}\Pi \tag{9}$$

$$\Pi_{,t} = \frac{1}{x^2}\left(\frac{x^2\alpha\Psi}{a}\right)_{,x} - a\alpha(\Phi + \Lambda\Phi^3) \tag{10}$$

$$\Psi_{,t} = \left(\frac{\alpha\Pi}{a}\right)_{,x} \tag{11}$$

Equations (6-11) are the total set of evolution equations that we solve numerically for different values of the Λ parameter.

3. Boundary conditions

Note that the set of equations is singular at $x = 0$. We take a spatial grid of the form $x_i = (i - 1/2)\Delta x$. to avoid the singularity at the origin. From the evolution equation of Φ, see Eq. (9), we can see that it not necessary to apply boundary condition for Φ, because the evolution equation can be integrated all the way from the boundary point $x_0 = -\Delta/2$ to the outer boundary point.

As in Ref.[4], we use the fictitious point $x_0 = -\Delta x/2$ to impose appropriate parity conditions: Π is even and Ψ is odd. At the outer boundary, we assume that Π behaves as an outgoing wave pulse of the form:

$$\Pi = u(x - t)/x \tag{12}$$

where u is an arbitrary function. In differential form this becomes

$$\partial_x \Pi + \partial_t \Pi + \Pi/x = 0 \tag{13}$$

By using finite difference in (13), we can solve it to find the unknown boundary value at the new time level. Because Π behaves as an outgoing wave at the boundary, so does Φ. Then, the outgoing wave boundary condition applied to Φ can imply that at the outer boundary

$$\Psi = -\Pi - \Phi/x. \tag{14}$$

We used this expression to obtain boundary values for Ψ once those of Φ and Π are known. Boundary condition for the metric functions are the following. Local flatness at the origin implies that $a(x = 0) = 1$ and $\partial_x a(x = 0) = 0$, and this two conditions imply that $a(x_0) = a(x_1) = 1 + \mathcal{O}(x^3)$. The Hamiltonian constraint (6) is then integrated outwards to obtain $a(x)$.

For the lapse function we impose $\alpha = 1/a$ as an outer boundary condition, this is because in a vacuum our slicing condition implies that we are in Schwarzschild coordinates. Then, we are assuming that our boundary conditions are sufficiently far away as to be always in vacuum. Thus, the slicing condition is integrated inwards to obtain $\alpha(x)$.

4. Numerical Results

For our numerical experiments, we use an equilibrium configuration found in[5] as the initial condition for $\Phi(0, x)$, and from this we obtain $\Psi(0, x) = \Phi'(0, t)$ and $\Pi(0, x) = 0$ through Eqs.(8).

We apply a transformation to the evolution equation corresponding to Π, as Eq. (10) could have some problems if discretized without care, because it is not second order accurate due to the presence of the factor $1/x^2$ in the principal part. The transformation that guarantees the accuracy of our simulations at the origin is

$$\Pi_{,t} = 3\frac{\partial}{\partial x^3}\left(\frac{x^2 \alpha \Psi}{a}\right) - a\alpha\Phi \tag{15}$$

and this is the equation which we discretized. In Fig. 1, we show the values obtained for the masses and fundamental frequencies of equilibrium configurations as functions of the central value of the first scalar field coefficient $\phi_1(0)$ for the equilibrium configuration with $\Lambda = 1$[2, 4, 14, 5]. The critical mass of the configurations, given by the maximum value of the mass curve, is $M_{\Phi c} = 0.694(m_{Pl}^2/m_\Phi)$, which is reached at $\phi_{1c}(0) = 0.48$; the corresponding fundamental frequency is $\Omega_\Phi = 0.859$. Before the critical mass value (left side) there are the values of masses that represent the S-branch, and after critical mass value (right side) there are the values masses that represent the U-branch.

Also in Fig. 1, we show the function $x^2\rho$ at $x = 100$ for three different times. We can observe that the different evolution coincide very well up to $x = 20$, but they differ significantly for large radii where $x^2\rho < 10^{-5}$. This means that part of the scalar field has to be reflected from the boundary, because our boundary conditions correspond, strictly, to the massless case. In any case, the error is not significant and we were able to keep it under control in the numerical evolutions.

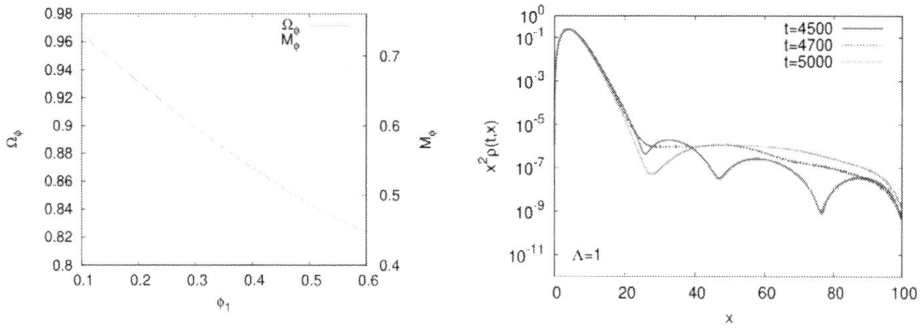

Figure 1. Plots of M_Φ as functions of the central value $\phi_1(0)$ for $\Lambda = 1$(left). And function $x^2\rho$ at $x = 100$ for three different times and $\Lambda = 1$

On the other hand, in Fig. 2 we show the maximum value of the radial metric $a(t, x)$ as read from the numerical evolution up to $t = 5000$. We see that it maintain the same oscillatory pattern in all time, and we take this as evidence of the stability of this equilibrium configuration. To verify the convergence of our numerical method, we use the momentum constrain equation

$$\beta := a_{,t} - \frac{1}{2}x\alpha\Psi\Pi = 0 \tag{16}$$

Its evolution is also shown in Fig. 2, in which we plot the $L2$ norm of the value of β across the grid as a function of the time for three different resolutions. This shows that our numerical code is second order convergent.

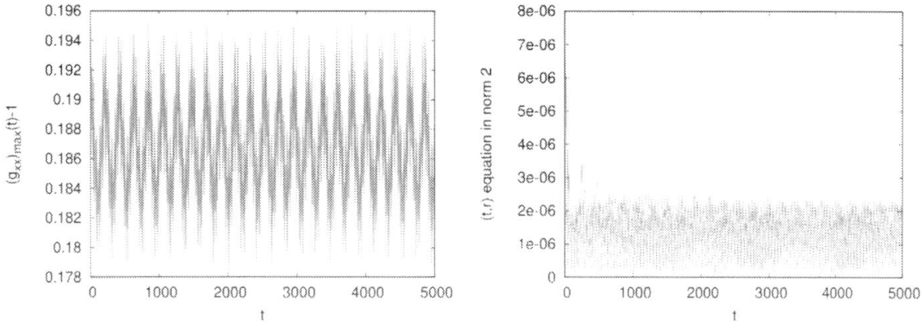

Figure 2. (Left) Maximum values as function of the time of the radial metric function $g_{rr} = a^2(t, r)$ for a stable equilibrium configuration. The numerical outer boundary is located at $x = 100$, the resolution is $\Delta x = 0.01$, and the quartic parameter is $\Lambda = 1$. (Right) The $L2 - norm$ of the momentum constrain, see Eq. (16), for three different spatial resolutions: $\Delta x = 0.01$ (fine), $\Delta x = 0.02$ (medium) and $\Delta x = 0.04$ (coarse). The numerical code is then second order convergent

We present some evolutions of different perturbed S-oscillatons in Fig. 3. We can see that when the perturbation is strong, the oscillaton migrate back to the line of equilibrium configurations if the initial mass is less than the critical value, $M_c = 0.694$, see Fig. 1.

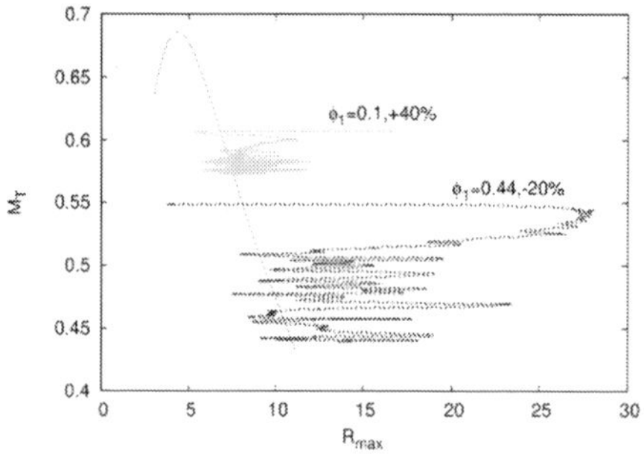

Figure 3. Evolution of different strongly perturbed oscillatons. The perturbated oscillatons migrate and settles down onto another equilibrium configuration (represented by de solid line) when the initial mass is less than the critical one, $M_c = 0.694$.

5. Final Comments

Results obtained for oscillatons with quartic self-interaction indicate that their behavior is similar tho those obtained in Refs.[15, 4] for the case of a quadratic potential only.

It is known that massive oscillatons may be classified into stable and unstable configurations, and that this classification can be read off from the mass function similar to that shown in Fig. 1. We have only shown the existence of stable quartic oscillatons under small and strong (radial) perturbations.

More work is needed to have a complete classification of the stability and, in particular, to determine the existence of unstable configurations. This is research under progress that we expect to report elsewhere.

Acknowledgments

SV-A acknowledges support from CONACyT, México. This work was partially supported PROMEP, DAIP,and by CONACyT México under grants 56946, 83825, and I0101/131/07 C-234/07 of the Instituto Avanzado de Cosmologí (IAC) collaboration (http://www.iac.edu.mx/)

References

[1] Seidel E and Suen W M 1991 *Phys. Rev. Lett.* **66** 1659–1662
[2] Urena-Lopez L A, Matos T and Becerril R 2002 *Class. Quant. Grav.* **19** 6259–6277
[3] Schunck F E and Mielke E W 2003 *Class. Quant. Grav.* **20** R301–R356 (*Preprint* 0801.0307)
[4] Alcubierre M *et al.* 2003 *Class. Quant. Grav.* **20** 2883–2904 (*Preprint* gr-qc/0301105)
[5] Valdez-Alvarado S, Becerril R and Urena-Lopez L A 2010 *AIP Conf. Proc.* **1256** 357–361
[6] Sahni V and Wang L M 2000 *Phys. Rev.* **D62** 103517 (*Preprint* astro-ph/9910097)
[7] Hu W, Barkana R and Gruzinov A 2000 *Phys. Rev. Lett.* **85** 1158–1161 (*Preprint* astro-ph/0003365)
[8] Matos T and Urena-Lopez L A 2000 *Class. Quant. Grav.* **17** L75–L81 (*Preprint* astro-ph/0004332)
[9] Matos T and Urena-Lopez L A 2001 *Phys. Rev.* **D63** 063506 (*Preprint* astro-ph/0006024)
[10] Matos T and Urena-Lopez L A 2007 *Gen. Rel. Grav.* **39** 1279–1286
[11] Bernal A 2008 *AIP Conf. Proc.* **1083** 20–27
[12] Bernal A, Barranco J, Alic D and Palenzuela C 2009 (*Preprint* 0908.2435)
[13] Urena-Lopez L A and Bernal A 2010 *Phys. Rev.* **D82** 123535 (*Preprint* 1008.1231)
[14] Urena-Lopez L A 2002 *Class. Quant. Grav.* **19** 2617–2632 (*Preprint* gr-qc/0104093)

[15] Seidel E and Suen W M 1994 *Phys. Rev. Lett.* **72** 2516–2519 (*Preprint* gr-qc/9309015)

Journal of Physics: Conference Series **315** (2011) 012014 doi:10.1088/1742-6596/315/1/012014

Z' and Dark Matter

Yann Mambrini

Laboratoire de Physique Théorique, Université Paris-Sud, F-91405 Orsay, France

E-mail: yann.mambrini@th.u-psud.fr

Abstract. Several constructions motivate the existence of a dark $U(1)_D$ gauge boson which interacts with the Standard Model only through its kinetic mixing or loop induced processes. We describe two typical examples with specific signatures in particular we show that a region with relatively light WIMPS, $M_{Z_D} < 40$ GeV and a kinetic mixing $10^{-4} < \delta < 10^{-3}$ is not yet excluded by the last experimental data and seems to give promising signals in a near future. We also show that conditions from anomaly cancelation generate tri-vector couplings $Z_D Z \gamma$ leading to a specific gamma ray line observable by FERMI telescope.

1. Introduction

Neutral gauge sectors with an additional dark $U(1)_D$ symmetry in addition to the Standard Model (SM) hypercharge $U(1)_Y$ and an associated Z_D are among the best motivated extensions of the SM, and give the possibility that a dark matter candidate lies within this new gauge sector of the theory [1]. The new vector boson Z_D can interact with the SM, even if no SM fermions are directly charged under the additional gauge symmetry. This interaction can occurs via mixed kinetic terms between the SM's hypercharge field strength and the new abelian field strength [2, 3, 4, 5] or through couplings generated by counter-term to preserve the anomaly cancelation condition [6, 7]. Whereas the former couplings can give significant signals in direct detection experiment even fitting the last DAMA [8] or COGENT [9] excesses [10, 11, 12, 13] or INTEGRAL 511 keV line [14], the latter can give rise to a gamma-ray line observable in satellite telescopes [15, 16, 17, 18]. A summary of gamma-ray line constraints can be found in [19], and within a supersymmetric [20] in extra-dimensional framework in [21]. A light Z' can be justified in leptophobic constructions [22] and even be an interbretation of the Wjj anomaly observed by CDF [23]. Using LEP/Tevatron constraints, a higher dimensional approach can be found in [25] whereas a summary of all the constraints has been studied in [26]

2. The dark kinetic mixing

The matter content of any *dark* $U(1)_D$ extension of the SM can be decomposed into three families of particles:

- The *Visible sector* is made of particles which are charged under the SM gauge group $SU(3) \times SU(2) \times U(1)_Y$ but not charged under $U(1)_D$ (hence the *dark* denomination for this gauge group)
- the *Dark sector* is composed by the particles charged under $U(1)_D$ but neutral with respect of the SM gauge symmetries. The dark matter (ψ_0) candidate is the lightest particle of the *dark sector*

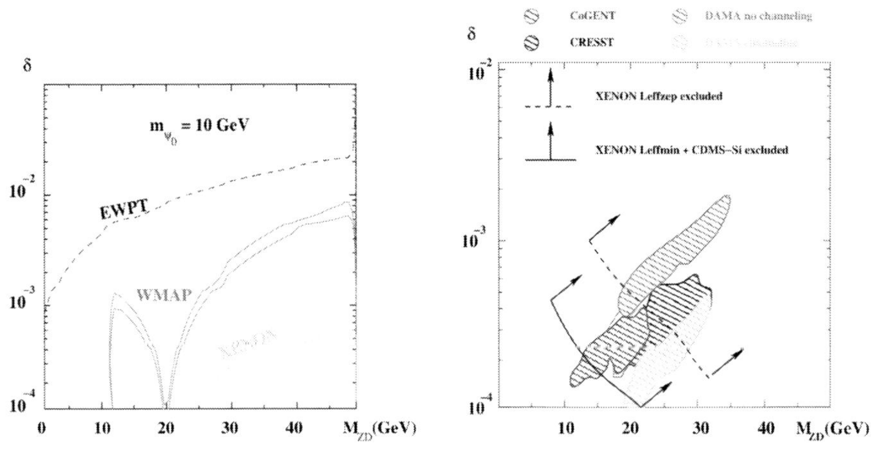

Figure 1. Left : example of allowed parameter space for m_{ψ_0} = 10 GeV in the (M_{Z_D}, δ) plane (left). The points between the full-red region respect the 5σ WMAP constraint, the points below the dashed-black line do not exceed accelerator data on precision tests, and the points above the dotted green line are excluded by XENON100 data. Right: parameter space allowed within 90 % of C.L. for the CoGeNT signal (blue), DAMA without channeling (red), with channeling (green), CRESST (black), and the exclusion region depending on the hypothesis concerning L_{eff}.

- The *Hybrid sector* contains states with SM *and* $U(1)_D$ quantum numbers. These states are fundamental because they act as a portal between the two previous sector through the kinetic mixing they induce at loop order.

From these considerations, it is easy to build the effective lagrangian generated at one loop :

$$\begin{aligned}
\mathcal{L} &= \mathcal{L}_{\rm SM} - \frac{1}{4}\tilde{B}_{\mu\nu}\tilde{B}^{\mu\nu} - \frac{1}{4}\tilde{X}_{\mu\nu}\tilde{X}^{\mu\nu} - \frac{\delta}{2}\tilde{B}_{\mu\nu}\tilde{X}^{\mu\nu} \\
&+ i\sum_i \psi_i\gamma^\mu D_\mu \psi_i + i\sum_j \Psi_j\gamma^\mu D_\mu \Psi_j \qquad (1)
\end{aligned}$$

B_μ being the gauge field for the hypercharge, X_μ the gauge field of $U(1)_D$ and ψ_i the particles from the hidden sector, Ψ_j the particles from the hybrid sector, $D_\mu = \partial_\mu - i(q_Y\tilde{g}_Y\tilde{B}_\mu + q_D\tilde{g}_D\tilde{X}_\mu + gT^aW^a_\mu)$, T^a being the $SU(2)$ generators, and

$$\delta = \frac{\tilde{g}_Y\tilde{g}_D}{16\pi^2}\sum_j q_Y^j q_D^j \log\left(\frac{m_j^2}{M_j^2}\right) \qquad (2)$$

with m_j and M_j being hybrid mass states [4]. Notice that the sum is on all the hybrid states, as they are the only ones which can contribute to the $Y_\mu X_\mu$ propagator. After diagonalization of the current eigenstates, one makes the gauge kinetic terms of Eq.(1) diagonal and canonical.

We show in Fig.1 (left) the points that fulfill the WMAP 5σ bound [27] on $\Omega_{\rm DM}$ for $m_{\psi_0} = 10$ GeV in the (M_{Z_D}, δ) plane. One can clearly see the Z_D−pole region when $M_{Z_D} \sim m_{\psi_0}$. One important point is that for a given M_{Z_D} and m_{ψ_0}, there exists a unique solution δ (up to the very small uncertainties at 5σ) fulfilling WMAP constraints : from 3 parameters $(m_{\psi_0}, M_{Z_D}, \delta)$, the WMAP constraints reduce it to two (M_{Z_D}, δ).

We show in Fig.1 the points respecting WMAP, and the DAMA/LIBRA (with and without channeling) CoGeNT and CRESST[1] results at 90 % of CL. All the constraints have been calculated for a standard Maxwellian velocity distribution (with mean velocity $v_0 = 230$ km/s and an escape velocity $v_{esc} = 600$ km/s). One can observe in Fig.1 that for all experiments, the regions are quite surprisingly near and correspond to 10 GeV $\lesssim M_{Z_D} \lesssim$ 30 GeV and $10^{-4} \lesssim \delta \lesssim 10^{-3}$, which is in complete agreement with the measurement of electroweak precision tests. Moreover, such values of δ are typical of one loop-order corrections and can easily be generated by heavy-fermions loops in the $Z - Z_D$ propagator.

3. Anomalies and gamma-ray line

It is well known that any extension of the SM which introduces chiral fermions with respect to gauge fields suffers from anomalies, a phenomenon of breaking of gauge symmetries of the classical theory at one-loop level. Anomalies are responsible for instance for a violation of unitarity and make a theory inconsistent [29, 30]. For this reason if any construction introduces a new fermionic sector to address the DM issue of the SM, it is vital to check the cancelation of anomalies and its consequences on the Lagrangian and couplings. The idea is to add to the Lagrangian local gauge non-invariant terms in the effective action whose gauge variations cancel the anomalous triangle diagrams. There exist two kinds of term which can cancel the mixed $U(1)_D \times G_A^{\rm SM}$ anomalies, with $G_A^{\rm SM}$ being one of the SM gauge group $SU(3) \times SU(2) \times U_Y(1)$: the Chern Simons (CS) term which couples the $G_A^{\rm SM}$ to the $U(1)_D$ gauge boson, and the Peccei-Quinn (PQ, or Wess-Zumino (WZ)) term which couples the $G_A^{\rm SM}$ gauge boson to an axion. In the effective action, these terms are sometimes called Generalized Chern–Simons (GCS) terms [6]:

$$\begin{aligned}
\mathcal{L}_{inv} &= -\frac{1}{4g'^2}F^{Y\mu\nu}F^{Y}_{\mu\nu} - \frac{1}{4g_X^2}F^{X\mu\nu}F^{X}_{\mu\nu} - \frac{1}{2}(\partial_\mu a_X - M_X X_\mu)^2 - i\bar{\psi}\gamma^\mu D_\mu \psi \\
\mathcal{L}_{var} &= \frac{C}{24\pi^2} a_X \epsilon^{\mu\nu\rho\sigma} F^{Y}_{\mu\nu} F^{Y}_{\rho\sigma} + \frac{E}{24\pi^2}\epsilon^{\mu\nu\rho\sigma} X_\mu Y_\nu F^{Y}_{\rho\sigma}.
\end{aligned} \tag{3}$$

The Stueckelberg axion a_X ensures the gauge invariance of the effective Lagrangian and g_X and $F^X_{\mu\nu} = \partial_\mu X_\nu - \partial_\nu X_\mu$ are the gauge coupling and field strength of $U(1)_D$. The axion has a shift transformation under $U(1)_D$

$$\delta X_\mu = \partial_\mu \alpha \quad , \quad \delta a_X = \alpha \, M_X. \tag{4}$$

The $\mathcal{L}_{var}$ will generate after the $SU(2) \times U(1)_Y$ tri-vectorial couplings $Z_D Z Z$ and $Z_D \gamma Z$ ($Z\gamma\gamma$ coupling being forbidden by spin- momentum conservation). This generates new annihilation processes $\psi_0\psi_0 \to Z_D \to ZZ/Z\gamma$ which can be observable through the *only* monochromatic gamma–ray line with energy $E_\gamma = m_{\psi_0}\left(1 - \frac{M_Z^2}{4m_{\psi_0}^2}\right)$ [15, 16]. Other models predicts several lines [31, 17], but none of them just one line.

As an illustrative point, we show in the left panel of Fig.2 an example of spectrum from the centre annulus that could be observable by the FERMI telescope, generated by DM annihilation within the pole region respecting WMAP constraint($m_{\psi_0} = 258$ GeV and $M_{Z_D} = 591$ GeV). We can clearly distinguish a $\gamma-$ray line centered around $E_\gamma = m_{\psi_0}\left(1 - \frac{M_Z^2}{4m_{\psi_0}^2}\right)$ above the continuous flux produced by the annihilation process $\psi_0\psi_0 \to ZZ/Z\gamma$. The expected sensitivity of FERMI telescope after 5 years of data taking is presented in the right panel of Fig.2.

[1] For the CRESST estimation, we used an extrapolation given in the talk of T. Schwetz and the CRESST collaboration [28].

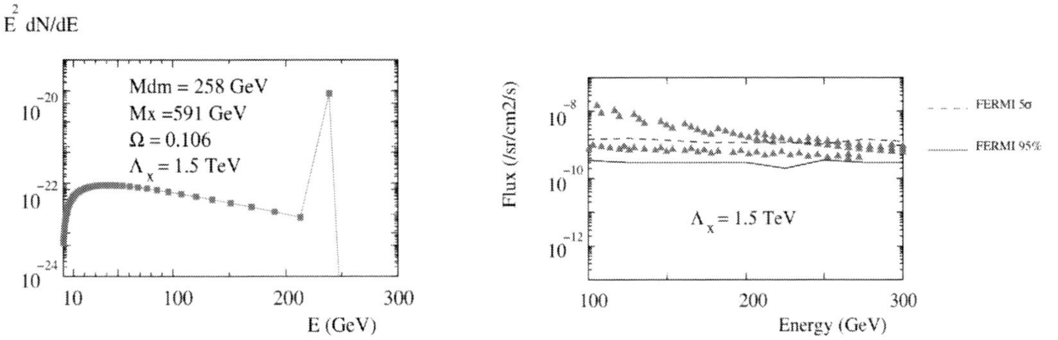

Figure 2. Left: example of gamma–ray flux respecting WMAP constraint for a DM mass of 258 GeV. Right: monochromatic γ−ray fluxes generated by anomaly-cancelation mechanism in comparison with expected 5σ and 95% CL sensitivity contours (5 years of FERMI operation) for the conventional background and unknown WIMP energy, for an effective scale $\Lambda_X = 1.5$ TeV

We clearly see in the right panel of Fig.2 that for an effective scale $\Lambda_X = 1.5$ TeV (scale of the "new physics" corresponding to the fermions generating the anomalies), all the parameter space would be observable by FERMI at 95% CL. Indeed, the points that respect the WMAP constraints lie around the pôle $M_{Z_D} \sim 2m_{\psi_0}$ where $\sim 60\%$ of the annihilation rate is dominated by the $Z\gamma$ final state. This proportion still holds for annihilating DM in the Galactic halo and gives a monochromatic line observable by FERMI.

4. Conclusion

We showed that the existence of a *dark* $U(1)_D$ gauge sector which interacts with the Standard Model only through its kinetic mixing or anomaly-generated couplings possesses a valid dark matter candidate respecting accelerator, cosmological and the more recent direct detection constraints. Moreover, considering the latest results of DAMA/LIBRA, CoGENT and CRESST, we demonstrated that a specific range of the kinetic mixing ($\delta \sim 10^{-4} - 10^{-3}$) can explain all these excesses for a dark boson mass $M_{Z_D} \sim 10-20$ GeV, whereas anomaly cancelationconditions generate a monochromatic γ−ray line from DM annihilation into $Z\gamma$. Such a signature would be a smoking gun signal for these types of constructions It is interesting to notice that other constraints, coming from synchrotron radiation [32] or difuse gamma-ray emission [33] can give more restriction to the analysis.

Acknowledgements

Y.M. wants to thank particularly E. Dudas, T. Schwetz, G. Belanger, N. Fornengo and A. Romagnoni for useful discussions. The work was supported by the french ANR TAPDMS **ANR-09-JCJC-0146** and the spanish MICINNs Consolider-Ingenio 2010 Programme under grant Multi- Dark **CSD2009-00064** and the E.C. Research Training Networks under contract **MRTN-CT-2006-035505**. The author also wants to send all his regards to the "magic" organizing comitee.

References

[1] See e.g. P. Langacker, arXiv:0801.1345 [hep-ph]; M. Pospelov, A. Ritz and M. B. Voloshin, Phys. Lett. B **662** (2008) 53 [arXiv:0711.4866 [hep-ph]]; M. Pospelov, Phys. Rev. D **80** (2009) 095002 [arXiv:0811.1030 [hep-ph]]; M. Pospelov, A. Ritz and M. B. Voloshin, Phys. Lett. B **662** (2008) 53 [arXiv:0711.4866 [hep-ph]]; J. Kumar and J. D. Wells, Phys. Rev. D **74**, 115017 (2006) [arXiv:hep-ph/0606183]; B. Batell,

arXiv:1007.0045 [hep-ph]; K. M. Belotsky, M. Y. Khlopov, K. I. Shibaev, D. Fargion, R. V. Konoplich and M. G. Ryskin, Grav. Cosmol. **11**, 3 (2005).

[2] D. Feldman, B. Kors and P. Nath, Phys. Rev. D **75** (2007) 023503 [arXiv:hep-ph/0610133]; D. Feldman, Z. Liu and P. Nath, Phys. Rev. D **75** (2007) 115001 [arXiv:hep-ph/0702123].

[3] B. Holdom, Phys. Lett. B **166**, 196 (1986); F. del Aguila, M. Masip and M. Perez-Victoria, Nucl. Phys. B **456** (1995) 531 [arXiv:hep-ph/9507455]; T. Cohen, D. J. Phalen, A. Pierce and K. M. Zurek, arXiv:1005.1655 [hep-ph].

[4] M. Baumgart, C. Cheung, J. T. Ruderman, L. T. Wang and I. Yavin, JHEP **0904** (2009) 014 [arXiv:0901.0283 [hep-ph]].

[5] E. J. Chun, J. C. Park and S. Scopel, arXiv:1011.3300 [hep-ph].

[6] P. Anastasopoulos, M. Bianchi, E. Dudas and E. Kiritsis, JHEP **0611** (2006) 057 [arXiv:hep-th/0605225].

[7] J. Kumar, A. Rajaraman and J. D. Wells, Phys. Rev. D **77** (2008) 066011 [arXiv:0707.3488 [hep-ph]]; I. Antoniadis, A. Boyarsky, S. Espahbodi, O. Ruchayskiy and J. D. Wells, arXiv:0901.0639 [hep-ph].

[8] R. Bernabei *et al.*, Riv. Nuovo Cim. **26N1** (2003) 1 [arXiv:astro-ph/0307403]; R. Bernabei *et al.* [DAMA Collaboration], Eur. Phys. J. C **56** (2008) 333 [arXiv:0804.2741 [astro-ph]].

[9] C. E. Aalseth *et al.* [CoGeNT collaboration], arXiv:1002.4703 [astro-ph.CO].

[10] S. Andreas, C. Arina, T. Hambye, F. S. Ling and M. H. G. Tytgat, Phys. Rev. D **82** (2010) 043522 [arXiv:1003.2595 [hep-ph]]; S. Andreas, T. Hambye and M. H. G. Tytgat, JCAP **0810** (2008) 034 [arXiv:0808.0255 [hep-ph]].

[11] Y. Mambrini, JCAP **1009** (2010) 022 [arXiv:1006.3318 [hep-ph]].

[12] M. R. Buckley, D. Hooper and T. M. P. Tait, arXiv:1011.1499 [hep-ph].

[13] T. Schwetz, J. Zupan, [arXiv:1106.6241 [hep-ph]]; D. Hooper, C. Kelso, [arXiv:1106.1066 [hep-ph]].

[14] J. M. Cline, A. R. Frey and F. Chen, arXiv:1008.1784 [hep-ph].

[15] E. Dudas, Y. Mambrini, S. Pokorski and A. Romagnoni, JHEP **0908** (2009) 014 [arXiv:0904.1745 [hep-ph]].

[16] Y. Mambrini, JCAP **0912** (2009) 005 [arXiv:0907.2918 [hep-ph]].

[17] C. B. Jackson, G. Servant, G. Shaughnessy, T. M. P. Tait and M. Taoso, JCAP **1004** (2010) 004 [arXiv:0912.0004 [hep-ph]]; G. Bertone, C. B. Jackson, G. Shaughnessy, T. M. P. Tait and A. Vallinotto, arXiv:1009.5107 [astro-ph.HE].

[18] J. Goodman, M. Ibe, A. Rajaraman, W. Shepherd, T. M. P. Tait and H. B. P. Yu, Nucl. Phys. B **844** (2011) 55 [arXiv:1009.0008 [hep-ph]].

[19] G. Vertongen, C. Weniger, JCAP **1105** (2011) 027. [arXiv:1101.2610 [hep-ph]]; J. Goodman, M. Ibe, A. Rajaraman, W. Shepherd, T. M. P. Tait, H. -B. Yu, Nucl. Phys. **B844** (2011) 55-68. [arXiv:1009.0008 [hep-ph]]; A. A. Abdo, M. Ackermann, M. Ajello, W. B. Atwood, L. Baldini, J. Ballet, G. Barbiellini, D. Bastieri *et al.*, Phys. Rev. Lett. **104** (2010) 091302. [arXiv:1001.4836 [astro-ph.HE]]; J. Conrad, [arXiv:1103.5638 [astro-ph.CO]].

[20] N. Chen, D. Feldman, Z. Liu, P. Nath, G. Peim, Phys. Rev. **D83** (2011) 035005. [arXiv:1011.1246 [hep-ph]].

[21] G. Bertone, C. B. Jackson, G. Shaughnessy, T. M. P. Tait, A. Vallinotto, [arXiv:1009.5107 [astro-ph.HE]].

[22] P. Gondolo, P. Ko, Y. Omura, [arXiv:1106.0885 [hep-ph]].

[23]

[24] K. Cheung, J. Song, Phys. Rev. Lett. **106** (2011) 211803. [arXiv:1104.1375 [hep-ph]]; M. R. Buckley, D. Hooper, J. L. Rosner, [arXiv:1106.3583 [hep-ph]]; M. Buckley, P. Fileviez Perez, D. Hooper, E. Neil, [arXiv:1104.3145 [hep-ph]].

[25] Y. Mambrini, B. Zaldivar, [arXiv:1106.4819 [hep-ph]].

[26] Y. Mambrini, JCAP **1107** (2011) 009. [arXiv:1104.4799 [hep-ph]]; M. T. Frandsen, F. Kahlhoefer, S. Sarkar, K. Schmidt-Hoberg, [arXiv:1107.2118 [hep-ph]]; Y. Mambrini, PoS **QFTHEP2010** (2010) 027. [arXiv:1012.0447 [hep-ph]].

[27] D. N. Spergel *et al.* [WMAP Collaboration], Astrophys. J. Suppl. **170** (2007) 377 [arXiv:astro-ph/0603449]; E. Komatsu *et al.* [WMAP Collaboration], arXiv:0803.0547 [astro-ph].

[28] see talk by W. Seidel, WONDER 2010 Workshop, Laboratory Nazionali del Gran Sasso, Italy, March 22-23, 2010 and MPIK seminar by T. Schwetz, june21, 2010; T. Schwetz, arXiv:1011.5432 [hep-ph].

[29] D. J. Gross and R. Jackiw, Phys. Rev. D **6** (1972) 477; C. Bouchiat, J. Iliopoulos and P. Meyer, Phys. Lett. B **38** (1972) 519.

[30] C. Coriano, N. Irges and S. Morelli, JHEP **0707** (2007) 008 [arXiv:hep-ph/0701010].

[31] L. Bergstrom, P. Ullio and J. H. Buckley, Astropart. Phys. **9** (1998) 137 [arXiv:astro-ph/9712318]; T. Bringmann, L. Bergstrom and J. Edsjo, JHEP **0801**, 049 (2008) [arXiv:0710.3169 [hep-ph]]. L. Bergstrom, T. Bringmann, M. Eriksson and M. Gustafsson, Phys. Rev. Lett. **94** (2005) 131301 [arXiv:astro-ph/0410359]; A. Birkedal, K. T. Matchev, M. Perelstein and A. Spray, arXiv:hep-ph/0507194; J. F. Beacom, N. F. Bell and G. Bertone, Phys. Rev. Lett. **94** (2005) 171301 [arXiv:astro-ph/0409403]; M. Gustafsson, E. Lundstrom, L. Bergstrom and J. Edsjo, Phys. Rev. Lett. **99** (2007) 041301 [arXiv:astro-

ph/0703512].
[32] C. Boehm, J. Silk and T. Ensslin, arXiv:1008.5175 [astro-ph.GA]; D. Hooper and T. Linden, arXiv:1011.4520 [astro-ph.HE].
[33] C. Arina and M. H. G. Tytgat, arXiv:1007.2765 [astro-ph.CO].

Earth matter effect on active-sterile neutrino oscillations

Mario A. Acero, Alexis A. Aguilar-Arevalo and J. C. D'Olivo

Instituto de Ciencias Nucleares, Departamento de Física de Altas Energías, Universidad Nacional Autónoma de México (ICN-UNAM).
Apdo. Postal 70-543, México, D.F. 04510, México.

E-mail: mario.acero@nucleares.unam.mx

Abstract. Oscillations between active and sterile neutrinos remain as an open possibility to explain some experimental observations. In a four-neutrino mixing scheme, we use the Magnus expansion of the evolution operator to study the evolution of neutrino flavor amplitudes within the Earth. We apply this formalism to calculate the transition probabilities from active to sterile neutrinos taking into account the matter effect for a varying terrestrial density.

1. Introduction

Experimental results from solar, atmospheric, reactor and accelerator neutrino experiments, have proved the existence of neutrino mixing, showing also that neutrinos are massive particles. Although the dominant behavior of neutrino oscillations is well described by a three–neutrino picture with two mass splitting parameters and three mixing angles (plus a CP–violating phase), some anomalous experimental results can be interpreted as an indication of the existence of an additional (sterile) neutrino.

It has been proved that the Magnus expansion [1] is an useful tool to obtain an approximate analytical description of matter effects on neutrino oscillations in the framework of two or three neutrinos [2, 3, 4, 5]. Here we apply this formalism to describe the propagation of neutrinos through the Earth, in the case of four neutrinos: the three standard active neutrinos plus and additional sterile neutrino.

2. Evolution of the four neutrino system

The evolution of the four-neutrino system can be written as

$$\Psi(t) = \mathcal{U}(t, t_0)\Psi(t_0), \tag{1}$$

where $\Psi(t)$ is the neutrino state vector at time t and $\mathcal{U}(t, t_0)$ is the evolution operator, which satisfies the Schrödinger-like equation ($\hbar = c = 1$)

$$i\,\frac{d\mathcal{U}}{dt}(t, t_0) = H(t)\,\mathcal{U}(t, t_0), \tag{2}$$

with the initial condition $\mathcal{U}(t_0, t_0) = I$. In the flavor basis $\{|\nu_\alpha\rangle, \alpha = e, \mu, \tau, s\}$

$$H(t) = UH_0U^\dagger + V(t), \tag{3}$$

where $H_0 = \text{diag}(0, 0, \Delta_{31}, \Delta_{41})$, $V(t) = \text{diag}(V_{CC}(t), 0, 0, -V_{NC}(t))$, and U is the mixing matrix that relates the neutrino flavor states to the mass states: $|\nu_\alpha\rangle = \sum_{i=1}^{4} U^*_{\alpha i}|\nu_i\rangle$. Here, $\Delta_{ij} = \Delta m^2_{ij}/2E = (m_i^2 - m_j^2)/2E$, and V_{CC} and V_{NC} are the charge current and neutral current contributions to the matter effective potential for the active neutrinos. Since we are interested in neutrinos with an energy E of a few GeV and higher, the effect of the Δm^2_{21} mass splitting can be safely discarded and we put $\Delta m^2_{21} = 0$. We also subtracted a common term $V_{NC}(t)$, which has no observable effect on the oscillations.

Hereafter, we assume that neutrinos are of Dirac type and parametrize the mixing matrix as follows:

$$U = R_{23}\Gamma_3 R_{34}\Gamma_3^* R_{14}\Gamma_2 R_{24}\Gamma_2^*\Gamma_1 R_{13}\Gamma_1^* R_{12}, \tag{4}$$

which helps to simplify the calculations. In this formula, R_{ij} are 4×4 rotation matrices expressed in terms of the vacuum mixing angles θ_{ij} and the diagonal matrices Γ_i include the three CP-violating phases δ_i $(i = 1, 2, 3)$ that appear in the present scheme. For $\theta_{i4} = 0$, U reduces to the standard 3-neutrino mixing matrix [6].

It is convenient to introduce a new vector $\hat{\Psi} \equiv \hat{U}^\dagger\Psi$, with $\hat{U} = R_{23}\Gamma_3 R_{34}\Gamma_3^* R_{14}$. Its evolution is driven by the Hamiltonian $\hat{H} = \hat{U}^\dagger H \hat{U}$ which, under the assumption $\Delta m^2_{41} \gg \Delta m^2_{31}, V_{CC}, V_{NC}$, becomes

$$\hat{H}(t) \cong \begin{pmatrix} D_1 & 0 & Z_1^* & 0 \\ 0 & D_2 & 0 & Z_2^* \\ Z_1 & 0 & D_3 & 0 \\ 0 & Z_2 & 0 & D_4 \end{pmatrix}, \tag{5}$$

with

$$D_1 = \Delta_{31}s^2_{13} + V_{CC}c^2_{14} - V_{NC}s^2_{14}c^2_{34}, \qquad D_2 = \Delta_{41}s^2_{24},$$
$$D_4 = \Delta_{41}c^2_{24} + V_{CC}s^2_{14} - V_{NC}c^2_{14}c^2_{34}, \qquad D_3 = \Delta_{31}c^2_{13} - V_{NC}s^2_{34},$$
$$Z_1 = \Delta_{31}c_{13}s_{13}e^{i\delta_1} - V_{NC}s_{14}c_{34}s_{34}e^{i\delta_3}, \qquad Z_2 = \Delta_{41}c_{24}s_{24}e^{i\delta_2}.$$

For brevity we defined $c_{ij} \equiv \cos\theta_{ij}$ and $s_{ij} \equiv \sin\theta_{ij}$.

The matrix in Eq. (5) can be diagonalized by a time-dependent unitary transformation

$$V_m = \Gamma_2 R_{24}(\theta^m_{24})\Gamma_2^*\Gamma_1 R_{13}(\theta^m_{13})\Gamma_1^* R_{12}(\pi/2), \tag{6}$$

such that $V_m \hat{H} V_m^\dagger = H_D = \text{diag}(\lambda_1, \lambda_2, \lambda_3, \lambda_4)$, with the energy eigenvalues in matter $\lambda_k(t)$ ordered in such a way that $\lambda_1 < \lambda_2 < \lambda_3 < \lambda_4$. The mixing angles in matter are given by formulas similar to the familiar ones for two neutrinos:

$$\cos 2\theta^m_{13} = \frac{D_3 - D_1}{\sqrt{(D_3 - D_1)^2 + 4|Z_1|^2}}, \tag{7}$$

$$\cos 2\theta^m_{24} = \frac{D_4 - D_2}{\sqrt{(D_4 - D_2)^2 + 4|Z_2|^2}}. \tag{8}$$

The left panel of Figure 1 shows the mixing angles in matter and the energy eigenvalues λ_i as functions of the potential. We see that λ_2 and λ_3 almost cross and the angle θ^m_{13} exhibit the characteristic resonant behavior for the values of the terrestrial density.

The evolution operator in the flavor basis can be expressed as $\mathcal{U}(t, t_0) = U_m(t)\mathcal{U}_A(t, t_0)U_m^\dagger(t_0)$, where $U_m(t) = \hat{U}V_m(t)$ is the mixing matrix that relates the flavor states with the set $\{|\nu^m_k(t)\rangle, k = 1, \ldots, 4\}$ of the (instantaneous) energy eigenstates in matter. The latter define the adiabatic basis and in it the evolution operator $\mathcal{U}_A(t, t_0)$ obeys Eq. (2) with the Hamiltonian

$$H_A(t) = H_D(t) - iV_m^\dagger(t)\,\dot{V}_m(t). \tag{9}$$

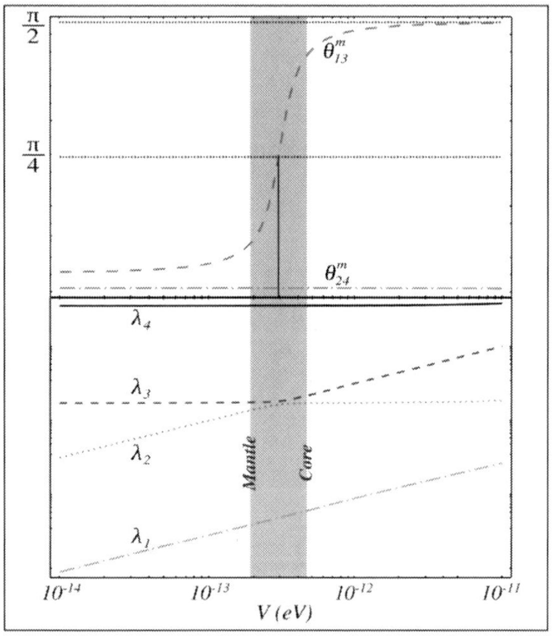

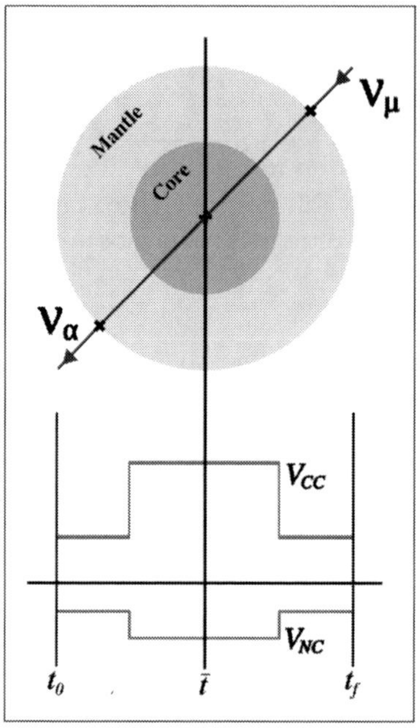

Figure 1. *Left panel.* Evolution of the matter-mixing angles and the (approximated) eigenvalues of the Hamiltonian vs. the potential. *Right panel.* Mantle–core–mantle potential modelling the Earth's interior.

where dot means differentiation with respect to time. Discarding the second term in Eq. (9) corresponds to solving the problem in the adiabatic approximation. In any case, the time dependence generated by H_D can be integrated exactly by a change of the representation accomplished by means of the unitary transformation $\mathcal{P}(t) = \exp[-\int_{t_0}^{t} dt' H_D(t')]$.

For the remaining part of Eq. (9), in general it is not possible to find an analytical solution in a close form and one has to rest on some approximation to determine it. As mentioned in the introduction, an appropriate procedure, which preserves unitarity, is based in the exponential expansion of the evolution operator. Accordingly, next we write $\mathcal{U}_A(t, t_0) = \mathcal{P}(t) \exp \Omega(t, t_0)$ and evaluate Ω in terms of the first two terms of its Magnus expansion: $\Omega \cong \Omega_1 + \Omega_2$. Proceeding in this manner, after some algebraic manipulations we arrive at:

$$\mathcal{U}_A(t_f, t_0) = \mathcal{I}_{32}^* \begin{pmatrix} \mathcal{I}_{32} e^{-i\alpha_1} & 0 & 0 & 0 \\ 0 & (c_\xi - i s_\xi \frac{\xi_{(2)}}{\xi}) e^{i\phi_{\bar{t} \to t_f}} & i e^{i\delta_1} s_\xi \frac{\xi_{(1)}}{\xi} & 0 \\ 0 & i e^{-i\delta_1} s_\xi \frac{\xi_{(1)}}{\xi} & (c_\xi + i s_\xi \frac{\xi_{(2)}}{\xi}) e^{-i\phi_{\bar{t} \to t_f}} & 0 \\ 0 & 0 & 0 & \mathcal{I}_{32} e^{-i\alpha_4} \end{pmatrix} \quad (10)$$

where

$$\mathcal{I}_{32} = \exp\Big\{ -\frac{i}{2} \int_a^b dt' \big[\lambda_3(t') + \lambda_2(t')\big] \Big\}, \quad (11)$$

$$\phi_{a \to b} = \int_a^b dt' \big[\lambda_3(t') - \lambda_2(t')\big], \quad (12)$$

and we have introduced the notations $c_\xi = \cos\xi$, $s_\xi = \sin\xi$, and $\xi = \sqrt{\xi_{(1)}^2 + \xi_{(2)}^2}$. The

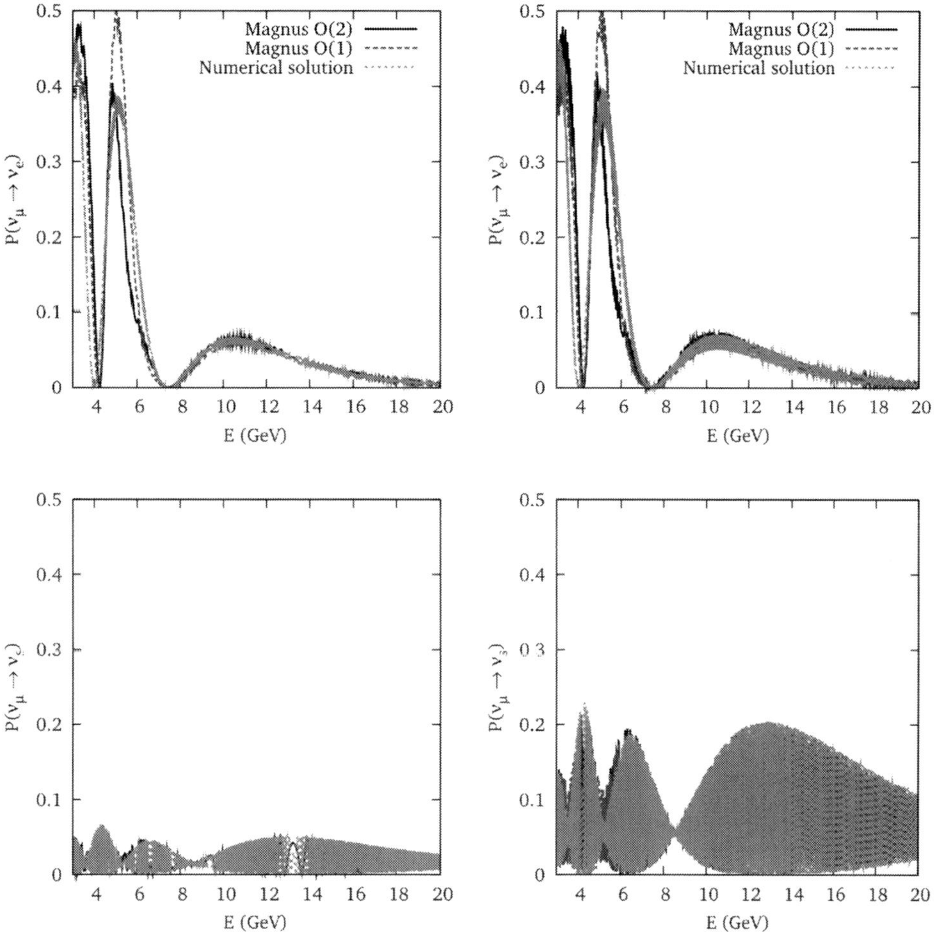

Figure 2. $P(\nu_\mu \to \nu_{e,s})$ as a function of the energy for a neutrino crossing the center of the Earth. Parameters: Common: $\Delta m^2_{31} = 2.5 \times 10^{-3}$ eV^2, $\Delta m^2_{41} = 1.0$ eV^2, $\theta_{12} = 34°$, $\theta_{13} = 7°$, $\theta_{23} = 45°$. Changed: *Left:* $\theta_{i4} = 5°$; *Right:* $\theta_{14} = 5°$, $\theta_{24} = \theta_{34} = 10°$.

quantities

$$\xi_{(1)} = 2\int_{\bar{t}}^{t_f} dt' \dot{\theta}^m_{13}(t') \sin\phi_{\bar{t}\to t'}, \tag{13}$$

$$\xi_{(2)} = \int_{t_0}^{t_f} dt' \int_{t_0}^{t'} dt''\, \dot{\theta}^m_{13}(t')\dot{\theta}^m_{13}(t'') \sin\phi_{t'\to t''}, \tag{14}$$

come from $\Omega_1(t_f, t_0)$ and $\Omega_2(t_f, t_0)$, respectively [4]. In the above expressions we take into account that the potential is symmetric with respect to the middle point of the neutrino trajectory $\bar{t} = (t_f + t_0)/2$ (Fig. 1).

In Figure 2, we show the behavior of the transition probabilities in terms of the neutrino energy for different values of the parameters involved in the analysis. We consider a simplified model for the electron density inside the Earth, the so-called mantle-core-mantle [7] (see Fig. 1). One can see that the Magnus approximation agrees very well with the numerical calculation for $E \gtrsim 7$ GeV. As the plots shows, the effect of the inclusion of the sterile neutrino generates noticeable changes on the $\nu_\mu \to \nu_\alpha$ ($\alpha = e, \mu, \tau$) transition probabilities. Moreover, the $\nu_\mu \to \nu_s$ transition remains as an open possibility. These effects increase for larger values of the three additional mixing angles θ_{i4} ($i = 1, 2, 3$). This is particularly relevant for θ_{24} and θ_{34}, on which no stringent limits have been placed directly, contrary to the case of θ_{14} that has been constrained by the Bugey results [8].

3. Conclusions

We used the Magnus expansion to calculate the flavor transition of a system composed by the three active neutrinos plus a sterile neutrino under the effect of the terrestrial potential. This approximation coincides very well with the numerical computations for high neutrino energies ($E \gtrsim 7$ GeV). The inclusion of the fourth neutrino has a significative effect on $P(\nu_\mu \to \nu_\alpha)$ which depends strongly on θ_{i4} and exhibits a rapid oscillatory behavior superimposed to the standard oscillations for the active neutrinos.

Acknowledgments

M. A. A. is supported by a postdoctoral grant from the UNAM. He also thanks the organizers for the support to participate in DSU2010. The authors acknowledge the support of PAPIIT-UNAM through Grant No. IN117210 and CONACYT through grant 83534.

References

[1] W. Magnus, *Commun. Pure Appl. Math.*, **7**, 649 (1954).
[2] J. C. D'Olivo, *Phys. Rev. D* **45**, 924 (1992).
[3] J. C. D'Olivo and J. A. Oteo,
[4] A.D. Supanitsky, J.C. D'Olivo and G. Medina-Tanco, *Phys. Rev. D* **78**, 0455024 (2008)
[5] A. Ioannisian and A.Yu. Smirnov, *Nucl. Phys.* **B916**, 94 (2009).
[6] C. Amsler *et al.* [Particle Data Group], *Phys. Lett. B* **667**, 1 (2008).
[7] Stacey, F. D. 1977, *Physics of the earth*, John Wiley & Sons, New York, NY (USA) 414 p.
[8] Y. Declais *et al.*, *Nucl. Phys. B* **434**, 503 (1995).

Journal of Physics: Conference Series **315** (2011) 012016 doi:10.1088/1742-6596/315/1/012016

Lepton Flavor Violation Signals from GUT Theories [1]

M. E. Gómez[a], S. Lola[b], P. Naranjo[a], J. Rodríguez-Quintero[a]

[a]Departamento de Física Aplicada, Facultad de Ciencias Experimentales, Universidad de Huelva, 21071 Huelva, Spain.

[b] Department of Physics, University of Patras, 26500 Patras, Greece

E-mail: mario.gomez@dfa.uhu.es

Abstract. We show how quantum corrections above the GUT scale can result in supersymmetric models where lepton flavour violation is naturally suppressed, in agreement with all experimental bounds. Within this framework, a soft term structure as predicted by an Abelian flavour symmetry combined with $SU(5)$ RGEs for scales above M_{GUT}, results to an efficient suppression of the off-diagonal terms in the scalar soft matrices, particularly for $m_0 < 100$ GeV.

1. Introduction

The purpose of this presentation is to show the predictions for lepton flavor violation (LFV) in models where SU(5) unification is combined with flavour symmetries, taking into account the RGE evolution above the GUT scale from Ref. [1] and focusing on the WMAP preferred area presented in [2]. We show that, by postulating the mass matrices at a high scale M_X and evolving them down to M_{GUT}, the pathological situation encountered in the conventional models [3, 4] may be remedied. For regions of the parameter space with a low m_0, the pattern we end up with exhibits a sizable suppression in the off-diagonal terms as compared to the textures at M_X, yielding acceptable LFV predictions. We show that this is true, even in the case of maximal mixing in the charged lepton sector, which is the most dangerous one as far as LFV is concerned.

2. LFV in SUSY-SU(5) with see-saw neutrinos

In SUSY theories, charged lepton flavour violation may be generated at the loop-level, even in models with universal soft terms at a high scale. This is also true for the MSSM, when extended with a see-saw mechanism to generate small neutrino masses [5]. In this case, LFV terms are generated radiatively, since it is not possible to simultaneously diagonalise neutrino, charged lepton and slepton mass matrices [6].

In this section we consider a SUSY-SU(5) theory with see-saw neutrinos. Our starting point is the following SUSY $SU(5)$ superpotential:

$$\mathcal{W}_X = T_1^T\, \mathcal{Y}_u^\delta\, T_1\, H + T_1^T\, \bar{\mathcal{Y}}_d\, \bar{F}_1\, \bar{H} + \bar{F}_1^{\,T}\, \mathcal{Y}_\nu^\delta\, S_1\, H + S_1^T\, \bar{M}_R\, S_1, \tag{1}$$

where $\mathcal{Y}_\alpha$ ($\alpha = u, d, \nu$) are the Yukawa matrices for the up-type quarks, down-quarks/charged-leptons and Dirac neutrinos, respectively. M_R is the heavy Majorana mass matrix. The symbol

[1] Presented by M.E. Gómez, based on Ref. [1].

δ stands for *diagonal*, the original fields rotated as

$$T = U_{10}T_1, \qquad \bar{F} = U_{\nu L}\bar{F}_1, \qquad S = U_{\nu R}S_1, \tag{2}$$

with the rotating matrices defined as

$$\mathcal{Y}_u = U_{10}\mathcal{Y}_u^\delta U_{10}^T, \qquad \mathcal{Y}_d = U_{5L}^*\mathcal{Y}_d^\delta U_{5R}^\dagger, \qquad \mathcal{Y}_\nu = U_{\nu L}^*\mathcal{Y}_\nu^\delta U_{\nu R}^\dagger, \tag{3}$$

and

$$\not{\mathcal{Y}}_d = V_{CKM}^*\mathcal{Y}_d^\delta V_E^\dagger, \tag{4}$$

Here, $V_{CKM} = U_{10}^\dagger U_{5L}$ and $V_E = U_{\nu L}^\dagger U_{5R}$ denote the mixing in the quark and lepton sectors, while $\bar{M}_R = U_{\nu R}^T M_R U_{\nu R}$.

The off-diagonal contributions to slepton mass matrices, when the superfields are rotated so that charged leptons become diagonal, can be understood through three rotations at different energy scales:

- M_X. The rotations in the superpotential fields lead to the following transformation of the soft terms:
$$\bar{m}_{10}^2 = U_{10}^\dagger m_{10}^2 U_{10} \qquad \bar{m}_5^2 = U_{\nu L}^\dagger m_5^2 U_{\nu L} \tag{5}$$
- M_{GUT}. Assuming that $SU(5)$ is broken down to the MSSM gauge group, the superpotential becomes
$$\mathcal{W}_{MSSM} = Q^T\mathcal{Y}_u^\delta UH_2 + Q^T(V_{CKM}^*\mathcal{Y}_d^\delta)DH_2 + L^T(V_E^*\mathcal{Y}_d^\delta)EH_2 + L^T\mathcal{Y}_\nu^\delta SH_2 + S^T\bar{M}_R S \tag{6}$$
where we have absorbed the matrices V_E^* and V_{CKM}^* in the definitions of the superfields E and D respectively. The scalar soft masses then become:
$$m_E^2 = V_{CKM}^\dagger \bar{m}_{10}^2 V_{CKM}, \qquad m_L^2 = \bar{m}_5^2, \tag{7}$$
$$m_Q^2 = m_U^2 = \bar{m}_{10}^2, \qquad m_D^2 = V_E^\dagger \bar{m}_5^2 V_E, \tag{8}$$
- M_N. M_N is the scale at which the heavy right handed neutrinos decouple. Below this scale, the particle content is just the one of the MSSM complemented with the neutrino mass operator resulting from the see-saw mechanism. Consequently, the superpotential can be written in a basis where the charged-lepton mass matrix becomes diagonal and the left slepton mass matrix becomes
$$\bar{m}_L^2 = V_E^\dagger m_L^2 V_E \simeq U_{5R}^\dagger m_5^2 U_{5R} \tag{9}$$

3. Runs above M_{GUT}: *Umbrella Effect*.

The introduction of a non-trivial flavour structure for the slepton soft terms at M_{GUT}, as predicted by the family symmetry that also generates Yukawa couplings, typically results to a large violation of the bounds on $l_j \to l_i\gamma$ [3, 4, 7]. This picture may be remedied by taking into account RGE effects from a scale $M_X > M_{GUT}$. In this case, the cosmological requirement of having a neutral particle as the LSP imposes low values on m_0, such that $m_{\tilde{\tau}} > m_\chi$ [2, 8, 9] (diagonal terms in the soft mass matrices have a large RGE growth, while non-diagonal elements remain almost unaffected by the runs). Thus, even assuming non-diagonal soft terms with matrix elements of the same order of magnitude at M_X, the corresponding matrix at M_{GUT} exhibits dominant diagonal elements.

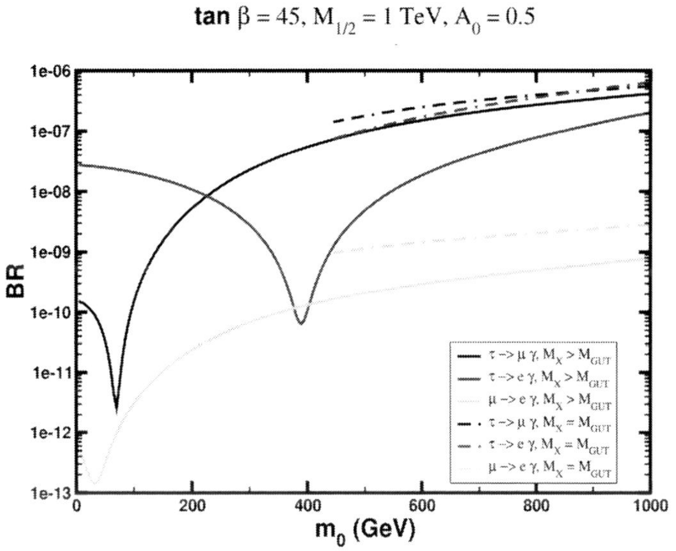

Figure 1. *Prediction for the charged-lepton flavour violating branching ratios showing the difference of taking either M_X or M_{GUT} as the starting point of the runs.*

To some extent, the RGE effect is similar to the action of closing an umbrella: the general non- universal soft terms at M_X resemble an open umbrella that approaches a diagonal matrix at the GUT scale. The sleptons tend then to be aligned by the flavour blind gauge interaction, driven by $\alpha_{\rm GUT}$ above the GUT scale. An analogous effect is well known to take place in the squark sector [10] where the strong radiative corrections to the squark masses, proportional to the gluino mass, are family independent and counteract the splitting effect on the top Yukawa coupling (the effect is commonly referred to as the "gluino focussing effect").

In Fig. 1, we show the differences between the following: i) SU(5) RGE evolution of the soft terms from a high scale M_X down to M_{GUT} and then to the MSSM with see-saw neutrinos (solid lines), and ii) Soft SUSY breaking terms given at M_{GUT} and then the MSSM with see-saw neutrinos (dash-lines). In case ii) we stop the lines at the value of m_0 below which $m_{\tilde{\tau}}$ becomes the LSP. In contrast, m_0 can even vanish at M_X in case i). The textures and soft terms we use are similar to Ref. [4]. However, unlike these authors, we decouple the right-handed neutrinos below M_{GUT}. As a result, the predicted BR's do not vanish in the limit $m_0 = 0$.

4. $SU(5)$ textures

Let us consider an specific example of $SU(5)$ Yukawa textures that match the fermion data and may also predict the pattern of soft terms to be expected.

Following the $U(1)_F$ charge assignment in [11], the Yukawa matrices have the form

$$\mathcal{Y}_u \propto \begin{pmatrix} \varepsilon^6 & \varepsilon^5 & \varepsilon^3 \\ \varepsilon^5 & \varepsilon^4 & \varepsilon^2 \\ \varepsilon^3 & \varepsilon^2 & 1 \end{pmatrix}, \quad \mathcal{Y}_\ell^T \propto \mathcal{Y}_d \propto \begin{pmatrix} \varepsilon^4 & \varepsilon^3 & \varepsilon^3 \\ \varepsilon^3 & \varepsilon^2 & \varepsilon^2 \\ \varepsilon & 1 & 1 \end{pmatrix}, \quad \mathcal{Y}_\nu \propto \begin{pmatrix} \varepsilon^{|1\pm n_1|} & \varepsilon^{|1\pm n_2|} & \varepsilon^{|1\pm n_3|} \\ \varepsilon^{|n_1|} & \varepsilon^{|n_2|} & \varepsilon^{|n_3|} \\ \varepsilon^{|n_1|} & \varepsilon^{|n_2|} & \varepsilon^{|n_3|} \end{pmatrix} \tag{10}$$

where n_i stand for the heavy Majorana neutrino charges. We will assume that the entire lepton mixing is arising from the charged-lepton sector, which is potentially the most dangerous case as far as LFV is concerned.

Journal of Physics: Conference Series **315** (2011) 012016 doi:10.1088/1742-6596/315/1/012016

In addition, flavour symmetries generally imply non-universal soft terms [12], since the structure of the soft terms is linked to the family charges. For the Yukawa textures in Eq.(10) the soft mass matrices m_{10}^2 and $m_{\bar{5}}^2$ become

$$m_{10}^2 \propto \begin{pmatrix} 1 & \varepsilon & \varepsilon^3 \\ \varepsilon & 1 & \varepsilon^2 \\ \varepsilon^3 & \varepsilon^2 & 1 \end{pmatrix} m_0^2, \quad m_{\bar{5}}^2 \propto \begin{pmatrix} 1 & \varepsilon & \varepsilon \\ \varepsilon & 1 & 1 \\ \varepsilon & 1 & 1 \end{pmatrix} m_0^2, \tag{11}$$

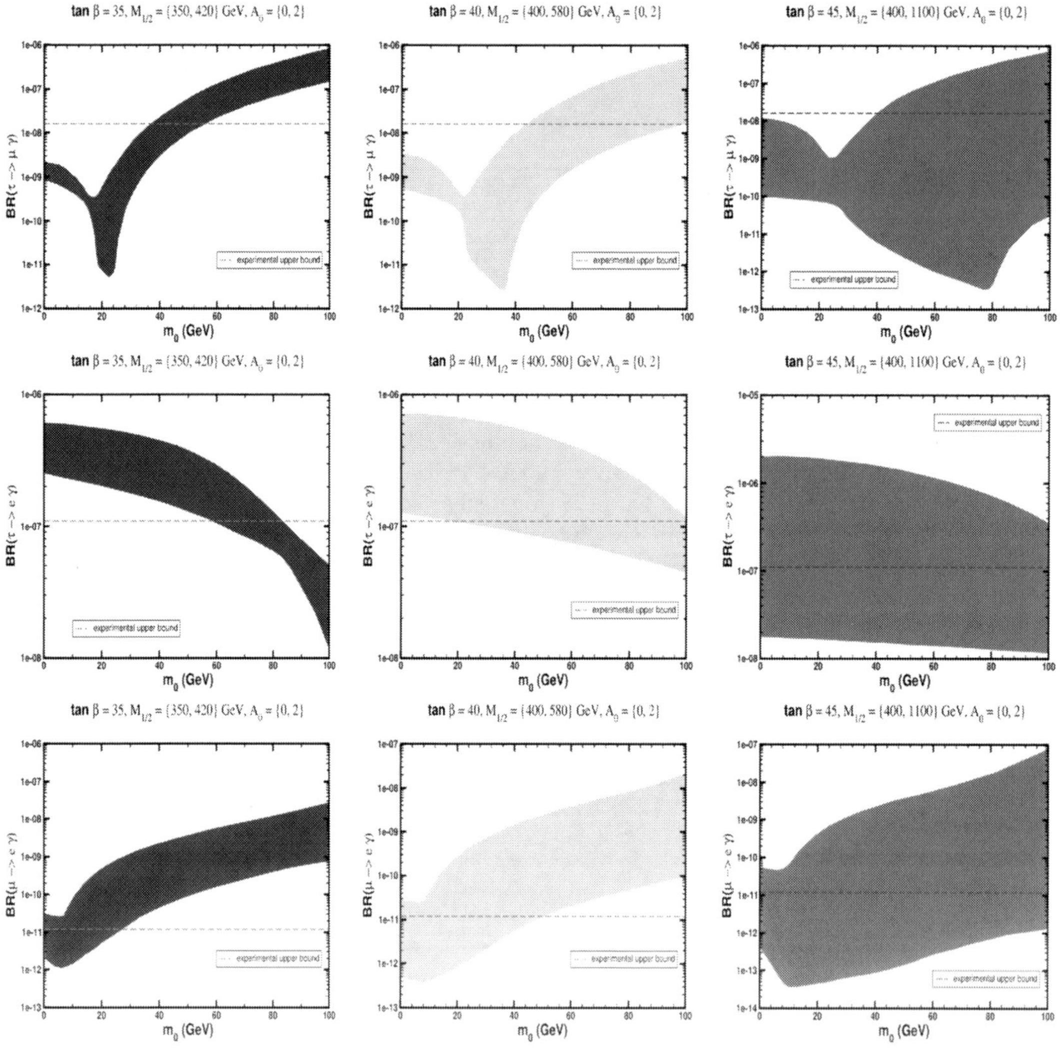

Figure 2. *Prediction for the $\tau \to \mu\gamma$ (top), $\tau \to e\gamma$ (middle) and $\mu \to e\gamma$ branching ratios for the cosmologically preferred area of values of $M_{1/2}$, A_0, for three different values of $\tan\beta$.*

Some LFV predictions of the above model are displayed in Fig. 4. We should stress that the depicted ranges for both m_0 and $M_{1/2}$ are the cosmologically preferred parameter space, as found in [2]. We can see that the case $\tan\beta = 35$ is ruled out, as there is no overlapping region for the

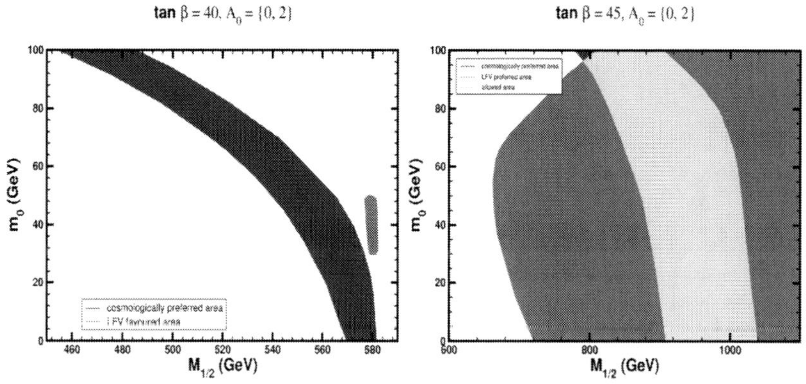

Figure 3. *Allowed parameter space when both LFV and cosmological constraints are taken into account for* $\tan\beta = 40$ *and* 45.

three decays. This is because of the RR sector-induced cancellations mentioned above, which arises for much larger values of m_0 for $\tau \to e\gamma$ than for the other two processes. The case $\tan\beta = 40$ does possess a common area for $20 < m_0 < 50$ GeV. However, such an area lies outside the cosmologically preferred region, as shown on the left panel of Fig. 4. Thus, we conclude that the case $\tan\beta = 40$ is only marginally allowed. Finally, for $\tan\beta = 45$ the whole range for $m_0 < 100$ GeV is allowed, when suitable values for $M_{1/2}$, A_0 are chosen. Moreover, as shown on the right panel of Fig. 4, there exists an overlapping region when considering the cosmologically favoured parameter space (values of any parameters involved in these plots (m_0, $M_{1/2}$, A_0) beyond the ranges shown lead to tachyonic soft masses. Thus, GUT runs efficiently suppress the off-diagonal entries, yielding charged-LFV rates that render the $SU(5)$ model *compatible* with current experimental bounds.

5. Conclusions

We have shown that $SU(5)$ runs above the GUT scale, naturally suppress the off-diagonal entries of the soft matrices, through what has been called the *umbrella effect*, leading to a nearly flavour-independent contribution for $m_0 < 100$ GeV. Such low values of m_0 are discarded if $SU(5)$ runs are not taken into account, since in this case the lightest stau becomes the LSP. Within this framework, $SU(5)$ runs lead to acceptable LFV rates for the cosmologically preferred parameter space found in [2], favoring values of $\tan\beta$ around 45. Significant deviations from this value, in either way, are harder to reconcile with cosmological data. Our results indicate that even in this case, the umbrella effect leads to suppressions to LFV rates, leading to a very significant enhancement of the available parameter space, as compared to the conventional schemes, with runs below M_{GUT}.

Acknowledgments

The research of S. Lola and P. Naranjo has been funded by the FP6 Marie Curie Excellence Grant MEXT-CT-2004-014297. M.E. Gómez and J. Rodríguez-Quintero acknowledge support from the project P07FQM02962 funded by "Junta de Andalucia", the Spanish MICINN projects FPA2009-10773 and MULTIDARK Consolider-Ingenio: CSD2009-00064.

References

[1] M. E. Gomez, S. Lola, P. Naranjo and J. Rodriguez-Quintero, JHEP **1006** (2010) 053 [arXiv:1003.4937 [hep-ph]].

[2] M. E. Gomez, S. Lola, P. Naranjo and J. Rodriguez-Quintero, JHEP **0904** (2009) 043 and AIP Conf. Proc. **1115** (2009) 273; E. Carquin et al., PoS IDM2008 (2008) 090.

[3] G. K. Leontaris and N. D. Tracas, Phys. Lett. B **431** (1998) 90; M. E. Gomez, G. K. Leontaris, S. Lola and J. D. Vergados, Phys. Rev. D **59** (1999) 116009.

[4] P. H. Chankowski, K. Kowalska, S. Lavignac and S. Pokorski, arXiv:hep-ph/0507133, proceedings of 40th Rencontres de Moriond on Electroweak Interactions and Unified Theories, La Thuile, Aosta Valley, March 2005.

[5] J. Hisano, T. Moroi, K. Tobe and M. Yamaguchi, Phys. Rev. D **53** (1996) 2442; J. Hisano and D. Nomura, Phys. Rev. D **59** (1999) 116005.

[6] F. Borzumati and A. Masiero, Phys. Rev. Lett. **57** (1986) 961; L. J. Hall, V. A. Kostelecky and S. Raby, Nucl. Phys. B **267** (1986) 415.

[7] R. Barbieri, L. J. Hall and A. Strumia, Nucl. Phys. B **445** (1995) 219; M. E. Gomez and H. Goldberg, Phys. Rev. D **53** (1996) 5244; G. Barenboim, K. Huitu and M. Raidal, Phys. Rev. D **63**(2001) 055006; L. Calibbi, A. Faccia, A. Masiero and S. K. Vempati, Phys. Rev. D **74** (2006) 116002.

[8] L. Calibbi, Y. Mambrini and S. K. Vempati, JHEP **0709** (2007) 081.

[9] J. Ellis, A. Mustafayev, K. Olive, arXiv:hep-ph/1003.3677.

[10] M. Dine, A. Kagan and S. Samuel, Phys. Lett. B **243** (1990) 250.

[11] J. R. Ellis, M. E. Gomez and S. Lola, JHEP **0707** (2007) 052.

[12] M. Raidal et al., Report of Working Group 3 of the CERN Workshop Flavor in the Era of the LHC, Eur.Phys.J. **C57** (2008) 13 and references therein.

VI International Workshop on the Dark side of the Universe (DSU 2010) IOP Publishing
Journal of Physics: Conference Series **315** (2011) 012017 doi:10.1088/1742-6596/315/1/012017

Background reduction and spin-dependent limits using DRIFT - a directionally sensitive dark matter detector

Mark Pipe on behalf of the DRIFT collaboration
Department of Physics and Astronomy, University of Sheffield, Sheffield, S3 7RH, UK
E-mail: m.pipe@sheffield.ac.uk

Abstract. The DRIFT (Directional Recoil Identification From Tracks) collaboration operates a $1\,m^3$ low pressure gaseous negative ion time projection chamber (NI-TPC) WIMP dark matter search experiment in the Boulby Underground Laboratory. Recent progress from the DRIFT collaboration is presented that includes a series of background reduction work and the addition of CF_4 to the target volume. A preliminary limit on the spin-dependent WIMP-proton interaction cross-section is presented from a non-blind analysis of 47.2 days live time with a $0.8\,m^3$ 30 Torr CS_2 - 10 Torr CF_4 target. The preliminary limit has a minimum of 1.1 pb for a 100 GeV WIMP.

1. Introduction

Weakly Interacting Massive Particles (WIMPs) are a particularly well-motivated candidate for cold dark matter. Dark matter halo models suggest that WIMPs may exist in the form of an isothermal sphere that encompasses our galaxy, resolving the problem of flat galactic rotation curves [1]. WIMPs are predicted to interact with baryonic matter such that sufficiently sensitive particle detectors should be able to detect the rare (<1 event/kg/day), low energy ($\sim$keV) nuclear recoil that would result. The nuclear recoil energy spectrum expected from WIMP interactions is exponential and featureless making it very difficult to discern above the many backgrounds that exist at this energy level. However, WIMPs arrive from a fixed point in the galactic rest frame (roughly from the constellation Cygnus) and this directionality offers a uniquely powerful signature for a directionally sensitive dark matter detector on Earth [2]. The DRIFT collaboration is attempting to utilise this powerful discriminant against terrestrial backgrounds with the development of a directionally sensitive dark matter detector.

The DRIFT-II detector (described in detail in [3]) is a $1\,m^3$ low pressure gaseous negative ion time projection chamber (NI-TPC). WIMP-nucleon interactions in the 40 Torr CS_2-CF_4 target gas are expected to create tracks of ionisation a few mm in length, and with an orientation biased in the direction opposite to that of the incoming WIMPs [4]. The DRIFT detectors use a novel negative ion drift concept to reduce diffusion of this track for drift distances up to 50 cm. Electronegative CS_2 gas is used in the target volume so that primary ionisation electrons quickly attach to the CS_2 molecules creating a track of negative CS_2 ions. These heavy ions drift with far less diffusion than electrons drifted alone, reducing diffusion to thermal levels and preserving the track's directional information.

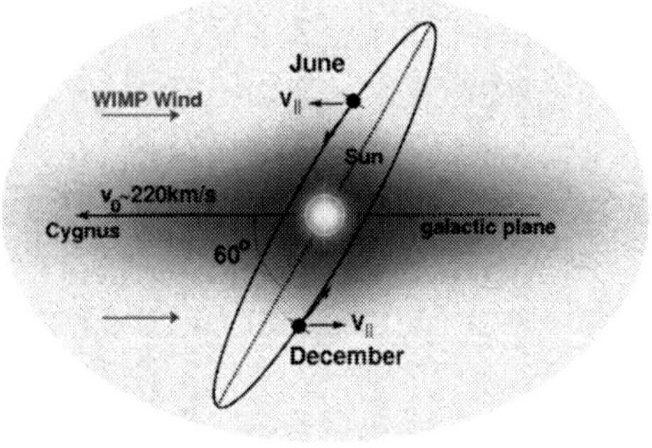

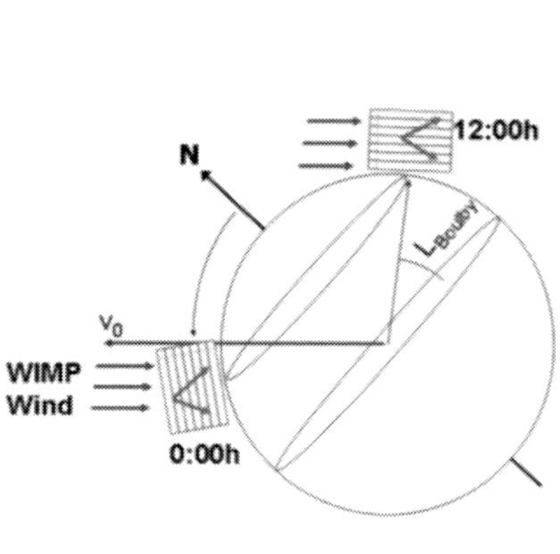

Figure 1. A detector on Earth is subject to a 'wind' of WIMPs with a velocity determined by the motion of the Earth through the essentially static WIMP halo. The $220\,\mathrm{km\,s^{-1}}$ mean WIMP velocity modulates with annual (left) and sidereal cycles (right) making a distinct signal in a direction sensitive WIMP detector on Earth.

There are currently two, essentially identical, DRIFT-II modules in operation. DRIFT-IId is located at a depth of 2805 m.w.e. at the Boulby Underground Laboratory, UK. DRIFT-IIc is located in a surface laboratory at Occidental College, Los Angeles, CA, USA. Results presented here are from the DRIFT-IId detector.

2. Background suppression

The current dominating background in the underground DRIFT-IId detector is from a class of events known as radon progeny recoils (RPRs). A schematic showing the source of these events can be seen in Figure 2. Radon gas (^{222}Rn and ^{220}Rn) is known to be present in the DRIFT-IId detector target volume, emitted from trace levels of U and Th in the detector components. ^{222}Rn is unstable and decays to ^{218}Po via the emission of a 5.49 MeV alpha, producing an ionisation track of length ~400 mm in the detector volume. The long track length allows these events to be discriminated from WIMP interactions with high efficiency so this background has little affect on detector performance. However, ~80% of the time this decay results in an unstable, positively charged $^{218}\mathrm{Po}^{+}$ ion that, in the electric field, drifts to, and plates out on, the surface of the 20μm stainless steel central cathode wires [5]. This unstable ^{218}Po atom then decays by emission of an 6.11 MeV alpha, which has a range of only 14μm in stainless steel. The geometry of this decay on the surface of the wire means that there is a ~37% chance that the alpha will become completely embedded in the cathode wire leaving only the recoil of the ^{218}Po into the fiducial volume. The result is a detection of a nuclear recoil of energy ~100 keV, potentially mimicking the expected WIMP signal. Furthermore, this decay chain results in ^{210}Pb plating out on the central cathode wires. ^{210}Pb, unstable with a half life of 22.3 years, can then be a source of further RPR background events long after the radon has been eliminated from the detector volume.

2.1. Radon reduction

Eliminating radon from the detector volume is the first crucial step in reducing the RPR background. The ambient radon background in the Boulby Underground Laboratory is low ($< 3\,\mathrm{Bq\,m^{-3}}$), so it is known that the high levels of radon contamination in the DRIFT-IId detector must be emanating from the components of the detector itself. Results from extensive radon emanation tests of all detector components are shown in Table 2. It is seen that the RG58

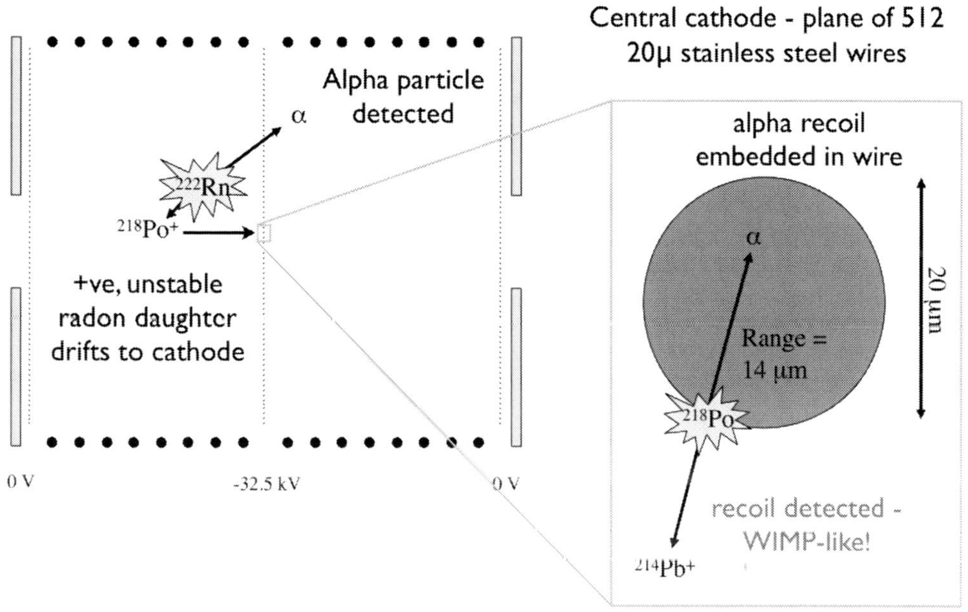

Figure 2. Schematic of the radon progeny recoil (RPR) background in the DRIFT-II detectors. Radon decay in the detector volume can result in an unstable, positively charged ion that drifts to, and plates out on, the surface of the $20\mu m$ stainless steel wires of the $1\,m^2$ central cathode plane. The unstable radon daughter on the wire surface then decays by alpha emission, which, if the alpha is oriented into the wire, can result in detection of a WIMP-like nuclear recoil without the tell-tale alpha event.

coaxial cables (PVC coated) and ribbon cables (PVC coated) were the dominant source of radon emanation. PTFE coated coaxial signal cables and FEP coated ribbon cables were tested and found to be suitable replacements for these with significantly lower radon emanation. The radon measurements in Table 2 show the total number of Rn atoms per second emitted by all detector components is reduced from 0.95 ± 0.06 to 0.09 ± 0.03 by replacing these cables.

An analysis of data from the DRIFT detector before and after replacing the components with high radon emanation shows a reduction in the rate of fully contained alpha events by a factor of 10 ± 1, consistent with the expected reduction from the Rn emanation results. A reduction in the residual RPR rate by a factor of 12.5 ± 0.8 is also seen and is also consistent with the removal of radon from the detector volume.

In March 2008 the $1\,m^2$ central cathode from the DRIFT-IId detector was etched in nitric acid in an attempt to remove long lived radon progeny (e.g. ^{210}Pb) and any other contaminants from the 512 $20\,\mu m$ wires of the central cathode wire plane. The cathode was placed in a clean plastic lined container that was then filled with 3 molar dilute nitric acid until all of the wires were covered. It was left to soak for 30 minutes. The water was then drained from the container and replaced with triply de-ionised water in which the cathode was left to soak for 48 hours. The water was then drained and the cathode allowed to dry for one week before being put back in the detector. The cathode was allowed to outgas in the vacuum vessel for two weeks before data taking was started again.

Analysis of data from before and after this procedure showed that the level of background events was reduced further, by a factor of 6.7 ± 1.0, to the current level of 12 ± 1 background

Table 1. Table caption.

Sample (emanating into vacuum)	Emanation time (days)	Adjusted result (Rn atoms s^{-1})
RG58 coaxial cables (72m)	12.5	0.36±0.03
Electronic boxes	12	0.05±0.02
Ribbon cables	6.5	0.50±0.04
Electronics	10	<0.02*
Coax cables	7	0.04±0.02
Field cage parts	7	<0.03*
PTFE signal cables	20	<0.02*
FEP ribbon cables	12.5	<0.02*

Table 2. Radon emanation results of various components from DRIFT-IIa. Measurements marked with '*' are below the level of sensitivity of the radon emanation setup used.

events per day.

2.2. *Analysis techniques*

Despite the significant reduction in RPR events achieved with the techniques above, the remaining rate of 12±1 RPR events per day needs to be addressed. RPRs create ionisation tracks in principle indistinguishable from WIMP/neutron induced tracks. However, RPR events occur solely on the surface of the central cathode. Thus all RPR tracks will have drifted the full 50 cm drift distance to the MWPC readout plane and will have suffered maximum diffusion of ~0.7 mm. Although there is not sufficient resolution in the x and y dimensions of the detector to measure the 3-dimensional diffusion at this level, the high resolution in the z axis (from the ~60 m s^{-1} ion drift velocity and the 1 MHz digitisation rate) is sufficient to determine some measurement of the diffusion. This measurement is challenging because the charge arrival of a track is affected not only by diffusion but by the orientation of the original track. Nevertheless, the track range in the z direction will generally be greater for events that occurred on the central cathode because of the diffusion of the track. The RMS time (RMST) of the event waveform was found to be the best measure of this. We show in the left hand side plot of Figure 5 the RMST vs equivalent F recoil energy in keV for background data remaining after all cuts (blue points) compared to the neutron data that has been through the same set of cuts (red points). For neutrons we expect an isotropic distribution in the detector. A clear separation can be seen with the background RPR events in a region ~25-65 keV equivalent F recoil energy with RMST ~15μs and the neutron events spread out between ~10-20 μs. This separation in neutron data vs background data allows a signal region to be derived that has zero background and a neutron efficiency that can be calculated using Monte Carlo simulations. The left hand side plot of Figure 3 shows the background events remaining, after cuts, from 47.2 days live time with a 0.8 m^3 fiducial volume of a 30 Torr CS_2 - 10 Torr CF_4 gas mixture in the underground DRIFT-IId detector. A signal region has been marked (tan line) that contains zero background events but in which events are seen in neutron calibration data. The right hand side plot of Figure 3 shows data from a Monte Carlo simulation, which has been verified by comparison to real neutron data. In this simulation 10000 100 GeV WIMP interactions were produced in the DRIFT-IId detector. The output data from the simulation was processed by the same analysis package and with the same cuts as the real data. The plot shows the 2742 simulated WIMP events that remain after all cuts, 228 of which are in the zero background signal region. A range of WIMP masses are simulated and analysed in this way in order to determine the response of the detector to WIMPs of all relevant masses.

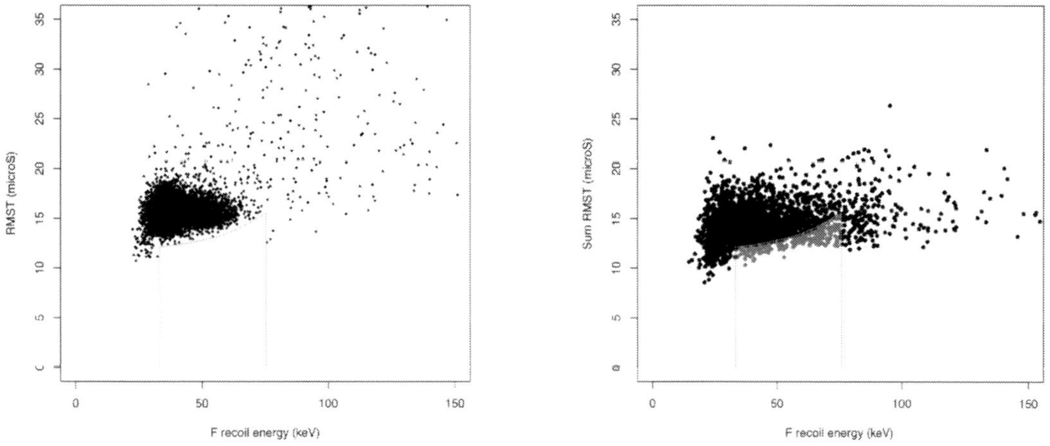

Figure 3. Plots of event RMST vs equivalent F recoil energy in keV. The left plot shows the remaining 6132 background events, after cuts, from 47.2 days of live time with a 30 Torr CS_2 - 10 Torr CF_4 gas mixture. A signal region has been chosen (tan line) that has zero background. The right plot shows the distribution of events from Monte Carlo data of 100 GeV WIMP recoils. From 10000 WIMP recoils generated, 2742 events pass all analysis cuts, 228 of which are in the zero background signal region. The Monte Carlo simulation has been verified by comparison to real neutron data.

3. Spin-dependent WIMP-proton limit

A target material of 30 Torr CS_2 - 10 Torr CF_4 was chosen to maximise fluorine content (where fluorine, with a nuclei of spin 1/2, is an ideal target material to detect spin-dependent WIMP interactions [6]) whilst maintaining negative ion drift and detector stability [7]. The detector was run for 47.2 days live time with a continuous flow of this gas mixture (to minimise radon buildup in the gas) giving 1.47 kg·days of fluorine fiducial mass. Preliminary limits on the SD WIMP-proton cross-section were produced using the zero background signal region and the WIMP detection efficiency determined in Section 2.2. These limits are shown in Figure 4. It should be noted that this first analysis is not blind, the signal region being chosen after analysis of the data and study of the background events. Further data is currently being taken that will be used in a fully blind analysis.

3.1. Thin film cathode

As described above, the WIMP-like RPR background is a result of unstable radon progeny on the surface of the central cathode wires recoiling into the fiducial volume, with the tell-tale alpha becoming embedded in the central cathode wire. A simple, short-term solution to drastically reduce this background could be to make the cathode more transparent to alphas so that the alpha particle does not get lost in the central cathode, but escapes and can be used to reject the event.

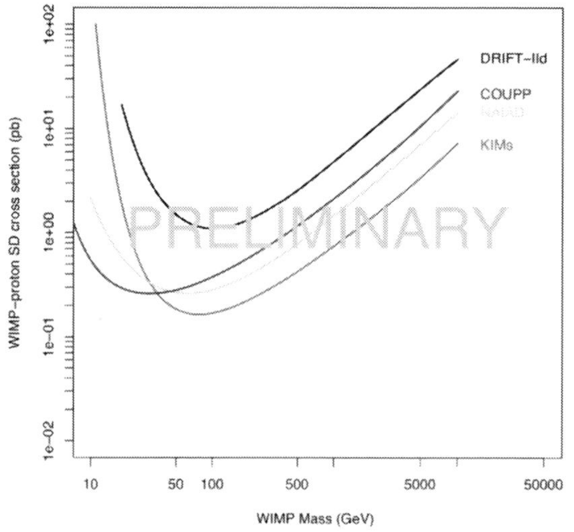

Figure 4. Preliminary limits on spin-dependent WIMP-proton coupling from 47.2 days live time with a $0.8\,\mathrm{m}^3$ fiducial volume of 30 Torr CS_2 - 10 Torr CF_4 in the DRIFT-IId detector underground at the Boulby Mine.

To study this possibility a number of potential cathodes were modelled and tested for structural integrity. $0.9\,\mu$m mylar sheet, evaporation coated with aluminium was found to be a suitable solution. The range of alphas in a sheet of this thickness is such that only ~1% of decays on the surface would result in a WIMP-mimicking RPR event compared to 37% of radon progeny decays on the wire central cathode. This suggests a potential reduction in background of a factor ~40.

A $1\,\mathrm{m}^2$ cathode constructed from $0.9\,\mu$m mylar sheet was tested at Occidental College on DRIFT-IIc and was shown not to have adverse effects on detector operation or stability. It was subsequently installed on the DRIFT-IId detector underground at the Boulby mine in March 2010 and the detector has been operating stably since the installation. Figure 5 shows a preliminary analysis of this data. The left hand side plot shows 12.25 days of background data (blue points) on top of 0.43 days of neutron data (red points) with the wire central cathode. The right hand side plot shows background and neutron data of similar live times with the new thin film central cathode. The rate and distribution of neutron events remains constant as expected, whereas the number of background events has reduced significantly, by a factor ~15. The expected full factor ~40 reduction is not seen suggesting there may be other backgrounds within the RPR region, or that there may be new backgrounds introduced by the thin film central cathode. These backgrounds will have to be understood and changes to cuts may be necessary to maximise WIMP efficiency and remove further of the background events. This job is underway as part of the preparation for the proposed blind analysis.

4. Summary

A 47.2 day run of the DRIFT-IId detector in the Boulby Underground Laboratory has been used to set a preliminary limit on SD WIMP-proton cross-sections ~3 orders of magnitude below published limits from other directional detectors and comparable to that achieved by

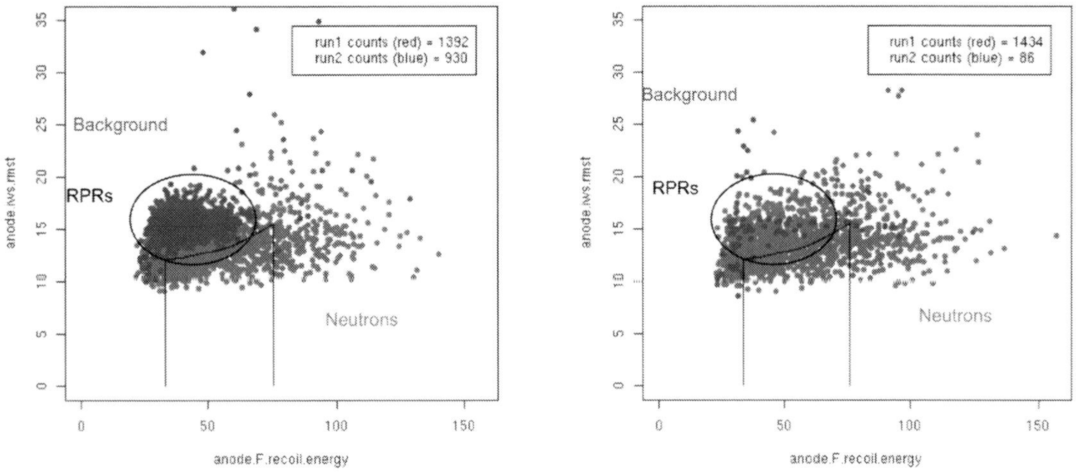

Figure 5. Plots of event RMST vs equivalent F recoil energy in keV. The left plot shows 0.43 days of neutron data (red points) that produce an isotropic distribution of tracks in the detector volume, and 12.25 days of background data (blue points) that occur on the surface of the wire central cathode and thus all suffer maximum diffusion. The right plot shows 0.44 days of neutron data (red points) and 12.22 days of background data with the new thin film central cathode. The neutron efficiency is seen to remain constant whilst a factor of ~15 reduction is seen in events in the RPR region.

non-directional detectors of much larger target mass.

The DRIFT-IId detector is currently running underground with the 20 μm wire plane cathode replaced by a 0.9 μm thin film cathode, which has been shown to reduce the RPR background by a factor of at least ~15. Preliminary analysis has shown a significant reduction in background but further work is required to fully understand the data with the new cathode.

The RMST parameter used here to reject RPRs in analysis has allowed competitive limits to be set but it is clear that a more accurate and efficient way of rejecting these events is required. To achieve this the current focus of the DRIFT collaboration is to develop a way of determining the absolute Z position of events in the detector by detecting the positive ions that drift to the central cathode from a particle interaction, as well as the negative charge at the MWPC. So far R&D work has succeeded in detecting ~1000 ion pairs (~45 keV equivalent F recoil energy) that arrive at the central cathode of a test device with similar conditions to the DRIFT detectors.

Acknowledgements
We would like to thank Cleveland Potash Ltd, operators of the Boulby Mine, NSF and STFC.

References
[1] J. D. Vergados, *Phys. Rev. D* **67** (2003) 103003
[2] B. Morgan, A. M. Green, and N. J. C. Spooner, *Phys. Rev. D* **71** (2005) 103507
[3] G. J. Alner, et al., *Nucl. Instr. and Meth. A* **555** (2005) 173
[4] D. P. Snowden-Ifft, C. J. Martoff, and J. M. Burwell, *Phys. Rev. D* **61** (2000) 101301
[5] S. Burgos, et al., *Nucl. Instr. and Meth. A* **584** (2008) 114
[6] J. Ellis, and R. A. Flores, *Phys. Lett. B* **263** (1991) 259
[7] K. Pushkin, and D. P. Snowden-Ifft., *Nucl. Instr. and Meth. A* **606** (2009) 569

VI International Workshop on the Dark side of the Universe (DSU 2010) IOP Publishing
Journal of Physics: Conference Series **315** (2011) 012018 doi:10.1088/1742-6596/315/1/012018

Internal bremsstrahlung in neutralino annihilation: revised impact on indirect detection from γ-rays.

M. Cannoni[1], M. E. Gómez[1], M. A. Sánchez-Conde[2,3], F. Prada[4] and O. Panella[5]

[1] Departamento de Física Aplicada, Facultad de Ciencias Experimentales, Universidad de Huelva, 21071 Huelva, Spain
[2] Instituto de Astrofísica de Canarias (IAC), E-38200 La Laguna, Tenerife, Spain
[3] Departamento de Astrofísica, Universidad de La Laguna (ULL), E-38205 La Laguna, Tenerife, Spain
[4] Instituto de Astrofísica de Andalucía (CSIC), E-18008, Granada, Spain
[5] Istituto Nazionale di Fisica Nucleare, Sezione di Perugia, Via Alessandro Pascoli, 06129, Perugia, Italy

E-mail: mirco.cannoni@dfa.uhu.es

Abstract. We revise the impact of internal bremsstrahlung photons in the context of the constrained minimal supersymmetric standard model on γ-ray dark matter annihilation searches. As an example, we review the γ-ray dark matter detection prospects from Draco dwarf spheroidal galaxy at the MAGIC stereoscopic system and the CTA project. We find that for a typical energy threshold of 100 GeV the flux of high energy photons is enhanced by an order of magnitude in the stau coannihilation region, where the signal remains still at least three orders of magnitude below the sensitivity of the instruments. However, the effect of internal bremsstrahlung is negligible or small in more optimistic scenarios for detection like the funnel and focus point regions.

1. γ-rays from neutralino annihilation

The detection of γ-rays coming from the annihilation of dark matter (DM) particles that should form the halo of galaxies is at present a very active field of research, which together with other indirect and direct search experiments [2], should give light to the nature of DM. The lightest neutralino of the minimal supersymmetric (SUSY) extension of the standard model (MSSM) is a natural candidate for dark matter (DM) [3]. There are three mechanisms producing photons in neutralino annihilation: (1) hadronization and decay of the annihilation products [4], mostly neutral pion decay, give *secondary photons* which show a continuous featureless energy spectrum decreasing towards m_χ, the maximum energy available. Typically this is the largest contribution over a wide portion of the parameter space; (2) at the one loop level neutralinos annihilate into photons through the processes [5] $\chi\chi \to \gamma\gamma$ and $\chi\chi \to Z\gamma$. Being neutralino highly non relativistic, annihilation is almost at rest, thus outgoing photons are almost monochromatic (*lines*) with energies $E_\gamma \sim m_\chi$ and $E_\gamma \sim m_\chi - m_Z^2/4m_\chi$, respectively. Though this gammas would give a very clear signal the cross section is $\mathcal{O}(\alpha^4)$ and loop suppressed; (3) finally *internal bremsstrahlung* (IB) [6, 7] which consists the emission of additional photons in neutralino pair annihilation into charged final states $\chi\chi \to X\bar{X}\gamma$ X being a charged lepton or a W boson. In the

Feynman diagrams these photons can be attached to the external legs representing final state charged particles or to the propagator of the virtual charged sparticle exchanged by neutralinos: the latter diagrams are at the origin of the hard photon spectrum of IB near m_χ. The cross section is $\mathcal{O}(\alpha^3)$, thus in principle intermediate between the two previous contribution, but as we will see, strongly dependent on the SUSY mass spectrum.

The expected flux of photons with energy above an energy threshold set by experiments is given by $F(E_\gamma > E_{\rm th}) = J(\Psi) \times \Phi_{PP}(E_\gamma > E_{\rm th})$. Here $J(\Psi)$ is the astrophysical factor and $\Phi_{PP}(E_\gamma > E_{\rm th})$ is the particle physics factor that includes all the particle physics informations. Given the three sources discussed above, the particle physics factor that in the following we call f_{susy}, explicitly reads

$$\begin{aligned} f_{SUSY} &= f_{cont} + f_{lines}, \\ f_{cont} &= \left(\sum_f \int_{E_{th}}^{m_\chi} \frac{dN_\gamma^f}{dE_\gamma} dE_\gamma \right) \frac{\langle \sigma_{\chi\chi}\, v \rangle}{2m_\chi^2} = f_{sec} + f_{IB}, \\ f_{lines} &= 2\frac{\langle \sigma_{\gamma\gamma}\, v \rangle}{2m_\chi^2} + \frac{\langle \sigma_{Z\gamma}\, v \rangle}{2m_\chi^2}. \end{aligned} \quad (1)$$

Here dN_γ^f/dE_γ is the differential yield of photons per annihilation to the final state f. The factor in parenthesis is thus $n_\gamma(E_\gamma > E_{th})$, the total number of photons per annihilation with energy greater than the threshold energy, $\langle \sigma_{\chi\chi}\, v \rangle$ is the thermal averaged total neutralino annihilation cross, $\langle \sigma_{\gamma\gamma}\, v \rangle$ and $\langle \sigma_{Z\gamma}\, v \rangle$ the cross sections for annihilation into lines and m_χ the neutralino mass.

For the numerical computation of IB effects we use *DarkSusy 5.0.5* [8]. In the contest of the constrained minimal supersymmetric standard model (CMSSM) the theory at the weak scale is determined by four parameters assigned at the unification scale: the common scalar mass m_0, the gauginos mass $m_{1/2}$, the trilinear couplings A_0, the ratio of the Higgs vacuum expectation values, $\tan\beta$ and the sign of μ, the Higgs mixing term, that we take positive. We require the neutralino relic abundance to be inside the cosmologically favored interval $0.09 < \Omega_\chi h^2 < 0.13$ (the most recent WMAP [9] interval at 3σ is $0.094 < \Omega_{DM} h^2 < 0.128$). We further require that SUSY models satisfy the LEP bounds on Higgs and chargino masses, $m_h > 114$ GeV and $m_{\chi^+} > 103.5$ GeV, and constraints from $b \to s\gamma$ as explained in Refs. [10].

After the imposition of the phenomenological constraints only few regions of the parameter space survive (see [?] for review). In these regions the relative weight of the three contributions to f_{SUSY} and to the flux changes drastically: to illustrate this behavior we select from our scan of the parameter space four points found in Table 1 where also the annihilation cross section and the distinct contributions to f_{SUSY} integrating the number of photons above $E_{th} = 100$ GeV can be read. The corresponding differential spectra of photons are shown in Fig. 1.

Point *(A)* is in the *stau coannihilation region* of the CMSSM parameter space: the mass of the lightest stau is $m_{\tilde\tau} = 195$ GeV very close to $m_\chi = 188$ GeV. Neutralino pair annihilation in $\tau^+\tau^-$ mediated by t-channel exchange of stau has the highest annihilation cross section. Here f_{IB} is the dominant contribution being 10 and 4.4 times greater than f_{sec} and f_{lines}.

Point *(B)*is in the *funnel or resonances region*: the mass of the CP-odd neutral Higgs is $m_A = 1211$ GeV while $m_\chi = 598$ GeV, thus $m_A \simeq 2m_\chi^0$: pair annihilation into $b\bar{b}$ through s-channel exchange of heavy neutral Higgs bosons is the dominant channel. In this case no photon line can be attached to the virtual particles in t channel exchange and the IB yield is negligible. f_{sec} is the only relevant contribution.

Point *(C)* is in the *focus point or hyperbolic branch region*. The mass of the lightest chargino is $m_{\chi^\pm} = 212$ GeV, not much bigger than $m_\chi = 163$ GeV and neutralino pairs annihilate into W^+W^- through t-channel chargino exchange. The IB yield is small because m_χ is not much

Table 1. CMSSM used in Fig. 1. The values of m_0, $m_{1/2}$, A_0, $m_{\tilde{\chi}}$ are in GeV, the sign of μ is positive. $\langle\sigma v\rangle$ is given in units of 10^{-29} cm^3 s^{-1}, the f's, defined in Eq. (1), are given in units of 10^{-32} GeV^{-2} cm^3 s^{-1}.

Model	$\tan\beta$	m_0	$m_{1/2}$	A_0	$\langle\sigma v\rangle$	f_{sec}	f_{lines}	f_{IB}	f_{SUSY}
(A)	18	127	459	−135	29	0.008	0.018	0.079	0.1
(B)	52	982	1377	725	2600	0.72	10^{-5}	10^{-5}	0.72
(C)	17	2200	430	805	2225	0.04	0.06	0.02	0.12
(D)	51	8940	2218	−4221	1203	0.3	0.003	0.017	0.32

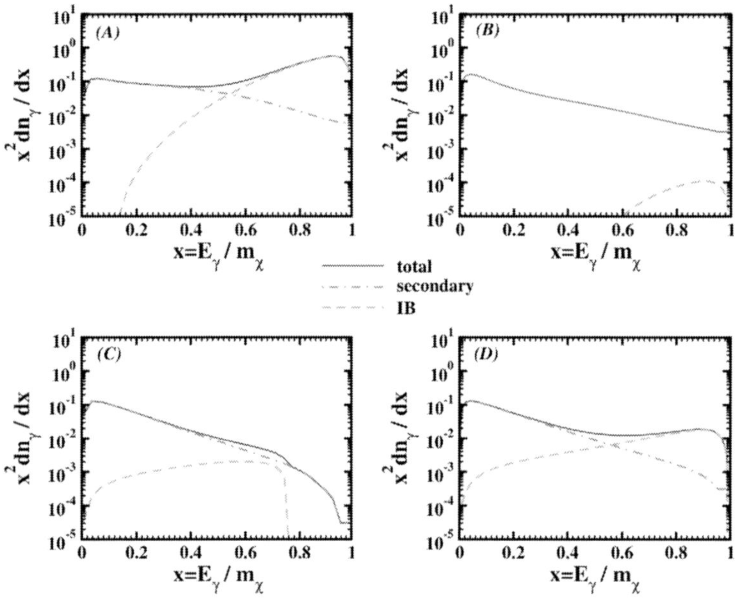

Figure 1. Continuum γ-rays spectra for the models in Table 1. The solid lines indicate the total yield, the dot-dashed lines the contribution of secondary photons, while dashed lines correspond just to the IB contribution.

greater than m_W and photons energy has a cut off which corresponds to the kinematic endpoint $x = 1 - m_W{}^2/m_\chi^2 \sim 0.75$. Here f_{lines} is bigger than f_{sec} and f_{IB}, and the three contribution are of the same order.

Point *(D)* is another example in the focus point region. The mass of the lightest chargino is $m_{\chi^\pm} = 954$ GeV, almost degenerate with $m_\chi = 918$ GeV. Neutralino pairs annihilate into W^+W^- through t-channel chargino exchange as in C but in this case $m_\chi \gg m_W$ thus IB photons contribution is more important and have endpoint at the neutralino mass: here f_{SUSY} is dominated by f_{sec} even if f_{IB} is greater than f_{lines}.

Note that although f_{IB} can be dominant, point *(A)*, or bigger or of the same order of f_{lines}, points *(D)* and *(C)*, the bigger f_{sec} is reached in point *(B)* where f_{IB} is irrelevant, thus from the point of view of the most promising scenario for detection there is no improvement: we discuss this point in a more general way in Section 3.

2. Astrophysical target: Draco dwarf galaxy

Gamma rays are detected by imaging air Cerenkov telescopes (IACT) like MAGIC [11], HESS [12], VERITAS [13] or satellites-based experiments like the FERMI satellite [14]. For these experiments, dwarf spheroidal (dSph) galaxies around the Milky Way represent a good alternative target option to e.g. the Galactic Center [15], already observed in γ-rays but with null DM detection so far [16]. dSphs are DM dominated systems with inferred very high mass-to-light ratios, and most of them are expected to be free from any other astrophysical source that might contribute to a possible γ-ray signal. Therefore, the detection of γ-rays from them would probably imply a successful DM annihilation detection. Here we consider Draco, located at 80 kpc, is one of the dwarfs with more observational constraints, which have helped to better determine its DM density profile and is one of the most studied dSph [17] [18]. Draco has already been observed by some of the quoted experiments:

The MAGIC telescope [19], found no gamma signal above an energy threshold of 140 GeV. As a consequence, an upper limit for the flux (2σ level) was set to be 1.1×10^{-11}photons cm^{-2} s^{-1}, assuming a power-law with spectral index -1.5 and a point-like source. This upper limit is $\mathcal{O}(10^3 - 10^9)$ above the values predicted by those SUSY models used in their analysis and therefore no constraints could be put on the parameter space. The FERMI collaboration has reported their upper limits for a possible γ-ray annihilation signal from Draco at lower energies [20], given that no significant gamma emission was detected above 100 MeV (see also A. Morselli contribution to these proceedings). Recently also the VERITAS collaboration [21] has published results of 18.38 hours of observation of Draco: finding no signal an upper limit at 95% on the total flux was set to be 0.49×10^{-12} photons cm^{-2} s^{-1} above 200 GeV.

The astrophysical factor $J(\Psi)$ represents the integral of the square of the dark matter density ρ_{DM} along the direction of observation Ψ relative to the center of the DM halo, and depends on the point spread function of the telescope (PSF). In the case of Draco we use the cuspy DM density profile given in Refs. [22]-[1] also used by the MAGIC collaboration in their analysis [11]. In particular we use here the value of $J(\Psi)$ integrated over the whole spatial extent of the source as the value of the astrophysical factor. This value, that does not depend on the PSF any longer, can be well approximated by $\overline{J} = \frac{1}{4\pi D^2} \int_V \rho^2_{DM}(r)\ dV$, with D the distance from the Earth to the center of the DM halo and r the galactocentric distance inside it: for Draco we take a value of $\overline{J} = 3.7 \times 10^{17}$ Gev^2 cm^{-5}. We remark that in literature others models for the halo profiles are used providing slightly different values for J: in particular both FERMI and VERITAS analysis are assume a two parameters Navarro-Frank-White profile [23]. As noted also in Ref. [18] the uncertainties on the astrophysical factor are of order of a factor 2-3 using different models: this uncertainty is however irrelevant with the present experimental sensitivity as we show in the next section.

3. Results

Armed with the previous ingredients we perform a scan on the parameter space of f_{SUSY} versus the threshold energy, which is the important quantity determining the flux once the astrophysical factor is fixed. We set $A_0 = 0$ and take two values of $\tan\beta$, 10 and 50 varying m_0 and $m_{1/2}$ such that the experimental constraints discussed in Section 1 are satisfied. For each value of $\tan\beta$ we separate the models with $m_0 > 2$ TeV and $m_0 < 2$ TeV in order to separate the focus point region from the stau coannihilation and the Higgs funnel ones. In Fig. 2 the shaded areas correspond to the total f_{SUSY}, the areas inside the dot-dashed lines correspond to f_{sec} and the areas inside the dashed lines give f_{lines}.

In panel (a) of Fig. 2 where points are in the stau coannihilation region we can appreciate the largest contribution of IB, as shown by the point (A). The absence of IB photons of the point (B) is evidenced by the panel (b) where points are in the funnel region, while the panels (c) and (d) have points mostly in the hyperbolic branch and share properties with the points (C) and (D).

To compare we the future experimental sensitivities we plot in these panels the sensitivity lines for Draco of the MAGIC telescopes in stereoscopic mode [24] and of the CTA project [25]. These lines are obtained dividing the Montecarlo simulated sensitivity for the flux (50 hours of

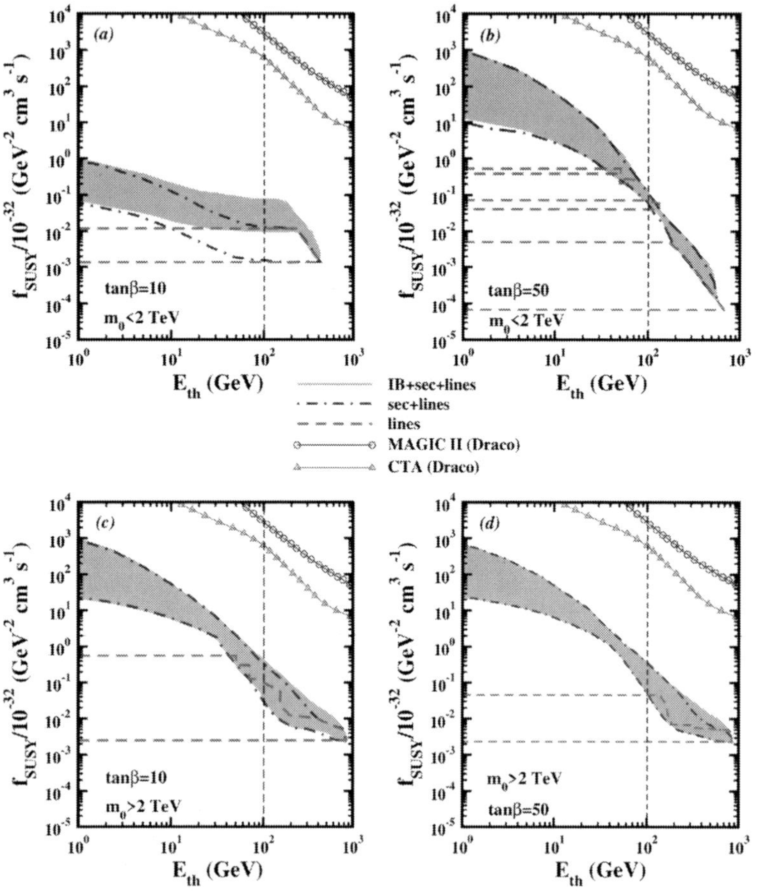

Figure 2. The particle physics factor f_{SUSY}, Eq. (1), versus E_{th}, energy threshold of the detector. We set $A_0 = 0$ and μ positive while m_0 and $m_{1/2}$ have values such that the CMSSM point predicts the neutralino relic density inside the WMAP bounds satisfying all the phenomenological constraints. Also plotted are the predicted sensitivity lines of MAGIC II and CTA for Draco corresponding to 50 hours of observation time and a 5σ detection level. The black vertical line corresponds to the typical threshold for the IACT.

observation time and a 5σ detection level) by the total astrophysical factor discussed in Section 2.

Let us focus at $E_{th} = 100$ GeV, the typical energy threshold for the telescopes. The IB contribution is seen to be important only in panel (a) corresponding to the stau coannihilation region, as expected. The order of magnitude effect has the net result of pushing f_{SUSY} to the value $\simeq 10^{-33}$ GeV^{-2} cm^3 s^{-1} as in panel (b): but this value is smaller than the value reached in panel (c) and (d), $f_{SUSY} \simeq 5 \times 10^{-33}$ GeV^{-2} cm^3 s^{-1} that is the most optimistic value. Thus we see that if from one side IB rises by an order of magnitude the flux in the stau cohannilation region, on the other side the sensitivity of the experiments to this region is more than three order of magnitude far away, that is the worst case scenario for γ-rays scenario remain the worst case. Note that the same is true at low threshold, say 1 GeV of interest for FERMI satellite, where f_{SUSY} is three orders of magnitude smaller than in panels (b), (c), (d).

4. Summary

Although IB has to be included for evaluation of fluxes of high energy photons from neutralino annihilation, its contribution is relevant only in the stau coannihilation region of the CMSSM parameter space: as a result, the most optimistic particle physics scenarios for DM detection, which typically correspond to those points of the parameter space where most of the flux is given by secondary photons, will not change substantially. As an example of the impact of the IB on DM search, we recalculated the DM detection prospects of the Draco dwarf galaxy for the MAGIC II and the future CTA telescopes: though the effect can rise the flux in the stau cohannilation region by an order of magnitude at E_{th} = 100 GeV, the predicted fluxes are still at least three orders of magnitude below the sensitivity of the IACTs, and the most optimistic scenario for detection are found in the funnel and focus point regions where internal bremsstrahlung is negligible or small.

Acknowledgments

M. C. acknowledges D. Delepine and all the members of the local organizing committee for the warm hospitality in Leon during the workshop. The authors acknowledges support from the project P07FQM02962 funded by "Junta de Andalucia", the Spanish MICINN-INFN(PG21) projects FPA2009-10773, FPA2008-04063-E and MULTIDARK project of Spanish MICINN Consolider-Ingenio: CSD2009-00064.

References

[1] Cannoni M, Gomez M E, Sanchez-Conde M A, Prada F and Panella O 2010 *Phys. Rev.* D **81**, 107303
[2] Munoz C 2004 *Int. J. Mod. Phys.* A **19** 3093
[3] Jungman G, Kamionkowski M and Griest K 1996 *Phys. Rept.* **267** 195; Bertone G, Hooper D and Silk J 2005 *Phys. Rept.* **405** 279
[4] Silk J and Srednicki M 1984 *Phys. Rev. Lett.* **53** 624
[5] Bergstrom L and Ullio P 1997 *Nucl. Phys.* B **504** 27; Bern Z, Gondolo P and Perelstein M 1997 *Phys. Lett.* B **411** 86; Ullio P and Bergstrom L 1998 *Phys. Rev.* D **57** 1962
[6] Bergstrom L 1989 *Phys. Lett.* B **225** 372; Flores R, Olive K A and Rudaz S 1989 *Phys. Lett.* B **232** 377; Bergstrom L, Bringmann T, Eriksson M and Gustafsson M 2005 *Phys. Rev. Lett.* **95** 241301
[7] Bringmann T, Bergstrom L and Edsjo J 2008 *J. High Energy Phys.* JHEP01(2008)049
[8] Gondolo P, Edsjo J, Ullio P, Bergstrom L, Schelke M and Baltz E A 2004 *J. Cosmol. Astropart. Phys.* JCAP07(2004)008
[9] Larson D *et al.* 2011 *Astrophys. J. Suppl.* **192** 16
[10] Gomez M E, Ibrahim T, Nath P and Skadhauge S, 2005 *Phys. Rev.* D **72** 095008; 2006 *ibid.* **74**, 015015; Gomez M E, Lazarides G and Pallis C 2002 *Nucl. Phys.* B **638** 165; 2003 *Phys. Rev.* D **67** 097701; Cannoni M and Panella O 2010 *Phys. Rev.* D **81** 036009
[11] Lorenz E 2004 *New Astron. Rev.* **48** 339
[12] Hinton J A 2004 *New Astron. Rev.* **48** 331
[13] Weekes T C *et al.* 2002 *Astropart. Phys.* **17** 221
[14] Gehrels N and Michelson P 1999 *Astropart. Phys.* **11** 277
[15] Bergstrom L, Ullio P and Buckley J H 1998 *Astropart. Phys.* **9** 137; Cesarini A, Fucito F, Lionetto A, Morselli A and Ullio P 2004 *Astropart. Phys.* **21** 267; Fornengo N, Pieri L and Scopel S 2004 *Phys. Rev.* D **70**, 103529
[16] Aharonian F *et al.* 2004 *Astron. Astrophys.* **425** L13; Albert J *et al.* 2006 *Astrophys. J.* **638** L101 Aharonian J *et al.* 2006 *Phys. Rev. Lett.* **97** 221102
[17] Tyler C 2002 *Phys. Rev.* D **66** 023509; Bergstrom L and Hooper D 2006 *Phys. Rev.* D **73** 063510; Profumo S and Kamionkowski M 2006, *J Cosmol. Astrop. Phys* JCAP03(0603)003; Colafrancesco S, Profumo S and Ullio P 2007 *Phys. Rev.* D **75** 023513; Strigari L E, Koushiappas S M, Bullock J S and Kaplinghat M 2007 *Phys. Rev.* D **75** 083526
[18] Bringmann T, Doro M and Fornasa M 2009 *J. Cosmol. Astropart. Phys.* JCAP01(0901) 016
[19] Albert J *et al.* 2008 *Astrophys. J.* **679** 428
[20] Abdo A A *et al.* 2010 *Astrophys. J.* **712** 147
[21] Acciari V A *et al.* 2010 *Astrophys. J.* **720** 1174
[22] Sanchez-Conde M A, Prada F, Lokas E L, Gomez M E, Wojtak R and Moles M 2007 *Phys. Rev.* D **76** 123509
[23] Navarro J F, Frenk C S and White S D M 1997 *Astrophys. J.* **490** 493
[24] P. Colin *et al.* 2009 Performance of the MAGIC telescopes in stereoscopic mode *Preprint* 0907.0960
[25] Doro M 2011 *Nucl. Instrum. Meth.* A **630** 285

Dark-Energy Equation-of-State parameter for high redshifts.

Ariadna Montiel and Nora Bretón

Physics Dep., Centro de Investigación y de Estudios Avanzados del I.P.N, Apdo. Post. 14-740 07300 México D.F., México.

E-mail: amontiel@fis.cinvestav.mx, nora@fis.cinvestav.mx

Abstract. Since the elucidation of the nature of dark energy depends strongly on redshift observations, it is desirable to measure them over a wider range, but supernovae cannot be detected out past redshift 1.7. Gamma-ray-bursts (GRBs) offer means to extend the analysis to at least redshifts of > 6. The reason is that GRBs are visible across much larger distances than supernovae. GRBs are now known to have several light-curve and spectral properties from which the luminosity of the burst can be calculated, and it might GRBs become into standard candles. We have used data of 69 GRB to study the behavior of the parameter of the dark energy equation of state as a function of redshift.

1. Introduction

Nowadays cosmological models can be confronted with real data. The observed cosmic microwave background (CMB) with a temperature of around 2.73 K with tiny temperature differences of about 10^{-5} between the different patches of the sky, sustains the assumption of homogeneity of the universe, at a certain scale. Moreover, in 1998, very refined observations of the brightness of distant supernovae seemed to hint the presence of a negative pressure component in the universe, which would make it to expand acceleratedly. In other words, some distant supernovae were fainter than expected and the most compeling explanation was that their light had travelled greater distances than assumed. To fit with the observed behaviour of supernovae redshifts, the cause of the accelerated expansion, the so called dark energy, should exert a negative pressure.

Therefore the more acurate knowledge of the dark-energy Equation of State (EoS) is of paramount importance to understand its nature: if it evolves with time, how much of it is there or if it is rather a manifestation of extra-dimensional physics. By making observations of distances over a wide range of redshifts we can place significant constraints on models of the universe. So a strong imperative in the quest for dark energy is to extend the analysis to high redshifts. However the range of observed redshifts of the supernovae is $0.1 < z < 1.7$. On the other hand, Gamma-ray bursts (GRB) are visible across much larger distances than supernovae, then GRBs offer the chance to extend the Hubble diagram (the plot of distance versus redshift) to at least redshifts of $z \approx 6$.

In this contribution we explore the dependence of the parameter $w(z)$ of the dark-energy (DE) equation-of-state ($p = w(z)\rho$), on the redshift derived from GRBs data. To this end we first find a reasonable calibration for the GRB in order to extract the luminosity distance d_L

as a function of the redshift. Then we proceed to calculate the Hubble function $H(z)$ to obtain $w(z)$. In the next section we briefly summarize the main features of the cosmological model.

2. The model with FRW spacetime

As we mention above, the homogeneity of the universe, to a certain scale, is strongly suggested by the uniformity of the observed CMB, with anisotropies of about 10^{-5}. This makes suitable to model the universe with a homogeneous and isotropic geometry, like the Friedmann-Robertson-Walker (FRW). It combined with the assumption that general relativity is the correct theory on cosmological scales leads to the Friedmann equations for the scale function $a(t)$, that governs the expansion of the universe.

General Relativity is the theory that models the gravitational interaction with the spacetime curvature, this quoted in the Einstein Eqs. $R_{\mu\nu} + Rg_{\mu\nu}/2 = 8\pi T_{\mu\nu}$. The right hand side of this equation is the energy-momentum tensor $T_{\mu\nu}$, that includes the matter content of the universe, that is assumed to be a perfect fluid, whose equation of state is put by hand. This perfect fluid is the contribution of several components among them: barionic matter and dark matter (the curvature component is assumed to vanish, according to the observations). The left hand side of the Einstein equations is related to the curvature quantities, that are derived from the assumed FRW geometry. From these assumptions the derived Friedmann equations are

$$H^2(a) = \left(\frac{\dot{a}}{a}\right)^2 = \frac{8\pi G}{3}\rho(a), \quad k = 0, \tag{1}$$

$$\frac{\ddot{a}}{a} = -\frac{4\pi G}{3}(\rho + 3p), \tag{2}$$

where $a(t)$ is the scale factor of the universe, a dimensionless quantity customarily chosen to be 1 at present; ρ is its average energy density and $H(a)$ is the Hubble parameter or the rate of expansion of the universe.

Both sides of the first Friedmann equation can be observationally probed directly. To fit with the supernovae redshift observations Eq. (1), demands another component besides the dust that models barionic matter, $\rho(a) = \rho_m(a) \sim a^{-3}$ even if a non-zero curvature is assumed. One possible resolution to this puzzle is to modify the right hand side of the Friedmann equation, Eq. (1), by introducing a new form of "dark" energy component, $\rho(a) = \rho_m(a) + \rho_X(a)$.

Before going on, we define the fractional density parameter as

$$\Omega_i = \frac{\rho_i}{\rho_{ic}} = \frac{8\pi G\rho_i}{3H^2}, \tag{3}$$

where ρ_{ic} is the critical density for which the spatial geometry is flat. The density in Eq. (3) is the fractional density for the i-component of the universe. The several components, tagged as ρ_i, mainly are the gravitational matter and the dark energy densities. Observations point to the flatness of the universe, i.e. $\Omega_{\text{total}} = \sum \Omega_i = 1$ (assuming cero curvature). The fact that the sum of densities is equal to one is called the *concordance model*, that can also be expressed as $\Omega_\Lambda + \Omega_m = 1$, where Ω_Λ and Ω_m are the critical density of dark energy and of gravitational matter, respectively; the latter includes dark matter and barionic matter.

3. $\omega(z)$ Parameter of the EoS from observations

The required gravitational properties of dark energy needed to induce the accelerated expansion are well described by its equation of state

$$\omega(z) = \frac{p_X(z)}{\rho_X(z)}, \tag{4}$$

which enters into the second Friedmann equation Eq. (2), implying that a negative pressure ($\omega < -1/3$) is necessary in order to induce accelerated expansion.

The parameter $\omega(z)$ determines not only the gravitational properties of dark energy but also its evolution. This evolution is easily obtained from the energy momentum conservation

$$d(\rho_X a^3) = -p_X d(a^3), \tag{5}$$

which leads to

$$\rho_X = \rho_{0X} e^{3\int_0^z \frac{dz'}{1+z'}(1+\omega(z'))}, \tag{6}$$

where the subscript '0' indicates the present value of the quantity. From Eq. (6) we see that the determination of $\omega(z)$ is equivalent to that of $\rho_X(z)$ which in turn is related to the Hubble parameter $H(z)$, that from the first Friedmann equation and using (6) can be expressed as

$$H(z)^2 = H_0(z)^2[\Omega_{0m}(1+z)^3 + \Omega_{0X} e^{3\int_0^3 \frac{dz'}{1+z'}(1+\omega(z'))}]. \tag{7}$$

Thus, the knowledge of Ω_{0m}, z and $H(z)$ suffices to determine $\omega(z)$ which is obtained from the previous equation as

$$\omega(z) = \frac{\frac{2}{3}(1+z)\frac{d\ln H}{dz} - 1}{1 - \frac{H_0^2}{H^2}\Omega_{0m}(1+z)^3}. \tag{8}$$

From observational data it is possible to extract $d_L(z)$ and then determine $H(z)$, since,

$$H(z) = \left\{\frac{d}{dz}\left(\frac{d_L(z)}{1+z}\right)\right\}^{-1}. \tag{9}$$

So the main quantity to find is the luminosity distance d_L as a function of the redshift z of distant objects. Measuring redshifts is straightforward from the detected spectrum of the luminous object; the hard part is determinig distances for objects of unknown intrinsic brightness. One of the most popular techniques is to try to find a standard candle. Standard candles are objects in the universe with a well calibrated intrinsic luminosity that can be used to determine distances on cosmological scales. Type Ia supernovae are these kind of objects [1], and it has to do with their origin, as a white dwarf that accreting matter (from a companion star, for instance) gets to the Chandrasekhar limit mass.

Therefore, to enlarge the range of redshifts in probing cosmological models using Gamma-ray-bursts (GRBs), as a first step we must establish under which parameters can GRB be considered as reliable standard candles.

4. Calibrating Gamma-ray burst.

There are several distance indicators to determine the cosmological luminosity distance $d_L(z)$. If a certain distance indicator is calibrated without any cosmological model, the indicator can be used to determine the cosmological parameters such as Ω_m, Ω_Λ and $\omega = p/\rho$.

In this section we show a calibration for a 69 GRBs sample [2], where the luminosity distance is a function of the redshift for two intervals, $z \leq 1.755$ and $z > 1.755$. Based in the data of Type Ia Supernovae, the calibration obtained by Kodama et al.[3] was,

$$\frac{d_L(z)}{10^{27}cm} = 14.57z^{1.02} + 7.16z^{1.76}. \tag{10}$$

With Eq. (10) and the 69 GRBs observational data, we can use Eq. (9) and in this way infer if the dark energy effective EoS parameter ω is close to -1 (cosmological constant). Any small deviation from this value could give a different theoretical scenario: if it is exactly equal (when of course referring to observational errors) to -1, we have a cosmological constant; if it is larger than -1, we have quintessence energy; while if it is smaller than -1 we have the so called phantom dark energy.

However with the Kodama calibration we obtained a plot for $w(z)$ that was physically unreasonable, diverging at certain redshift. This was not reported in [3].

We tried to repair this inconsistency by finding out another calibration for the GRBs extracted directly from data. In order to calibrate the GRB observed data and obtain the dependence rule between the luminosity distance d_L and the redshift z, we try the five posible relationships proposed in [2].

These relate GRBs isotropic luminosity, L, or the total burst energy in the gamma rays, E_γ, to the observables of the light curves and/or spectra: $\tau_{\rm lag}$, the time lag, V variability, the peak of the νF_ν spectrum $E_{\rm peak}$ and $\tau_{\rm RT}$, the minimum rise time, as follows,

$$log\left(\frac{L}{1ergs^{-1}}\right) = 52.26 - 1.01log\left[\frac{\tau_{lag}(1+z)^{-1}}{0.1s}\right], \tag{11}$$

$$log\left(\frac{L}{1ergs^{-1}}\right) = 52.49 + 1.77log\left[\frac{V(1+z)}{0.02}\right], \tag{12}$$

$$log\left(\frac{L}{1ergs^{-1}}\right) = 52.21 + 1.68log\left[\frac{E_{peak}(1+z)}{300keV}\right], \tag{13}$$

$$log\left(\frac{E_\gamma}{1ergs^{-1}}\right) = 52.57 + 1.63log\left[\frac{E_{peak}(1+z)}{300keV}\right], \tag{14}$$

$$log\left(\frac{L}{1ergs^{-1}}\right) = 52.54 - 1.21log\left[\frac{\tau_{RT}(1+z)^{-1}}{0.1s}\right]. \tag{15}$$

The isotropic luminosity L is given by

$$L = 4\pi d_L^2 P_{bolo}, \tag{16}$$

where P_{bolo} is a observational quantity reported in [2].

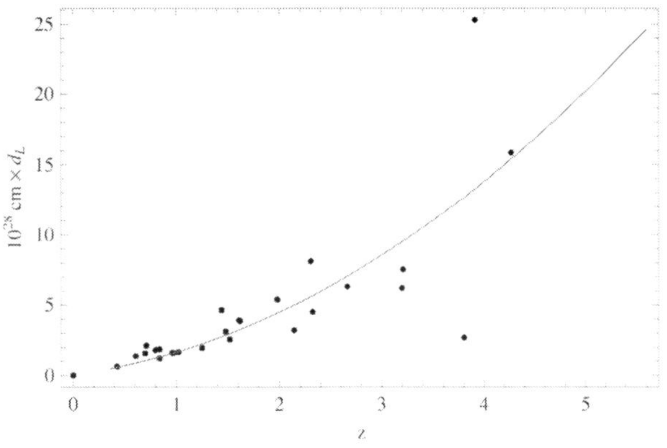

Figure 1. The luminosity distance as a function of the redshift of GRBs, Eq. (18) derived using the calibrated $E_\gamma - E_{peak}$ relation.

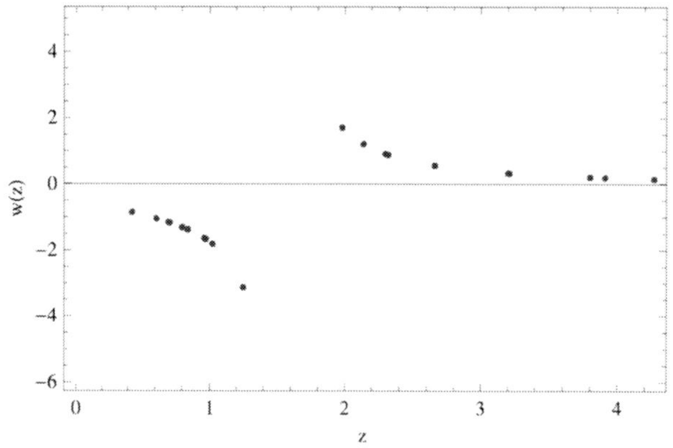

Figure 2. $w(z)$ parameter of the equation of state as a function of z, derived from 26 subset of GRBs given in [2].

The plot between E_γ vs E_{peak}, Eq. (14), turned out to be the best to find a tendency. To obtain $d_L(z)$ from this calibration one has the relation

$$d_L^2 = \frac{(1+z)}{4\pi S_{bolo} F_{beam}} 10^{\left[50.57+1.63log\left(\frac{E_{peak}}{300keV}(1+z)\right)\right]} cm^2, \tag{17}$$

where S_{bolo} and F_{beam} are observationally reported in [2].

The adjust of the data gives the following rule, see Fig. (1),

$$\frac{d_L(z)}{10^{27}cm} = 10.7z + 5.9z^2, \tag{18}$$

that is not far from the one proposed in [3], Eq. 10. We note that this calibration was obtained for a 26 subset of 69 GRBs data.

Using this extracted rule for the luminosity distance as function of z and using Eq. (8), we derived the dark-energy equation-of-state parameter $w(z)$, Fig. (2).

The plot of $w(z)$ is not continuos, but shows a divergence at certain z around 1.54. The reason of this divergence is that Eq. (8) blows up when $H^2(z) = H_0^2\Omega_{0m}(1+z)^3$, then Eq. (8) does not work for all the range of z, unless we consider values of Ω_{0m} smaller than ordinarily accepted. In other words, larger redshifts demand more abundance of dark energy and consequently less gravitational matter due to the concordance model. Our results point to the need of an alternative way to determine $w(z)$ appropriate to large redshifts.

5. Conclusions

The aim of this work is to determine the functional dependence of the dark-energy equation-of-state parameter in terms of the redshift, $w(z)$, from observational data coming from the GRBs. First we find the best calibration between the observational data of GRBs, obtaining the luminosity distance as function of redshift, $d_L(z)$, Eq. (18). Then we obtain the corresponding Hubble function and then $w(z)$. However we found an anomalous behavior in $w(z)$. The reason is an inconsistency of $w(z)$, Eq. (8), that has a limited range of validity, that we found to be for $z < 1.54$.

During the process of this research we learned of similar works, for instance [4], where the EoS is adopted and then tested. We remark that in our work we tried to extract $w(z)$ directly from data only assuming the FRW and concordance model.

With the improving of the observations, in particular with the recent launch of new satellites devoted to GRBs surveys, as Fermi-GLAST and AGILE, one should be able to expand the samples of GRBs and this will allow more robust analysis for $w(z)$ using GRBs.

Acknowledgments
A. M. acknowledges a M.Sc grant by Conacyt, México and also thanks to the organizers for partial support to attend the DSU Conference in León, Gto.

References

[1] S. Dodelson 2003 *Modern Cosmology* (USA: Academic Press)
[2] B. E. Schaefer 2007 *Astrophys. J.* **660** 16
[3] Y. Kodama et. al. 2008 *Mon. Not. R. Astron. Soc.* **391** L1-L4
[4] S. Capozziello, L. Izzo 2009 *Nuclear Physics B* **194** 206-211

The Fermi Large Area gamma ray Telescope and the current searches for dark matter in space

Aldo Morselli
INFN Roma Tor Vergata, Via della Ricerca Scientifica 1 00133 ROMA, ITALY
E-mail: aldo.morselli@roma2.infn.it

Abstract. Since its launch in the 2008, the Large Area Telescope, onboard of the Fermi Gamma-ray Space Telescope, has detected the largest amount of gamma rays, in the 20MeV 300GeV energy range and electrons + positrons in the 7 GeV- 1 TeV range. This impressive statistics allows one to perform a very sensitive indirect experimental search for dark matter. I will present the latest results on these searches

1. The Cosmic Ray Electron spectrum

Recently the experimental information available on the Cosmic Ray Electron (CRE) spectrum has been dramatically expanded as the Fermi-LAT Collaboration [1, 2] has reported a high precision measurement of the electron spectrum from 7 GeV to 1 TeV performed with its Large Area Telescope (LAT) [3, 4]. The spectrum shows no prominent spectral features and it is significantly harder than that inferred from several previous experiments. These data together with the PAMELA data on the rise above 10 GeV of the positron fraction [5] are quite difficult to explain with just secondary production [6],[7] [9]. The temptation to claim the discovery of dark matter is strong but there are competing astrophysical sources, such as pulsars, that can give a strong flux of primary positrons and electrons (see [10], [11], [12], [13] and references therein). At energies between 100 GeV and 1 TeV the electron flux reaching the Earth may be the sum of an almost homogeneous and isotropic component produced by Galactic supernova remnants and the local contribution of a few pulsars with the latter expected to contribute more and more significantly as the energy increases.

Two pulsars, Monogem, at a distance of d=290 pc and Geminga, at a distance of d=160 pc, can give a significant contribution to the high energy electron and positron flux reaching the Earth and with a set of reasonable parameters of the model of electron production we can have a nice fit of the PAMELA positron fraction [5] and Fermi data (see figures 1 and 2). However we have a lot of freedom in the choice of these parameters because we still do not know much about these processes, so further study on high energy emission from pulsars is needed in order to confirm or reject the pulsar hypothesis.

Nevertheless a dark matter interpretation of the Fermi-LAT and of the PAMELA data is still an open possibility. In Figure 3 is shown the parameter space of dark matter particle mass versus pair-annihilation rate, for models where dark matter annihilates into monochromatic $e^{\pm}$ [13]. The preferred range for the dark matter mass lies between 400 GeV and 1-2 TeV, with larger masses increasingly constrained by the H.E.S.S. results. The required annihilation rates, when employing a particular dark matter density profile imply typical boost factors ranging between

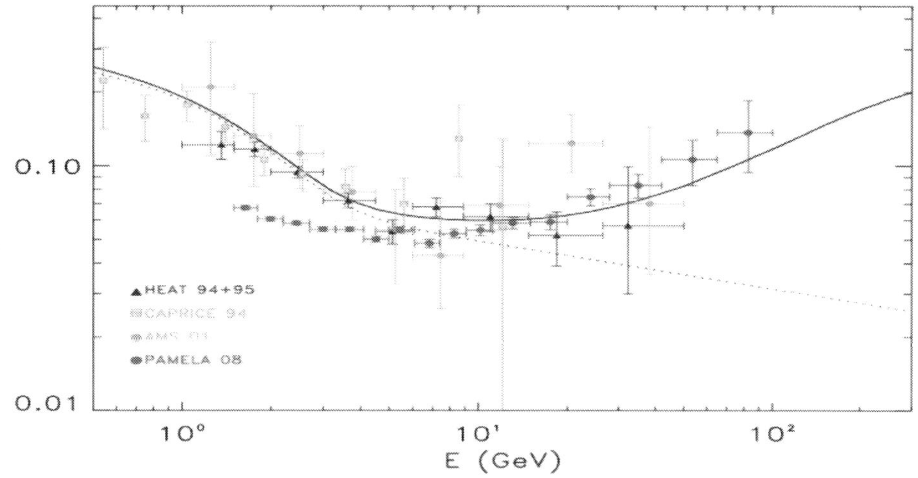

Figure 1. PAMELA data and a possible contribution from Monogem and Geminga pulsars [13]. Black-dotted line shows the background from secondary positrons in cosmic rays from GALPROP

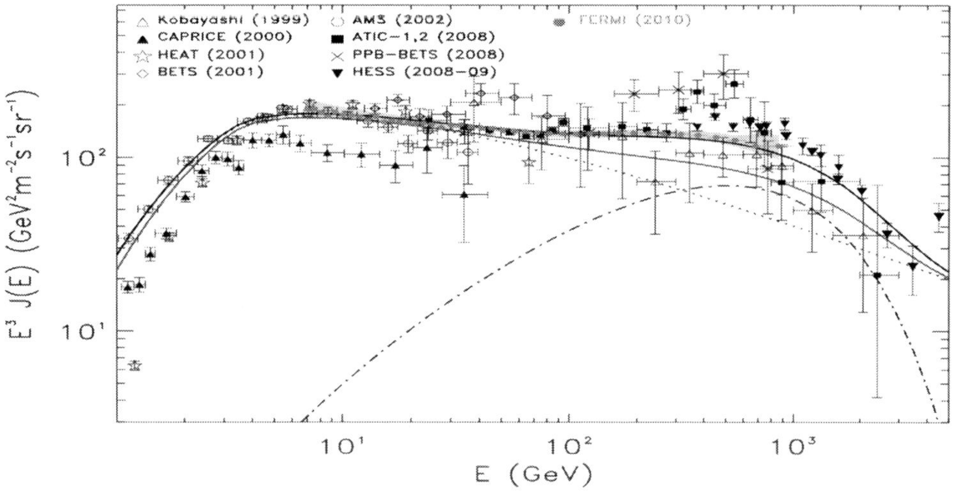

Figure 2. Electron-plus-positron spectrum (blue continuous line) for the same scenario as in figure 1. The gray band represents systematic errors on the Fermi-LAT data [3, 4]

20 and 100, when compared to the value $\langle \sigma v \rangle \sim 3 \times 10^{-26}$ cm^3/sec expected for a thermally produced dark matter particle relic.

How can one distinguish between the contributions of pulsars and dark matter annihilations? Most likely, a confirmation of the dark matter signal will require a consistency between different experiments and new measurements of the reported excesses with large statistics. The observed excess in the positron fraction should be consistent with corresponding signals in absolute positron and electron fluxes in the PAMELA data and all lepton data collected by Fermi. Fermi has a large effective area and long projected lifetime, 5 years nominal with a goal 10 years

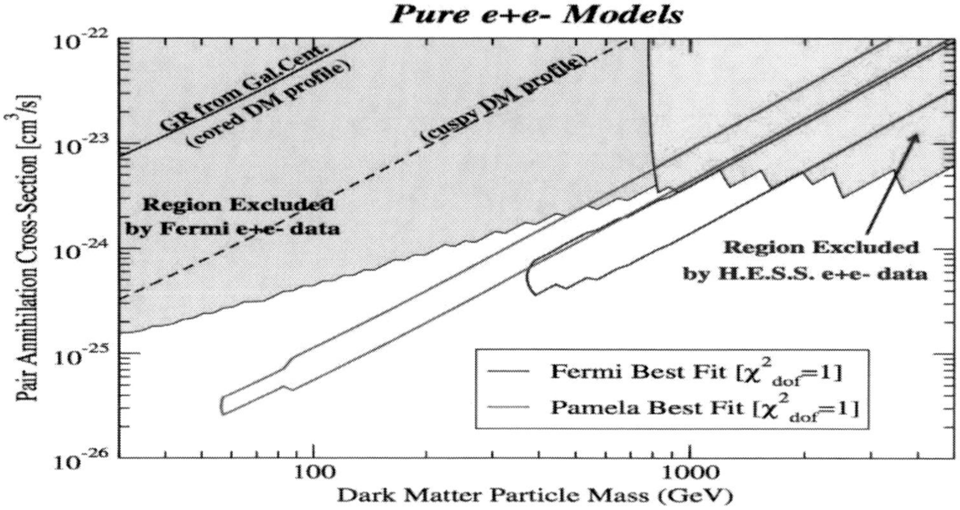

Figure 3. The parameter space of dark matter particle mass versus pair-annihilation rate, for models where dark matter annihilates into monochromatic $e^{\pm}$. Models inside the regions shaded in gray and cyan over-produce $e^{\pm}$ from dark matter annihilation with respect to the Fermi-LAT and H.E.S.S. measurements, at 2-σ level. The red and blue contours outline the regions where the χ^2 per degree of freedom for fits to the PAMELA and Fermi-LAT data is at or below 1.

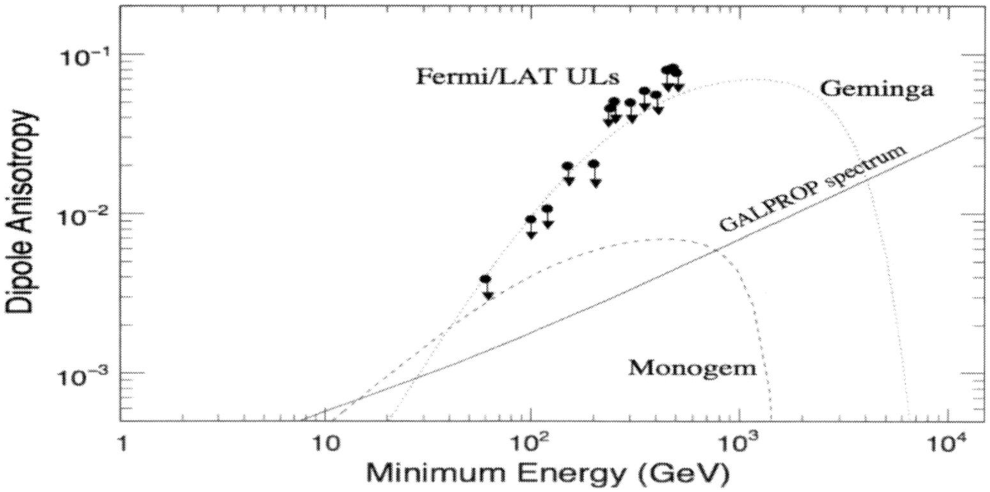

Figure 4. Dipole anisotropy δ versus the minimum energy for GALPROP (solid line), Monogem source (dashed line), and Vela source (dotted line). The 3σ confidence level from the data is also shown with circles. The solar modulation was treated using the force-field approximation with Φ=550 MV

mission, which makes it an excellent detector of cosmic-ray electrons up to ~1 TeV. Future Fermi measurements of the total lepton flux with large statistics make possible to distinguish a gradual change in slope with a sharp cutoff with high confidence [14]. The latter, can be an indication in favor of the dark matter hypothesis.

Another possibility is to look for anisotropies in the arrival direction of the electrons.

Journal of Physics: Conference Series **315** (2011) 012020 doi:10.1088/1742-6596/315/1/012020

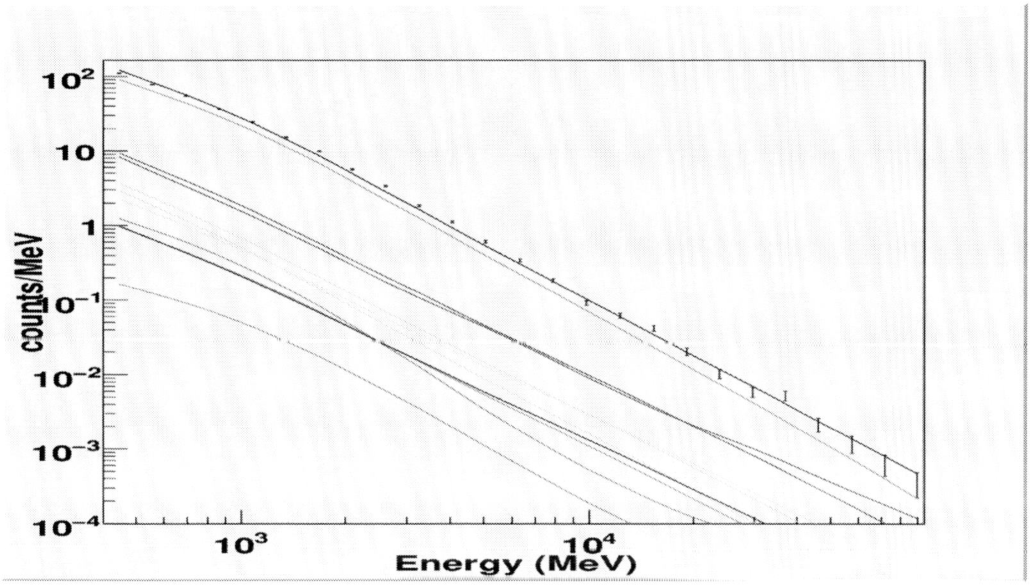

Figure 5. Spectra from the likelihood analysis of the Fermi/LAT data (number of counts vs reconstructed energy) in a 7°×7° region around the Galactic Center (number of counts vs reconstructed energy). Here in the plot, from above to below: -the black point are the observed data; -the black line is the sum of all the components; -the red line is the Galactic diffuse emission; - the lower black line is the isotropic extragalactic; -other components are the sources detected. These results are preliminary.

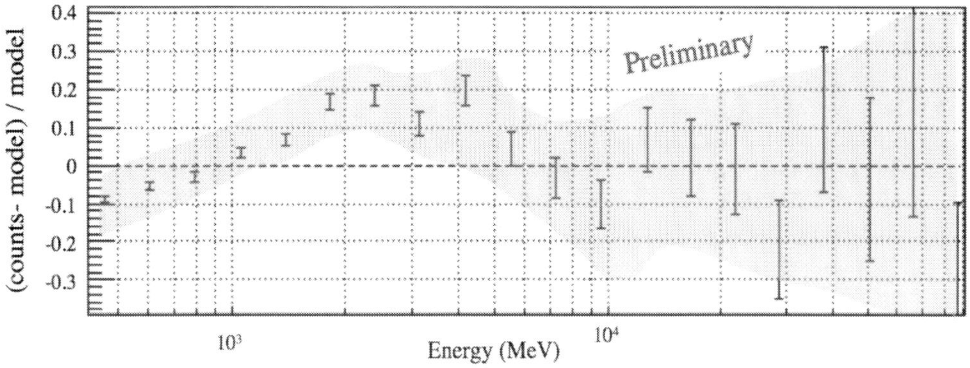

Figure 6. Residuals ((exp.data - model)/model) of the above likelihood analysis. The blue area shows the systematic errors on the effective area. These results are preliminary.

The Large Area Telescope on-board the *Fermi* satellite (*Fermi*-LAT) detected more than 1.6 million cosmic-ray electrons/positrons with energies above 60 GeV during its first year of operation. The arrival directions of these events were searched for anisotropies of angular scale extending from ~10° up to 90°, and of minimum energy extending from 60 GeV up to 480 GeV. Upper limit for the dipole anisotropy has been set to 0.5 - 5% depending on the energy [15].

The levels of anisotropy expected for Vela like and Monogem-like sources (i.e. sources with

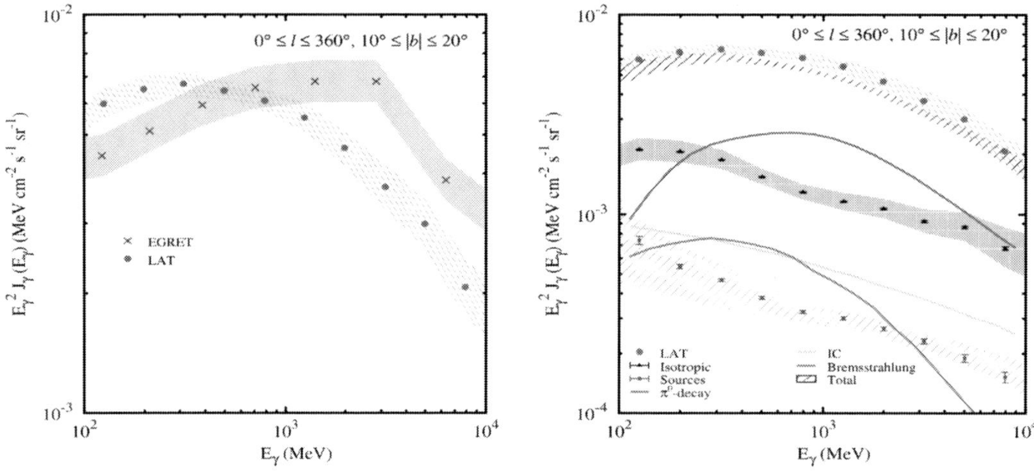

Figure 7. *Left:* Diffuse emission intensity averaged over all Galactic longitudes for latitude range $10° \leq |b| \leq 20°$. Data points: Fermi LAT, red dots; EGRET, blue crosses. Systematic uncertainties: Fermi LAT, red; EGRET, blue. *Right:* Fermi LAT data with model, source, and isotropic components for same sky region.

similar distances and ages) seem to be higher than the scale of anisotropies excluded by the results (see figure 4). However, it is worth to point out that the model results are affected by large uncertainties related to the choice of the free parameters.

2. The gamma-ray signals

A strong leptonic signal should be accompanied by a boost in the γ-ray yield providing a distinct spectral signature detectable by Fermi.

The Galactic center (GC) is expected to be the strongest source of γ-rays from DM annihilation, due to its coincidence with the cusped part of the DM halo density profile [16, 17]. A preliminary analysis of the data, taken during the first 11 months of the Fermi satellite operations, is shown in figures 5 and 6.

The reported results were obtained with a binned likelihood analysis, performed by means of the tools developed by the Fermi/LAT collaboration (gtlike, from the Fermi analysis tools [18]). For this analysis:

- a ROI of $7° \times 7°$ was considered. This ROI was used in order to minimize the background contribution and to avoid significative leakage of the gamma ray signal under study;
- the ROI was centered at the position RA = 266.46°, Dec=-28.97°, i.e. the position of the brightest source;
- the data taken during the first 11 months (8/2008-7/2009) have been used;
- the events were selected to have an energy between 400MeV and 100GeV;
- only events classified as "diffuse" class and which converted in the *front* part of the tracker have been selected for the analysis. The selection in energy, event classification and conversion provided us with events with very well reconstructed incoming direction and data have been binned into a 100×100bins map;
- the IRF and the events classification are those relative to the Pass6V3 version of the Fermi/LAT analysis software.

In order to perform the likelihood analysis for the LAT data, a model of the already known sources and the diffuse background should be built. The model in use for the presented analysis contains 11 sources in the Fermi 1 year catalog [19] which are located into or very close to the considered ROI. These sources have a point-like spatial model and a spectrum in the form of a power-law. The model also contain the diffuse gamma-ray background which is made of two components:

(i) the Galactic Diffuse gamma-ray background. The observed Galactic Diffuse emission was modeled by means of the GALPROP code [20] and [21], and the realization of the galactic emission named gll_iem_54_87Xexph7S.fit was used. During the likelihood maximization only the normalization of the GALPROP model is varied, not its components;

(ii) the Isotropic Background. This component should account for both the Extragalactic gamma-ray emission and residual charged particles. It is modeled as an isotropic emission with a template spectrum.

The diffuse gamma-ray backgrounds and discrete sources, as we know them today, can account for the large majority of the detected gamma-ray emission from the Galactic Center. Nevertheless a residual emission is left, not accounted for by the above models [22].

Improved modeling of the Galactic diffuse model as well as the potential contribution from other astrophysical sources (for instance unresolved point sources) could provide a better description of the data. Analyses are under-way to investigate these possibilities.

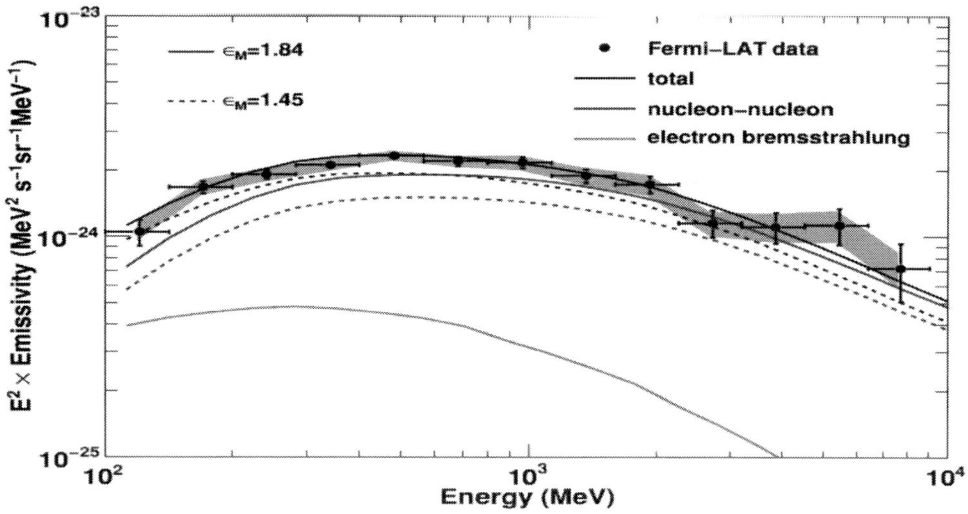

Figure 8. Differential γ-ray emissivity from the local atomic hydrogen gas compared with the calculated γ-ray production. The horizontal and vertical error bars indicate the energy ranges and 1 σ statistical errors, respectively. Estimated systematic errors of the LAT data are indicated by the shaded area. A nucleus enhancement factor ϵ_M of 1.84 is assumed for the calculation of the γ-rays from nucleon-nucleon interactions. Dotted lines indicate the emissivities for the case of $\epsilon_M = 1.45$, the lowest values in the referenced literature.

An excess in gamma-ray should also be seen in the Galactic diffuse spectrum. Figure 7 (left) shows the LAT data averaged over all Galactic longitudes and the latitude range $10^\circ \leq |b| \leq 20^\circ$. The hatched band surrounding the LAT data indicates the systematic uncertainty in the measurement due to the uncertainty in the effective area. Also shown on the right are the

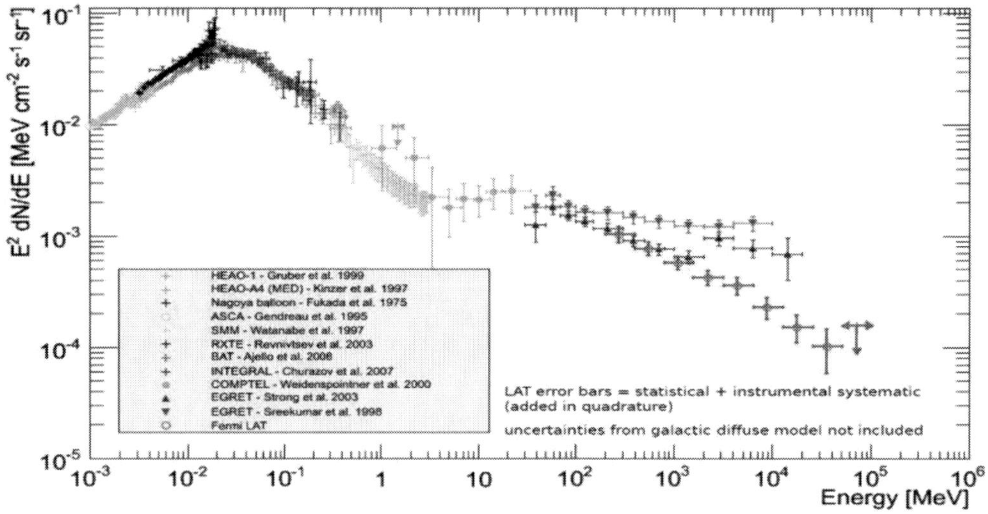

Figure 9. Fermi-LAT extragalactic intensity compared with EGRET-derived intensities. Fermi-LAT derived spectrum is compatible with a simple power-law with index $\gamma = 2.41 \pm 0.05$ and intensity $I(> 100\,\mathrm{MeV}) = (1.03 \pm 0.17) \times 10^{-5}$ cm^{-2} s^{-1} sr^{-1} where the uncertainties are systematics dominated.

EGRET data for the same region of sky where one can see that the LAT-measured spectrum is significantly softer than the EGRET measurement [23]. Figure 7 (right) compares the LAT spectrum with the spectra of an *a priori* diffuse Galactic emission (DGE) model. While the LAT spectral shape is consistent with the DGE model used in this paper, the overall model emission is too low thus giving rise to a $\sim 10 - 15\%$ excess over the energy range 100 MeV to 10 GeV. However, the DGE model is based on pre Fermi data and knowledge of the DGE. The difference between the model and data is of the same order as the uncertainty in the measured CR nuclei spectra at the relevant energies. Overall, the agreement between the LAT-measured spectrum and the model shows that the fundamental processes are consistent with our data, thus providing a solid basis for future work in understanding the DGE.

In figure 8 it is shown the diffuse γ-rays in a mid-latitude region in the third quadrant (Galactic longitude l from 200° to 260° and latitude $|b|$ from 22° to 60°). The region contains no known large molecular clouds and most of the atomic hydrogen is within 1 kpc of the solar system. The contributions of γ-ray point sources and inverse Compton scattering are estimated and subtracted. The residual γ-ray intensity exhibits a linear correlation with the atomic gas column density in energy from 100 MeV to 10 GeV The differential emissivity from 100 MeV to 10 GeV agrees with calculations based on cosmic ray spectra consistent with those directly measured, at the 10 % level. The results obtained indicate that cosmic ray nuclei spectra within 1 kpc from the solar system in regions studied are close to the local interstellar spectra inferred from direct measurements at the Earth within ~ 10 % [24].

The high-energy diffuse γ-ray emission is dominated by γ-rays produced by cosmic rays (CR) interacting with the Galactic interstellar gas and radiation fields, the so-called diffuse Galactic emission. A much fainter component, commonly designated as "extragalactic γ-ray background" (EGB) by definition has an isotropic sky distribution and is considered by many to be the superposition of contributions from unresolved extragalactic sources including active galactic nuclei, starburst galaxies and γ-ray bursts and truly-diffuse emission processes.

Figure 9 shows the spectrum of the EGB above 200 MeV derived with the data taken during the first 11 months [25] , and from EGRET data [26], [27]. Our intensity extrapolated to 100 MeV based on the power-law fit, $I(>100\,\mathrm{MeV}) = (1.03 \pm 0.17) \times 10^{-5}$ cm^{-2} s^{-1} sr^{-1}, is significantly lower than that obtained from EGRET data: $I_{\mathrm{EGRET}}(>100\,\mathrm{MeV}) = (1.45 \pm 0.05) \times 10^{-5}$ cm^{-2} s^{-1} sr^{-1}. Furthermore, our spectrum is compatible with a featureless power law with index $\gamma = 2.41 \pm 0.05$. This is significantly softer than the EGRET spectrum with index $\gamma_{\mathrm{EGRET}} = 2.13 \pm 0.03$. To check that the different spectra are not due to the instrumental point-source sensitivities, we adopt $F(>100\,\mathrm{MeV}) = 10^{-7}$ cm^{-2} s^{-1}, comparable to the average EGRET sensitivity, and attribute the flux of all detected LAT sources below this threshold to the EGB. We obtain an intensity $I_{res}(>100\,\mathrm{MeV}) = (1.19 \pm 0.18) \times 10^{-5}$ cm^{-2} s^{-1} sr^{-1} and a spectrum compatible with a power-law with index $\gamma_{res} = 2.37 \pm 0.05$. Therefore, the discrepancy cannot be attributed to a lower threshold for resolving point sources. Our EGB intensity is comparable to that obtained in the EGRET re-analysis by [27] with an updated DGE model, $I_{\mathrm{SMR}}(>100\,\mathrm{MeV}) = (1.11 \pm 0.1) \times 10^{-5}$ cm^{-2} s^{-1} sr^{-1}. However, our EGB spectrum does not show the distinctive harder spectrum above ≥ 1 GeV and peak at ~ 3 GeV found in the same EGRET re-analysis.

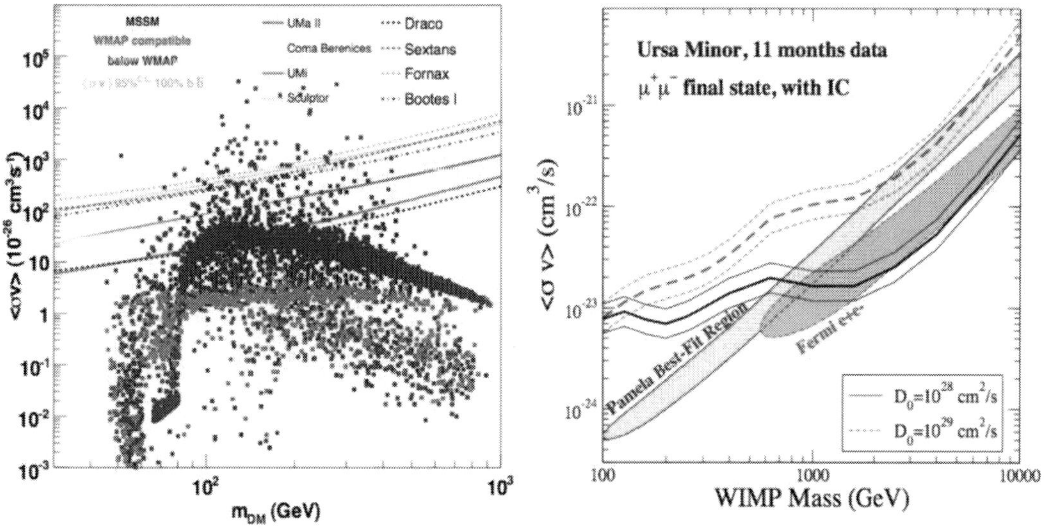

Figure 10. (Left) MSSM models in the ($m_{\mathrm{wimp}}, < \sigma v >$) plane consistent with all accelerator constraints. Red points have a neutralino thermal relic abundance corresponding to the inferred cosmological dark matter density (blue points have a lower thermal relic density, and we assume that neutralinos still comprise all of the dark matter in virtue of additional non-thermal production processes). The lines indicate the Fermi 95% upper limits obtained from likelihood analysis on the selected dwarfs. (Right) Constraints on the annihilation cross-section for a $\mu^+\mu^-$ final state based on the 95% confidence limits on the γ-ray flux compared to dark matter annihilation models which fit well either the PAMELA or Fermi $e^+ + e^-$ measurements. The constraints are for the Ursa Minor dwarf including both γ-ray emission from IC scattering and final state radiation. Here we consider two different diffusion coefficients and show the effect of the uncertainties in the Ursa Minor density profile.

3. Dwarf spheroidal galaxies and Clusters of galaxies

Local Group dwarf spheroidal galaxies, the largest galactic substructures predicted by the cold dark matter scenario, are attractive targets for dark matter indirect searches because they are

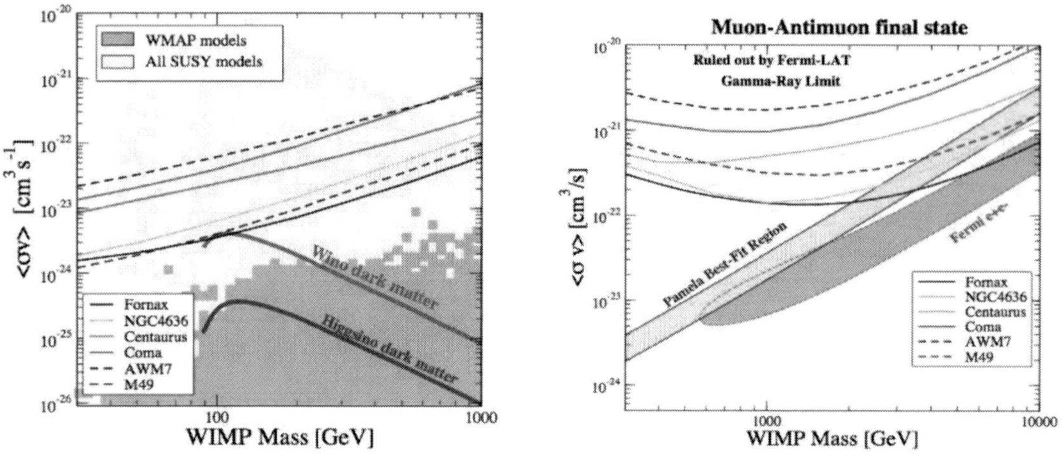

Figure 11. *Left:* The expected gamma-ray spectrum from dark matter annihilation into a $\mu^+\mu^-$ final state including both FSR and emission from IC scattering of CMB photons for different assumed particle masses. The spectral normalization is arbitrary, but the same effective J is used in all cases. *Right:* Flux upper limits as a function of particle mass for an assumed $\mu^+\mu^-$ final state, including the contributions of both FSR and IC gamma-ray emission, for the clusters. The upper limits reflect the 95% confidence level limits on the photon flux in the 100 MeV to 100 GeV energy range.

nearby and among the most extreme dark matter dominated environments. With the data taken during the first 11 months no significant γ-ray emission was detected above 100 MeV from the candidate dwarf galaxies. So we can determine upper limits to the γ-ray flux assuming both power-law spectra and representative spectra from WIMP annihilation. The resulting integral flux above 100 MeV is constrained to be at a level below around 10^{-9} photons $cm^{-2}s^{-1}$ [28] Using recent stellar kinematic data, the γ-ray flux limits can be combined with improved determinations of the dark matter density profile in 8 of the 14 candidate dwarfs to place limits on the pair annihilation cross-section of WIMPs in several widely studied extensions of the standard model, including its supersymmetric extension and other models that received recent attention. With the present data we are able to rule out large parts of the parameter space where the thermal relic density is below the observed cosmological dark matter density and WIMPs (neutralinos here) are dominantly produced non-thermally, e.g. in models where supersymmetry breaking occurs via anomaly mediation. These γ-ray limits also constrain some WIMP models proposed to explain the *Fermi* and PAMELA e^+e^- data, including low-mass wino-like neutralinos and models with TeV masses pair-annihilating into muon-antimuon pairs (see figure 10).

The same kind of analysis can be done with the clusters of galaxies. Nearby clusters and groups of galaxies are potentially bright sources of high-energy gamma-ray emission resulting from the pair-annihilation of dark matter particles. However, no significant gamma-ray emission has been detected so far from clusters in the first 11 months of observations with the Fermi Large Area Telescope. We interpret this non-detection in terms of constraints on dark matter particle properties. In particular for leptonic annihilation final states and particle masses greater than $\sim$ 200 GeV, gamma-ray emission from inverse Compton scattering of CMB photons is expected to dominate the dark matter annihilation signal from clusters, and our gamma-ray limits exclude large regions of the parameter space that would give a good fit to the recent anomalous Pamela and Fermi-LAT electron-positron measurements [29]. For an example see figure 11.

Finally a line at the WIMP mass, due to the 2γ production channel, could be observed as a feature in the astrophysical source spectrum [14]. Such an observation is a "smoking gun" for WIMP DM as it is difficult to explain by a process other than WIMP annihilation or decay and the presence of a feature due to annihilation into γZ in addition would be even more convincing.

Up to now however no lines was observed and we obtain γ-ray line flux upper limits in the range $0.6 - 4.5 \times 10^{-9}\mathrm{cm}^{-2}\mathrm{s}^{-1}$ [30] and corresponding DM annihilation cross-section and decay lifetime limits shown in figure 12 and in figure 13 .

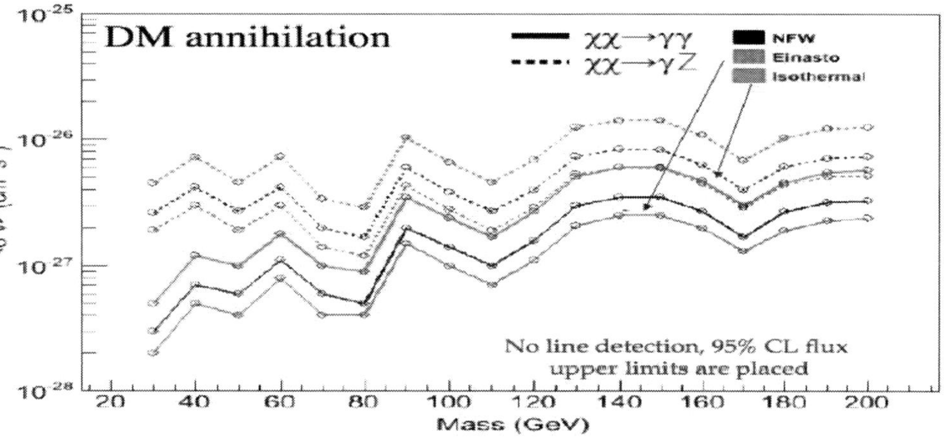

Figure 12. Cross-section limits for various dark matter halo profiles for the annihilation in monochromatic gamma-rays

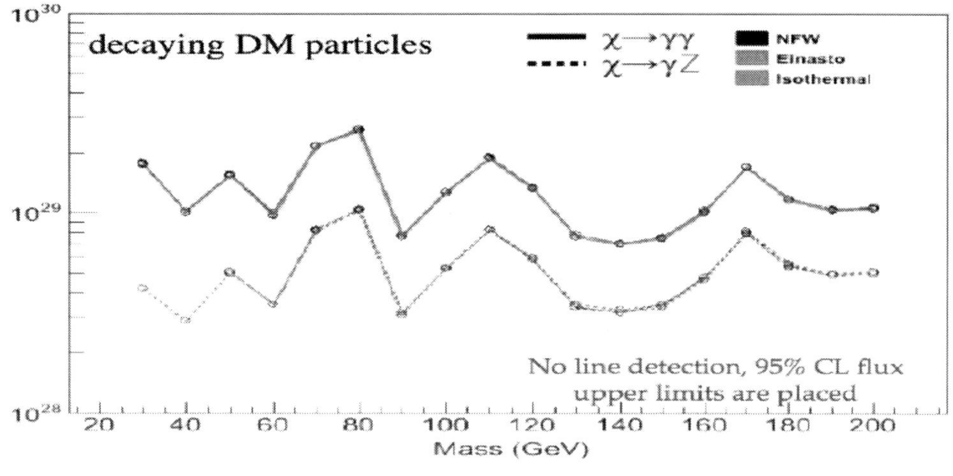

Figure 13. Lifetime limits for various dark matter halo profiles for the decay in monochromatic gamma-rays channel

4. Conclusion

Recent accurate measurements of cosmic-ray positrons and electrons by PAMELA, and Fermi have opened a new era in particle astrophysics. The CRE spectrum measured by Fermi-LAT is significantly harder than what was expected on the basis of previous data. Adopting the presence of an extra $e^{\pm}$ primary component with $\sim$ 2.4 spectral index and $E_{cut} \sim 1TeV$ allows to consistently interpret Fermi-LAT CRE data, HESS and PAMELA. Such extra-component can be originated by pulsars for a reasonable choice of relevant parameters or by annihilating dark matter for model with $M_{DM} \sim 1TeV$. Improved analysis and complementary observations (CRE anisotropy, spectrum and angular distribution of diffuse γ, DM sources search in γ) are required to possibly discriminate the right scenario. Their exotic origin has to be confirmed by complimentary findings in γ-rays by Fermi and atmospheric Cherenkov telescopes, and by LHC in the debris of high-energy proton destructions. A positive answer will be a major breakthrough and will change our understanding of the universe forever. On the other hand, if it happens to be a conventional astrophysical source of cosmic rays, it will mean a direct detection of particles accelerated at an astronomical source, again a major breakthrough. In this case we will learn a whole lot about our local Galactic environment. However, independently on the origin of these excesses, exotic or conventional, we can expect very exciting several years ahead of us.

[1] W.B.Atwood et al. [Fermi Coll.] ApJ 697 No 2 (2009 June 1) 1071-1102 [arXiv:0902.1089]
[2] A.A.Abdo et al. [Fermi Coll.] Astroparticle Physics 32 (2009) 193-219 [arXiv:0904.2226]
[3] A.A.Abdo et al. [Fermi Coll.], Phys. Review Letters, 102, 181101 (2009) [arXiv:0905.0025]
[4] M.Ackermann, et al. [Fermi Coll.], Phys. Physical Review D, accepted [arXiv:1008.3999]
[5] O. Adriani et al. [PAMELA Coll.] Nature 458, 607, 2009 [arxiv:0810.4995]
[6] A. W. Strong and I. V. Moskalenko, ApJ 509 (1998) 212, ApJ 493 (1998) 694
[7] A. Lionetto, A. Morselli, and V. Zdravkovic, JCAP09 (2005) 010 [astro-ph/0502406]
V. S. Ptuskin et al., ApJ 642 (2006) 902
[8] K. Abe et al. [BESS Coll.], arXiv:0805.1754
[9] A. Morselli, I.Moskalenko, PoS(idm2008)025 [arXiv:0811.3526]
[10] A.Boulares APJ 342 (1989) 807-813
[11] F. A. Aharonian, A. M. Atoyan and H. J. Völk, Astron. Astrophys. 294 (1995) L41
[12] S. Coutu et al., Astropart. Phys. 11 (1999) 429
[13] D.Grasso et al. Astroparticle Physics 32 (2009), pp. 140-151 [arXiv:0905.0636]
[14] E. Baltz et al. , JCAP07 (2008) 013 [arXiv:0806.2911]
[15] M.Ackermann et.al. [Fermi Coll.] , accepted for publication in Physical Review D [arXiv:1008.5119]
[16] A. Morselli at al., Nucl. Phys. 113B (2002) 213
[17] A.Cesarini, F.Fucito, A.Lionetto, A.Morselli, P. Ullio, Astropart. Phys. 21 (2004) 267 [astro-ph/0305075]
[18] http://fermi.gsfc.nasa.gov/ssc/data/analysis/software/
[19] A.A.Abdo et al. [Fermi Coll.], Astrophysical Journal Supplement Series 2010 188 405 [arXiv:1002.2280]
[20] A.Strong et al., 2004, ApJ 613, 962S
[21] A.Strong et al., 2007 Annu. Rev. Nucl. Part. Sci., 57, 285
[22] V. Vitale and A. Morselli for the Fermi/LAT Collaboration, 2009 Fermi Symposium [arXiv:0912.3828]
[23] A.Abdo et al. [Fermi Coll.], Physical Review Letters, 103, 251101 (2009) [arXiv:0912.0973]
[24] A.A.Abdo et al., [Fermi Coll.], ApJ 703 (2009) 1249-1256 [arXiv:0908.1171]
[25] A.A.Abdo et al. [Fermi Coll.], Phys.Rev.Lett.104, 101101-1-7 (2010) [arXiv:1002.3603]
[26] P. Sreekumar et al., Astrophys. J. 494, 523 (1998)
[27] A. W. Strong, I. V. Moskalenko, and O. Reimer, Astro- phys. J. 613, 956 (2004)
[28] A.A.Abdo et al., [Fermi Coll.], ApJ 712 (2010) 147-158 [arXiv:1001.4531]
[29] M.Ackermann et al. [Fermi Coll.] ,JCAP, Journal of Cosmology and Astroparticle Physics, 05, 025 (2010) [arXiv:1002.2239]
[30] A.A.Abdo et al., [Fermi Coll.], Phys.Rev.Lett.104, 091302-08 (2010) [arXiv:1001.4836]

Dark Matter in SuperGUT Unification Models

Keith A. Olive
William I. Fine Theoretical Physics Institute,
University of Minnesota, Minneapolis, MN 55455, USA

Abstract. After a brief update on the prospects for dark matter in the constrained version of the MSSM (CMSSM) and its differences with models based on minimal supergravity (mSUGRA), I will consider the effects of unifying the supersymmetry-breaking parameters at a scale above M_{GUT}. One of the consequences of superGUT unification, is the ability to take vanishing scalar masses at the unification scale with a neutralino LSP dark matter candidate. This allows one to resurrect no-scale supergravity as a viable phenomenological model.

1. Introduction

While often used synonymously, the constrained minimal supersymmetric standard model (CMSSM) [1, 2, 3, 4, 5, 6] differs in two important ways from supersymmetric models based on minimal supergravity (mSUGRA) [7, 8, 9, 10]. The latter class of theories is in fact a subset of the former and can be thought of as a very constrained version of the theory [11]. The often studied CMSSM, is a 4+ parameter theory. Starting with the superpotential,

$$W = \left[y_e H_1 L e^c + y_d H_1 Q d^c + y_u H_2 Q u^c\right] + \mu H_1 H_2, \tag{1}$$

we can obtain the soft supersymmetry-breaking part of the Lagrangian

$$\begin{aligned} \mathcal{L}_{\rm soft} &= -\frac{1}{2} M_\alpha \lambda^\alpha \lambda^\alpha - m_{ij}^2 \phi^{*i} \phi^j \\ &\quad - A_e y_e H_1 L e^c - A_d y_d H_1 Q d^c - A_u y_u H_2 Q u^c - B\mu H_1 H_2. \end{aligned} \tag{2}$$

In the CMSSM, the four parameters are: the gaugino mass, $M_\alpha = m_{1/2}$, unified at some high energy input scale, M_{in}, usually assumed to be the grand unified (GUT) scale, M_{GUT}; the scalar masses $m_{ij}^2 = \delta_{ij} m_0^2$; and the trilinear terms, $A_f = A_0$, are all unified at the same input scale, M_{in}; and the ratio of the two Higgs expectation values, $\tan\beta$. In the CMSSM, $|\mu|$ and B are determined by the electroweak symmetry-breaking (EWSB) conditions by fixing M_Z and $\tan\beta$. In mSUGRA models, there is an additional constraint, namely

$$B_0 = A_0 - m_0. \tag{3}$$

In this case, it is no longer possible to satisfy the EWSB conditions (i.e., the minimization of the Higgs potential) and choose $\tan\beta$ independently. As a consequence, we have a 3+ parameter theory which is now specified by $m_{1/2}, m_0$, and A_0. The "+" in the number of parameters refers to the sign of μ. While the magnitude of μ is fixed by the EWSB conditions, its sign is left undetermined. In true minimal supergravity models, there is also a relation between m_0 and the gravitino mass, namely $m_0 = m_{3/2}$. Thus, the gravitino mass is no longer independent and as we will see, often in mSUGRA models, the gravitino is found to be the lightest supersymmetric particle (LSP). The gravitino mass is generally not counted as one of the parameters in the CMSSM, as it can be set to a sufficiently large value so as to render it irrelevant.

No-scale supergravity models [12] are still more restrictive, as the boundary conditions on the scalar masses and bi- and trilinear supersymmetry-breaking terms become

$$m_0 = A_0 = B_0 = 0. \tag{4}$$

Thus, we are left with a 1+ parameter theory specified by $m_{1/2}$. Generally, the condition $m_0 = 0$ is largely incompatible with phenomenology except for a restricted set of values of $\tan\beta$, but might be more viable if the supersymmetry-breaking scale were pushed above the GUT scale [13].

Below, I will try to highlight some of the differences between the CMSSM and mSUGRA models and explore the phenomenological consequences of pushing the scale at which the supersymmetry-breaking parameters are unified above the GUT scale [14, 15]. After a brief review on the status of the CMSSM, and its differences with mSUGRA, I will demonstrate the difficulties with no-scale supergravity models. Then, I will describe the effects of raising the supersymmetry-breaking scale above the GUT scale and apply this to no-scale models.

2. The CMSSM

For given values of $\tan\beta$, A_0, and $sgn(\mu)$, the regions of the CMSSM parameter space that yield an acceptable relic density and satisfy other phenomenological constraints may be displayed in the $(m_{1/2}, m_0)$ plane. In Fig. 1a, the dark (blue) shaded region corresponds to that portion of the CMSSM plane with $\tan\beta = 10$, $A_0 = 0$, and $\mu > 0$ such that the computed relic density yields the WMAP value [16] of

$$\Omega h^2 = 0.111 \pm 0.006. \tag{5}$$

The bulk region at relatively low values of $m_{1/2}$ and m_0, tapers off as $m_{1/2}$ is increased. At higher values of m_0, annihilation cross sections are too small to maintain an acceptable relic density and $\Omega_\chi h^2$ is too large. At large $m_{1/2}$, co-annihilation processes between the LSP and the next lightest sparticle (in this case the $\tilde{\tau}$) enhance the annihilation cross section and reduce the relic density. This occurs when the LSP and NLSP are nearly degenerate in mass. The dark (red) shaded region has $m_{\tilde{\tau}} < m_\chi$ and is excluded. The effect of coannihilations is to create an allowed band about 25-50 GeV wide in m_0 for $m_{1/2} \lesssim 950$ GeV, or $m_\chi \lesssim 400$ GeV, which tracks above the $m_{\tilde{\tau}_1} = m_\chi$ contour [17]. Also shown in the figure are some phenomenological constraints from the lack of detection of charginos [18], or Higgses [19] as well as constraints from $b \to s\gamma$ [20] and $g_\mu - 2$ [21]. The locations of these constraints are described in the caption.

At larger $m_{1/2}, m_0$ and $\tan\beta$, the relic neutralino density may be reduced by rapid annihilation through direct-channel H, A Higgs bosons, as seen in Fig. 1(b) [1, 3]. Finally, the relic density can again be brought down into the WMAP range at large m_0 in the 'focus-point' region close the boundary where EWSB ceases to be possible and the lightest neutralino χ acquires a significant higgsino component [22]. The start of the focus point region is seen in the upper left of Fig. 1b.

A global likelihood analysis enables one to pin down the available parameter space in the CMSSM. One can avoid the dependence on priors by performing a pure χ^2-based fit as done in [6] which used a Markov-Chain Monte Carlo (MCMC) technique to explore efficiently the likelihood function in the parameter space of the CMSSM.

The best fit point is shown in Fig. 2, which also displays contours of the $\Delta\chi^2$ function in the CMSSM (solid curves outline the 68 and 95% CL regions). The parameters of the best-fit CMSSM point are $m_{1/2} = 310$ GeV, $m_0 = 60$ GeV, $A_0 = 130$ GeV, and $\tan\beta = 11$, yielding

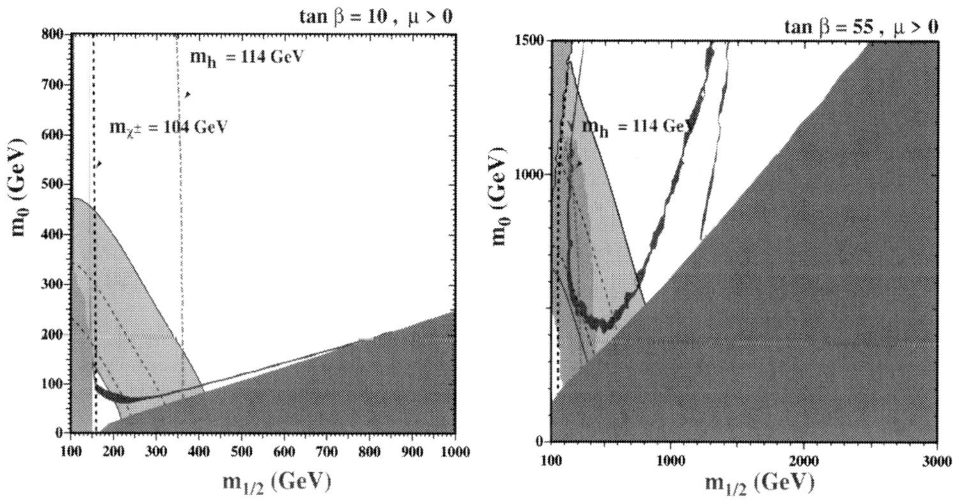

Figure 1. The $(m_{1/2}, m_0)$ planes for (a) $\tan\beta = 10$ and $\mu > 0$, assuming $A_0 = 0, m_t = 173.1$ GeV and $m_b(m_b)^{\overline{MS}}_{SM} = 4.25$ GeV. The near-vertical (red) dot-dashed lines are the contours $m_h = 114$ GeV, and the near-vertical (black) dashed line is the contour $m_{\chi^\pm} = 104$ GeV. The medium (dark green) shaded region is excluded by $b \to s\gamma$, and the dark (blue) shaded area is the cosmologically preferred region. In the dark (brick red) shaded region, the LSP is the charged $\tilde{\tau}_1$. The region allowed by the E821 measurement of $g_\mu - 2$ at the 2-σ level, is shaded (pink) and bounded by solid black lines, with dashed lines indicating the 1-σ ranges. In (b), $\tan\beta = 55$.

the overall $\chi^2/\mathrm{N_{dof}} = 20.6/19$ (36% probability) and nominally $M_h = 114.2$ GeV [6]. The best-fit point is in the coannihilation region of the $(m_0, m_{1/2})$ plane. The C.L. contours extend to somewhat large values of m_0. However, the qualitative features of the $\Delta\chi^2$ contours indicate a preference for small m_0 and $m_{1/2}$.

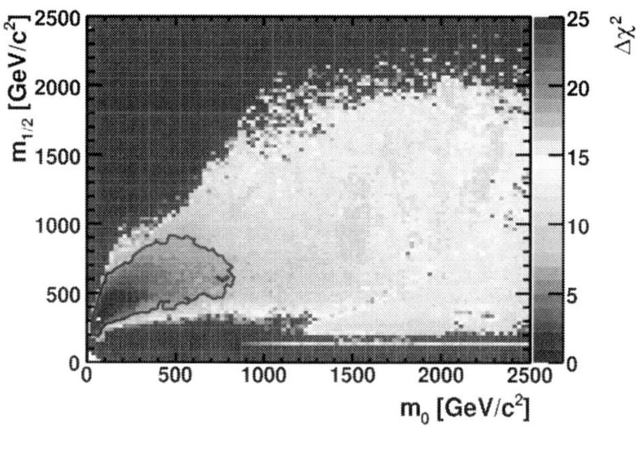

Figure 2. The $\Delta\chi^2$ function in the $(m_0, m_{1/2})$ plane for the CMSSM. We see that the coannihilation region at low m_0 and $m_{1/2}$ is favored.

The frequentist analysis described above can also be used to predict the neutralino-nucleon elastic scattering cross section [6]. The value of $\sigma_p^{\rm SI}$ shown in Fig. 3a is calculated assuming a π-N scattering σ term $\Sigma_{\pi N} = 64$ MeV. We see in Fig. 3 that values of the $\tilde{\chi}_1^0$-proton cross section $\sigma_p^{\rm SI} \sim 10^{-8}$ pb are expected in the CMSSM, and that much larger values seem quite unlikely. The 2D χ^2 function in the (m_χ, σ_p) plane is shown in Fig. 3b.

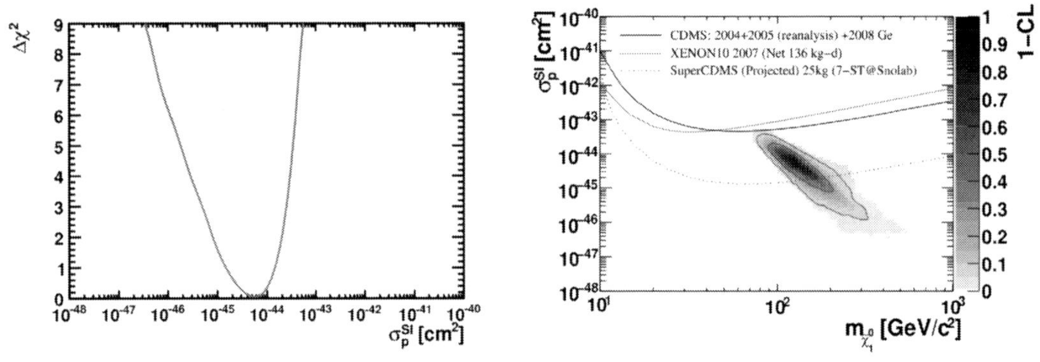

Figure 3. The likelihood function for the spin-independent $\tilde{\chi}_1^0$-proton scattering cross section $\sigma_p^{\rm SI}$ in the CMSSM (left panel). The correlation between the spin-independent DM scattering cross section $\sigma_p^{\rm SI}$ and $m_{\tilde{\chi}_1^0}$ in the CMSSM (right panel).

3. mSUGRA Models

As noted above, mSUGRA models are a very constrained version of the MSSM, as $\tan\beta$ is no longer a free parameter. In minimal $N = 1$ supergravity, the Kähler potential can be written as

$$G = K(\phi^i, \phi_i^*, z, z^*) + \ln(\kappa^6 |W|^2) \tag{6}$$

with

$$K = \kappa^2(\phi^i\phi_i^* + zz^*) \tag{7}$$

where $W = f(z) + g(\phi)$ is the superpotential, assumed to be separable in hidden sector fields, z, and standard model fields, ϕ. $\kappa^{-1} = M_P/\sqrt{8\pi}$ and the Planck mass is $M_P = 1.2 \times 10^{19}$ GeV. The scalar potential can be derived once the superpotential is specified. The simplest choice for a single hidden sector field is a superpotential [7] $f(z) = m(z + b)$. Using this in Eq. (6) and dropping terms inversely proportional to the Planck mass, we can write [10]

$$V = |\frac{\partial g}{\partial \phi}|^2 + m_{3/2}(\phi\frac{\partial g}{\partial \phi} - \sqrt{3}g + h.c.)) + m_{3/2}^2\phi\phi^*, \tag{8}$$

where the vacuum expectation value $\langle \kappa z \rangle = \sqrt{3} - 1$ (with $\kappa b = 2 - \sqrt{3}$, chosen to cancel the vacuum energy density at the minimum) has been inserted and the superpotential has been rescaled by a factor $e^{-\kappa b}$.

The first term in Eq. (8) is the ordinary F-term part of the scalar potential of global supersymmetry. The next term, proportional to $m_{3/2}$ represents a universal trilinear A-term. This can be seen by noting that $\sum \phi\partial g/\partial\phi = 3g$, so that in this model of supersymmetry breaking, $A_0 = (3 - \sqrt{3})m_{3/2}$. Note that if the superpotential contains bilinear terms, we would find $B_0 = (2 - \sqrt{3})m_{3/2}$. The last term represents a universal scalar mass of the type advocated in the CMSSM, with $m_0^2 = m_{3/2}^2$. The generation of such soft terms is a rather generic property of low energy supergravity models [23] and the relation $B_0 = A_0 - m_0$ is derived from the minimal form of the Kähler potential and does not depend on the specific form of $f(z)$.

The analogue of the CMSSM $(m_{1/2}, m_0)$ plane for mSUGRA models is shown in Fig. 4 for two fixed values of A_0/m_0 [11]. In the left panel, the Polonyi model choice of $A_0/m_0 = 3 - \sqrt{3}$ is made. As one can see, each point on the plane corresponds to a specific value of $\tan\beta$ as specified by the dot-dashed contours. Also seen in this panel is a solid (brown) diagonal line. Below this line, the gravitino is the LSP, since $m_\chi \approx 0.43 m_{1/2} < m_0 = m_{3/2}$. Below this line, there is a diagonal dotted (red) line, which separates the region for which the next lightest supersymmetric particle is the neutralino (above) or the stau (below). The area below this line is not shaded since in principle, this area could be allowed as the stau is not stable. However, constraints from big bang nucleosynthesis excluded much of gravitino LSP region in mSUGRA models [24].

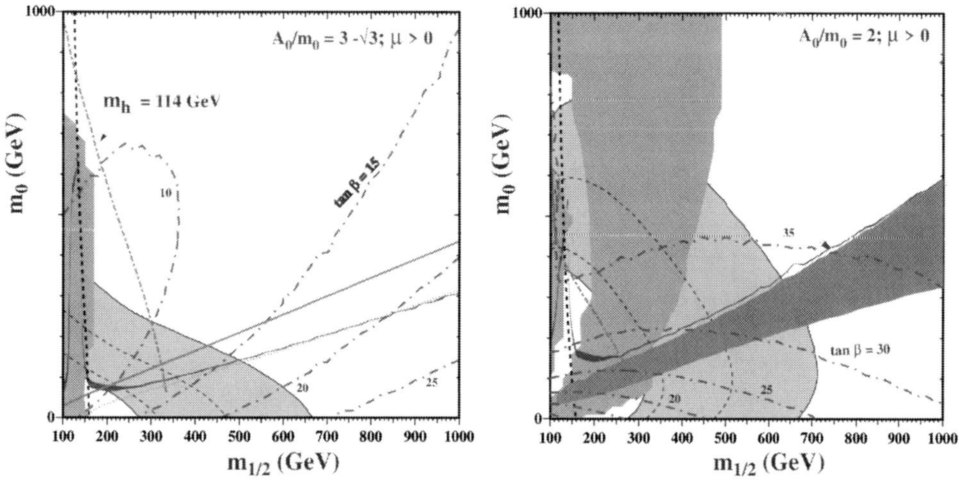

Figure 4. As in Fig. 1, the $(m_{1/2}, m_0)$ planes for (a) $A_0/m_0 = 3 - \sqrt{3}$, $B_0 = A_0 - m_0$, and $\mu > 0$. The dot-dashed contours show curves of constant $\tan\beta$. Below the dark (brown) line the gravitino is the LSP. In (b), $A_0/m_0 = 2$.

In the right panel of Fig. 4, $A_0/m_0 = 2$. In this case, the boundary of $m_{\tilde{\tau}} = m_\chi$ is above the boundary for $m_\chi = m_{3/2}$. As a consequence, there is now a viable co-annihilation strip with neutralino dark matter. Below that, there is an excluded region (shaded dark red) which has a stau LSP. Below the shaded region, the gravitino is again the LSP subject to BBN constraints [24]. Funnel and focus point regions are generally absent in mSUGRA models.

4. No-Scale Models

No-scale supergravity models [12] are characterized by a Kähler potential defined by

$$K = -3 \ln \left[\kappa(z + z^*) - \frac{1}{3}\kappa^2 \phi^i \phi_i^* \right]. \tag{9}$$

The scalar potential takes the simple form

$$V = e^{G - \frac{1}{3}K} \left| \frac{\partial g}{\partial \phi^i} \right|^2. \tag{10}$$

Thus one immediately finds that $m_0 = A_0 = B_0 = 0$, and the only source of supersymmetry breaking is transmitted through the gaugino masses (if the gauge kinetic function is a non-trivial function of z). In Fig. 5, an $(m_{1/2}, m_0)$ plane with $A_0/m_0 = B_0/m_0 = 0$ is shown. Like the mSUGRA model, specification of B_0 fixes $\tan\beta$ at each point on the plane. However, unlike mSUGRA models, the gravitino mass is rather arbitrary and can be set independently from other supersymmetry-breaking scales. No-scale models run along the $m_0 = 0$ axis of this plane. As one can see, along $m_0 = 0$, there is no way to achieve a sufficiently high relic density to match the WMAP determination and for $m_{1/2} \lesssim 150$ GeV, there is a problem with the branching ratio for $B_s \to X_s\gamma$, while at larger $m_{1/2}$, the LSP is either the stau or the gravitino. Furthermore, the LEP limit on the Higgs mass requires $m_{1/2} \gtrsim 340$ GeV. Thus it would appear that no-scale models are not phenomenologically viable.

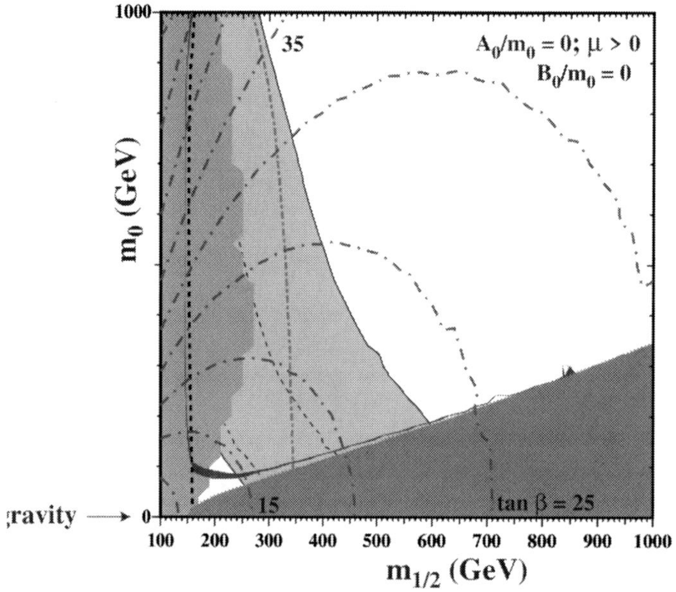

Figure 5. As in Fig. 1, the $(m_{1/2}, m_0)$ planes for (a) $A_0/m_0 = B_0/m_0 = 0$, and $\mu > 0$. The dot-dashed contours show curves of constant $\tan\beta$.

5. SuperGUT Models

In all of the models discussed above, it was assumed that the renormalization scale at which scalar and gaugino masses are unified, M_{in}, is the same scale at which gauge coupling unification occurs, M_{GUT}. This need not be the case. Indeed, since supersymmetry breaking may occur either below or above M_{GUT}, it is quite possible that $M_{in} \neq M_{GUT}$. For example, as M_{in} is decreased below M_{GUT}, the differences between the renormalized sparticle masses diminish and the regions of the $(m_{1/2}, m_0)$ planes that yield the appropriate density of cold dark matter move away from the boundaries [25]. Eventually, for small M_{in}, the coannihilation and focus-point regions of the conventional GUT-scale CMSSM merge and for very small M_{in} they disappear entirely.

On the other hand, increasing M_{in} increases the renormalization of the sparticle masses which tends to increase the splittings between the physical sparticle masses [26]. Furthermore, this in turn has the effect of increasing the relic density in much of the $(m_{1/2}, m_0)$ plane. As a consequence, the coannihilation strip is squeezed to lower values of $m_{1/2}$ [27, 14], particularly for $\tan\beta \sim 10$, and even disappears as M_{in} increases. At the same time, the focus-point strip often moves out to ever larger values of m_0. The allowed region of parameter space that survives longest is the rapid-annihilation funnel at large $m_{1/2}$ and $\tan\beta$. In the CMSSM with $M_{in} = M_{GUT}$, the funnel region also requires large m_0 and would make a contribution to $g_\mu - 2$ that is too small to explain the experimental discrepancy with Standard Model calculations based on low-energy e^+e^- data. However, for large M_{in}, the funnel region extends to low m_0 (including $m_0 = 0$) and in some cases will be compatible with the $g_\mu - 2$ measurements.

For superGUT models with $M_{in} > M_{GUT}$, one must specify in addition to Eq. (1), the GUT superpotential which for minimal SU(5) takes the form,

$$\begin{aligned} W_5 &= \mu_\Sigma Tr\hat{\Sigma}^2 + \frac{1}{6}\lambda' Tr\hat{\Sigma}^3 + \mu_H \hat{\mathcal{H}}_{1\alpha}\hat{\mathcal{H}}_2^\alpha + \lambda\hat{\mathcal{H}}_{1\alpha}\hat{\Sigma}^\alpha_\beta\hat{\mathcal{H}}_2^\beta \\ &\quad + (\mathbf{h_{10}})_{ij}\epsilon_{\alpha\beta\gamma\delta\zeta}\hat{\psi}_i^{\alpha\beta}\hat{\psi}_j^{\gamma\delta}\hat{\mathcal{H}}_2^\zeta + (\mathbf{h_{\overline{5}}})_{ij}\hat{\psi}_i^{\alpha\beta}\hat{\phi}_{j\alpha}\hat{\mathcal{H}}_{1\beta}, \end{aligned} \tag{11}$$

where the d^c and L superfields of the MSSM reside in the $\overline{\mathbf{5}}$ representation, $\hat{\phi}_i$, while the Q, u^c and e^c superfields are in the $\mathbf{10}$ representation, $\hat{\psi}_i$. $\hat{\Sigma}(\mathbf{24})$ is the SU(5) adjoint Higgs multiplet and the two Higgs doublets of the MSSM, H_1 and H_2 are extended to five-dimensional SU(5) representations $\hat{\mathcal{H}}_1(\overline{\mathbf{5}})$ and $\hat{\mathcal{H}}_2(\mathbf{5})$ respectively. There are now two new couplings: λ and λ'. λ affects directly the running of the soft Higgs masses, adjoint and Yukawa couplings, while λ' affects only the adjoint. Accordingly there are also new soft masses and μ terms associated with SU(5).

A superGUT version of the CMSSM based on SU(5) is now a 7+ parameter theory [14] specified by $m_{1/2}, m_0, A_0$, $\tan\beta$, and $sgn(\mu)$ as in the CMSSM, as well as M_{in}, λ, and λ'. At $M_{in} > M_{GUT}$, the universality conditions become,

$$\begin{aligned} m_{\overline{\mathbf{5}},1} = m_{\mathbf{10},1} = m_{\overline{\mathbf{5}}} = m_{\mathbf{10}} = m_{\mathcal{H}_1} = m_{\mathcal{H}_2} = m_\Sigma &\equiv m_0, \\ A_{\overline{\mathbf{5}}} = A_{\mathbf{10}} = A_\lambda = A_{\lambda'} &\equiv A_0, \\ M_5 &\equiv m_{1/2}. \end{aligned} \tag{12}$$

These parameters are evolved down to M_{GUT} and matched to their Standard Model CMSSM counterparts.

Some examples of $(m_{1/2}, m_0)$ planes in a superGUT model are shown in Fig. 6 for $\tan\beta = 10$ and two values of M_{in} and the specific choices $\lambda = 1, \lambda' = 0.1$ [14]. These results are quite insensitive to the value of λ' and can be compared with the CMSSM plane shown in Fig.1a. For $M_{in} = 2.5 \times 10^{16}$ GeV, we see two dramatic effects from the modest increase in M_{in}. One is the rapid disappearance of the stau LSP region, which has retreated to $m_0^2 < 0$. In the particular example shown, the coannihilation strip extends to $m_{1/2} \sim 450$ GeV, and there is a healthy portion compatible with the $g_\mu - 2$ constraint. Note that the chargino, m_h, $g_\mu - 2$ and $b \to s\gamma$ constraints are relatively stable in the $(m_{1/2}, m_0)$ plane with respect to changes in M_{in}. The other noticeable feature in panel (a) of Fig. 6 is the retreat of the EWSB constraint to smaller $m_{1/2}$ and larger m_0. This effect is quite sensitive to the value of λ, whereas the fate of the coannihilation region is relatively insensitive to its value.

For the choice $M_{in} = 10^{17}$ GeV, shown in panel (b) of Fig. 6, these effects are more pronounced: both the coannihilation and the focus-point strips have disappeared entirely. There is a small piece of the $(m_{1/2}, m_0)$ plane where the relic density falls within the WMAP range, but this is incompatible with m_h and gives too large a value of $g_\mu - 2$.

For $\tan\beta = 55$, we see in Fig. 7 that as M_{in} increases, the renormalization effects cause the stau LSP region to retreat as in the $\tan\beta = 10$ case, though more slowly, and it does not disappear entirely, even for $M_{in} = 2.4 \times 10^{18}$ GeV. Likewise, whilst the coannihilation strip shrinks with increasing M_{in}, it does not disappear, and much of it remains consistent with m_h,

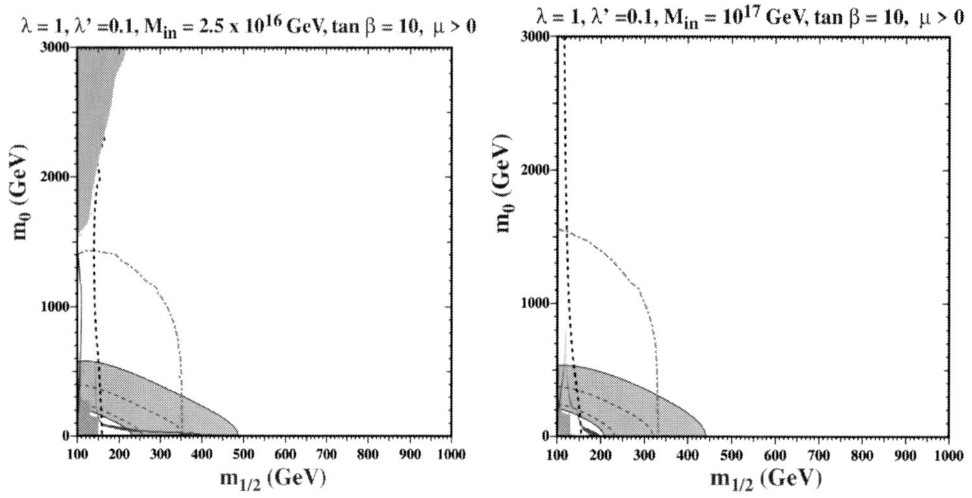

Figure 6. As in Fig. 1, the $(m_{1/2}, m_0)$ planes for $A_0 = 0, \tan\beta = 10, \lambda = 1$ and $\lambda' = 0.1$ for different choices of M_{in}: (a) 2.5×10^{16} GeV and (b) 10^{17} GeV.

$b \to s\gamma$ and $g_\mu - 2$. The rapid-annihilation funnel also persists as M_{in} increases, staying in a similar range of $m_{1/2}$, but shifting gradually to lower values of m_0. In particular, we note that for the case $M_{in} = 2.4 \times 10^{18}$ GeV, the no-scale possibility $m_0 = 0$ [12, 13] is allowed, on one or both flanks of the rapid-annihilation funnel. Finally, we note that the EWSB boundary disappears entirely for the displayed choices of $M_{in} > M_{GUT}$, as does the focus-point WMAP strip.

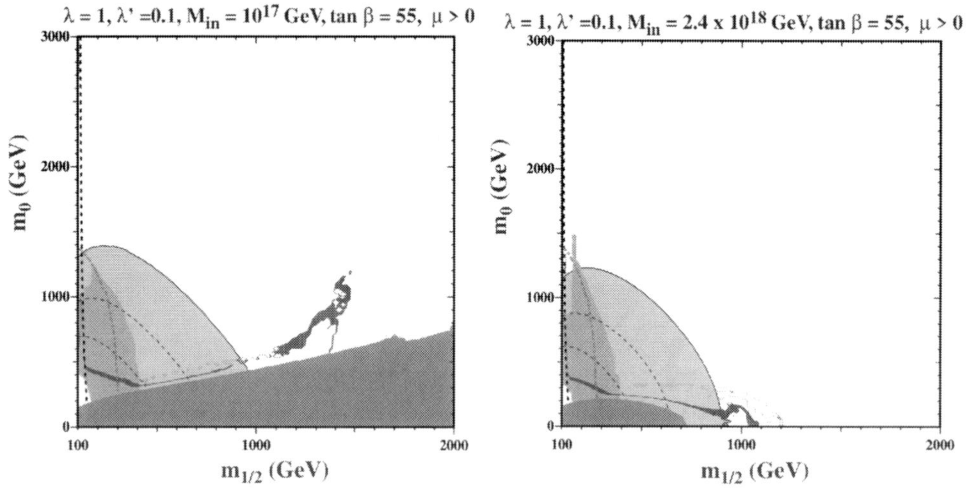

Figure 7. As in Fig. 6, for $A_0 = 0, \tan\beta = 55, \lambda = 1$ and $\lambda' = 0.1$ for different choices of M_{in}: (a) 10^{17} GeV and (b) 2.4×10^{18} GeV.

6. No-Scale Models Resurrected

As discussed above, no-scale models with GUT scale universality conditions are not phenomenologically viable. However, we have also seen when the universality scale in increased above the GUT scale, no-scale requirements such as $m_0 = 0$ may yield satisfactory models [13, 28, 15].

A no-scale superGUT model could be constructed from the same superpotential (11) with the additional requirements that $m_0 = A_0 = B_0 = 0$ at M_{in}. The latter condition translates to $B_\Sigma = B_H = B_0 = 0$ at M_{in}. And it would appear therefore that we now have a 4+ parameter theory specified by $m_{1/2}, M_{in}, \lambda, \lambda'$, and $sgn(\mu)$. However, the model really only depends on the ratio of the two Higgs couplings, λ/λ' and we are in fact left with a 3+ parameter theory. As in the mSUGRA models discussed above, $\tan\beta$ is determined from the EWSB conditions though the boundary condition is now $B_0 = 0$. Since this condition is applied at M_{in}, the MSSM (subGUT) B parameter must be matched to the GUT B parameters, B_Σ and B_H [29],

$$B = B_H - \frac{6\lambda}{\mu\lambda'}\left[(B_\Sigma - A'_\lambda)(2B_\Sigma - A'_\lambda) + m_\Sigma^2\right]. \tag{13}$$

In Fig. 8, some examples of $(m_{1/2}, M_{in})$ planes are shown for fixed $\lambda' = 2$ with $\lambda = 0, -0.14, -0.15$, and -0.16 [15]. We see that in the first panel of Fig. 8 the WMAP-compatible region takes the form of a thin L-shaped strip in the $(m_{1/2}, M_{in})$ plane with a rounded corner: points above and to the right of the L have values of $\Omega_\chi h^2$ that are too large. The near-horizontal part of the line is located at $M_{in} \sim 5 \times 10^{16}$ GeV and extends from $m_{1/2} \sim 200$ GeV to ~ 1000 GeV, larger values being excluded by the requirement that the LSP not be charged, and we find that $\tan\beta \in (16, 30)$. All the base strip is compatible with the LEP chargino constraint, and with $b \to s\gamma$. However, only the portion with $m_{1/2} \gtrsim 300$ GeV is compatible with the LEP lower limit on m_h. The near-vertical part of the no-scale strip is always incompatible with the LEP Higgs constraint and (mostly) $b \to s\gamma$.

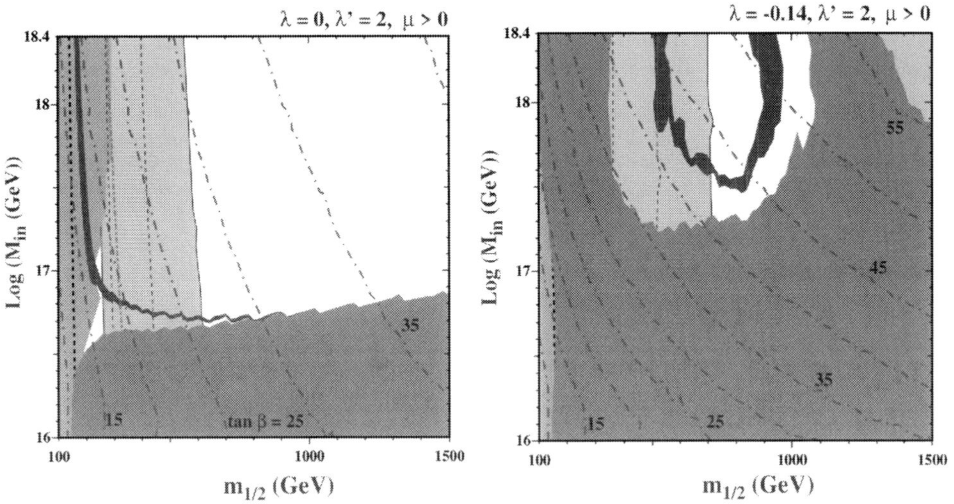

Panel (b) of Fig. 8 for $\lambda = -0.14$ has a rather different appearance, but is in fact a natural continuation of the trends seen in the previous panel. In particular, the near-vertical part of the WMAP-compatible strip has moved to larger $m_{1/2} \sim 400$ GeV, and is compatible with both m_h and $b \to s\gamma$, and the near-horizontal part of the strip has risen to $M_{in} \sim 3 \times 10^{17}$ GeV. More dramatically, the WMAP-compatible strip now becomes a loop, with a right side connecting

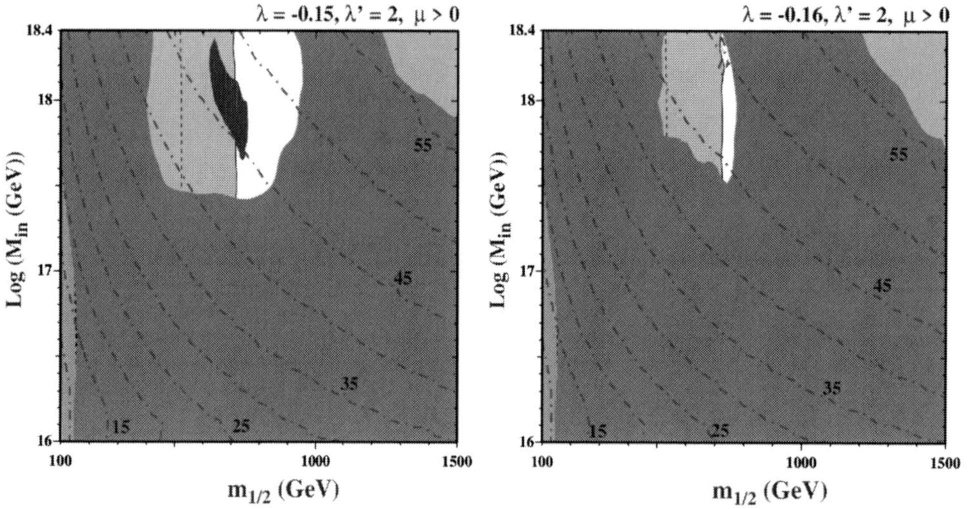

Figure 8. The $(m_{1/2}, M_{in})$ planes for the no-scale supergravity model with $\lambda' = 2$ and (a) $\lambda = 0$, (b) $\lambda = -0.14$, (c) $\lambda = -0.15$ and (d) $\lambda = -0.16$. Shading and line types are as in Fig. 1 with the exception that in the orange region we find no consistent solutions to the RGEs.

the previous two strips at relatively large $m_{1/2}$ and M_{in} and $\tan\beta \gtrsim 50$ (though the loop closes only when $M_{in} > M_P$). We emphasize that all of the loop has a neutralino LSP, and that the stau-LSP contour surrounds this loop: the region within the loop has too much dark matter.

In panel (c) of Fig. 8 for $\lambda = -0.15$ we see just a 'blob' with $m_{1/2} \in (500, 650)$ GeV and $M_{in} \in (5 \times 10^{17}, 3 \times 10^{18})$ GeV on the edge of the region preferred by $g_\mu - 2$. This remaining 'blob' disappears for larger $-\lambda$ as shown in panel (d) for $\lambda = -0.16$. Here, the area within the 'window' is phenomenologically allowed, though the relic density lies below the WMAP range.

In summary, we have seen that the CMSSM and mSUGRA are in fact different theories (despite their interchangeable use in the literature). In mSUGRA, $\tan\beta$ is generally small and funnel regions with the correct relic density do not appear. Focus point regions are also absent due to either a large value of A_0/m_0 or small $\tan\beta$. In addition, the gravitino is often the LSP in mSUGRA models. For the specific case, of no-scale supergravity with GUT scale universality, we have seen that the model is phenomenologically challenged. For $M_{in} > M_{GUT}$, no-scale supergravity models can however be resurrected. An upcoming challenge will be to differentiate between these models once data from the LHC or direct detection experiments become available.

Acknowledgments

This work was supported in part by DOE grant DE-FG02-94ER-40823.

References

[1] M. Drees and M. M. Nojiri, Phys. Rev. D **47** (1993) 376 [arXiv:hep-ph/9207234]; H. Baer and M. Brhlik, Phys. Rev. D **53** (1996) 597 [arXiv:hep-ph/9508321]; Phys. Rev. D **57** (1998) 567 [arXiv:hep-ph/9706509]; H. Baer, M. Brhlik, M. A. Diaz, J. Ferrandis, P. Mercadante, P. Quintana and X. Tata, Phys. Rev. D **63** (2001) 015007 [arXiv:hep-ph/0005027].

[2] J. R. Ellis, T. Falk, K. A. Olive and M. Schmitt, Phys. Lett. B **388** (1996) 97 [arXiv:hep-ph/9607292]; Phys. Lett. B **413** (1997) 355 [arXiv:hep-ph/9705444]; J. R. Ellis, T. Falk, G. Ganis, K. A. Olive and M. Schmitt, Phys. Rev. D **58** (1998) 095002 [arXiv:hep-ph/9801445]; V. D. Barger and C. Kao, Phys. Rev. D **57** (1998) 3131 [arXiv:hep-ph/9704403]. J. R. Ellis, T. Falk, G. Ganis and K. A. Olive, Phys. Rev. D **62** (2000) 075010 [arXiv:hep-ph/0004169].

[3] J. R. Ellis, T. Falk, G. Ganis, K. A. Olive and M. Srednicki, Phys. Lett. B **510** (2001) 236 [arXiv:hep-ph/0102098].
[4] J. R. Ellis, K. A. Olive, Y. Santoso and V. C. Spanos, Phys. Lett. B **565** (2003) 176 [arXiv:hep-ph/0303043]; H. Baer and C. Balazs, JCAP **0305**, 006 (2003) [arXiv:hep-ph/0303114]; A. B. Lahanas and D. V. Nanopoulos, Phys. Lett. B **568**, 55 (2003) [arXiv:hep-ph/0303130]; U. Chattopadhyay, A. Corsetti and P. Nath, Phys. Rev. D **68**, 035005 (2003) [arXiv:hep-ph/0303201]; C. Munoz, Int. J. Mod. Phys. A **19**, 3093 (2004) [arXiv:hep-ph/0309346]; J. Ellis and K. A. Olive, arXiv:1001.3651 [astro-ph.CO].
[5] J. R. Ellis, K. A. Olive, Y. Santoso and V. C. Spanos, Phys. Rev. D **69** (2004) 095004 [arXiv:hep-ph/0310356]; J. Ellis, S. Heinemeyer, K. Olive and G. Weiglein, *JHEP* **0502** 013, hep-ph/0411216; J. R. Ellis, S. Heinemeyer, K. A. Olive and G. Weiglein, JHEP **0605**, 005 (2006) [arXiv:hep-ph/0602220].
[6] O. Buchmueller *et al.*, Phys. Lett. B **657** (2007) 87 [arXiv:0707.3447 [hep-ph]]; O. Buchmueller *et al.*, JHEP **0809** (2008) 117 [arXiv:0808.4128 [hep-ph]]; O. Buchmueller *et al.*, Eur. Phys. J. C **64**, 391 (2009) [arXiv:0907.5568 [hep-ph]]
[7] J. Polonyi, *Generalization Of The Massive Scalar Multiplet Coupling To The Supergravity*, Hungary Central Inst Res - KFKI-77-93.
[8] E. Cremmer, B. Julia, J. Scherk, P. van Nieuwenhuizen, S. Ferrara and L. Girardello, Phys. Lett. B **79**, 231 (1978); E. Cremmer, B. Julia, J. Scherk, S. Ferrara, L. Girardello and P. van Nieuwenhuizen, Nucl. Phys. B **147**, 105 (1979).
[9] For reviews, see: H. P. Nilles, Phys. Rep. **110** (1984) 1; A. Brignole, L. E. Ibanez and C. Munoz, arXiv:hep-ph/9707209, published in *Perspectives on supersymmetry*, ed. G. L. Kane, pp. 125-148.
[10] R. Barbieri, S. Ferrara and C. A. Savoy, Phys. Lett. B **110**, 343 (1982).
[11] J. R. Ellis, K. A. Olive, Y. Santoso and V. C. Spanos, Phys. Lett. B **573** (2003) 162 [arXiv:hep-ph/0305212], and Phys. Rev. D **70** (2004) 055005 [arXiv:hep-ph/0405110].
[12] E. Cremmer, S. Ferrara, C. Kounnas and D. V. Nanopoulos, Phys. Lett. B **133**, 61 (1983); J. R. Ellis, C. Kounnas and D. V. Nanopoulos, Nucl. Phys. B **247**, 373 (1984).
[13] J. R. Ellis, D. V. Nanopoulos and K. A. Olive, Phys. Lett. B **525**, 308 (2002) [arXiv:hep-ph/0109288].
[14] J. Ellis, A. Mustafayev and K. A. Olive, arXiv:1003.3677 [hep-ph].
[15] J. Ellis, A. Mustafayev and K. A. Olive, arXiv:1004.5399 [hep-ph].
[16] E. Komatsu *et al.*, arXiv:1001.4538 [astro-ph.CO].
[17] J. Ellis, T. Falk, and K.A. Olive, Phys. Lett. **B444** (1998) 367 [arXiv:hep-ph/9810360]; J. Ellis, T. Falk, K.A. Olive, and M. Srednicki, *Astr. Part. Phys.* **13** (2000) 181 [Erratum-ibid. **15** (2001) 413] [arXiv:hep-ph/9905481].
[18] Joint LEP 2 Supersymmetry Working Group, *Combined LEP Chargino Results, up to 208 GeV*, `http://lepsusy.web.cern.ch/lepsusy/www/inos_moriond01/charginos_pub.html`.
[19] R. Barate *et al.* [ALEPH, DELPHI, L3, OPAL Collaborations: the LEP Working Group for Higgs boson searches], Phys. Lett. B **565**, 61 (2003) [arXiv:hep-ex/0306033]; D. Zer-Zion, *Prepared for 32nd International Conference on High-Energy Physics (ICHEP 04), Beijing, China, 16-22 Aug 2004*; LHWG-NOTE-2004-01, ALEPH-2004-008, DELPHI-2004-042, L3-NOTE-2820, OPAL-TN-744, `http://lephiggs.web.cern.ch/LEPHIGGS/papers/August2004_MSSM/index.html`.
[20] S. Chen *et al.* [CLEO Collaboration], Phys. Rev. Lett. **87** (2001) 251807 [arXiv:hep-ex/0108032]; P. Koppenburg *et al.* [Belle Collaboration], Phys. Rev. Lett. **93** (2004) 061803 [arXiv:hep-ex/0403004]. B. Aubert *et al.* [BaBar Collaboration], arXiv:hep-ex/0207076; E. Barberio *et al.* [Heavy Flavor Averaging Group (HFAG)], arXiv:hep-ex/0603003.
[21] [The Muon g-2 Collaboration], *Phys. Rev. Lett.* **92** (2004) 161802, hep-ex/0401008; G. Bennett et al. [The Muon g-2 Collaboration], *Phys. Rev.* **D 73** (2006) 072003 [arXiv:hep-ex/0602035].
[22] J. L. Feng, K. T. Matchev and T. Moroi, *Phys. Rev. D* **61** (2000) 075005 [arXiv:hep-ph/9909334].
[23] H.-P. Nilles, M. Srednicki, and D. Wyler, Phys. Lett. **120B** (1983) 345;
L.J. Hall, J. Lykken and S. Weinberg, Phys. Rev. **D27** (1983) 2359.
[24] R. H. Cyburt, J. R. Ellis, B. D. Fields, K. A. Olive and V. C. Spanos, JCAP **0611**, 014 (2006) [arXiv:astro-ph/0608562].
[25] J. R. Ellis, K. A. Olive and P. Sandick, Phys. Lett. B **642**, 389 (2006) [arXiv:hep-ph/0607002], JHEP **0706**, 079 (2007) [arXiv:0704.3446 [hep-ph]], and JHEP **0808**, 013 (2008) [arXiv:0801.1651 [hep-ph]].
[26] N. Polonsky and A. Pomarol, Phys. Rev. Lett. **73**, 2292 (1994) [arXiv:hep-ph/9406224], and Phys. Rev. D **51** (1995) 6532 [arXiv:hep-ph/9410231].
[27] L. Calibbi, Y. Mambrini and S. K. Vempati, JHEP **0709**, 081 (2007) [arXiv:0704.3518 [hep-ph]]; L. Calibbi, A. Faccia, A. Masiero and S. K. Vempati, Phys. Rev. D **74**, 116002 (2006) [arXiv:hep-ph/0605139]; E. Carquin, J. Ellis, M. E. Gomez, S. Lola and J. Rodriguez-Quintero, JHEP **0905** (2009) 026 [arXiv:0812.4243 [hep-ph]].
[28] M. Schmaltz and W. Skiba, Phys. Rev. D **62**, 095005 (2000) [arXiv:hep-ph/0001172].

[29] F. Borzumati and T. Yamashita, arXiv:0903.2793 [hep-ph].

Four Generation MSSM to Generate the Universe Matter-Antimatter Asymmetry

Benjamín Raziel Jaramillo Avila, David Delepine

Departamento de Física, Universidad de Guanajuato, Campus Léon, C.P. 37150, León, Guanajuato, México

E-mail: raziel@fisica.ugto.mx, delepine@fisica.ugto.mx

Abstract. A four generation extension to the MSSM is considered in the context of the electroweak phase transition. This model could be consistent with the generation of the baryonic and leptonic asymmetry of the universe during such phase transition. The phase transition strength is briefly discussed for this model and one way to study the CP violation during the electroweak phase transition is sketched.

1. Introduction

The electroweak phase transition is an important candidate to explain the baryonic and leptonic asymmetry of the universe. A strong phase transition is one of the necessary requirements to generate and maintain the baryonic asymmetry of the universe up to the present era.

The Standard Model does not provide an electroweak phase transition strong enough to generate and maintain this asymmetry up to present date. However it does have features that are necessary to generate this asymmetry. That is the reason why extensions to the Standard Model are considered as candidates to generate this asymmetry. The MSSM has been considered with this purpose [8, 9], however it requires the existence of a light stop quark, an scenario ruled out now. The introduction of a fourth generation within a supersymmetric framework strengthens the phase transition even beyond that of the MSSM since it introduces new bosons, the fourth generation superpartners, which strengthen the phase transition through their nonzero temperature contributions to the effective potential.

2. Effective potential and a phase transition

To study the v.e.v. of Higgs fields for a theory the Higgs effective potential can be introduced [1, 2]. The Higgs effective potential comes from considering the Higgs fields dependent part of the effective action when the classical fields, on which the effective action depends, are considered to be coordinate independent. Since Higgs v.e.v. are coordinate independent only coordinate independent classical fields need to be considered. Coordinate independent classical Higgs fields, on which the effective potential depends, will be denoted by ϕ.

The effective potential can be calculated by adding the classical potential (the zero loop potential in the lagrangian) higher order loop terms. There is a diagrammatic formula to add this higher loop terms. Loop orders higher than two are usually not calculated for an effective potential. In this work we only use one loop corrections.

In general one loop Higgs contributions to the effective potential for a field with mass that depends on ϕ are:

$$N\left(\frac{m^4(\phi)}{64\pi^2}\left[\log\left(\frac{m^2(\phi)}{\Lambda^2}\right)-\frac{3}{2}\right]\right.$$
$$\left.+\frac{T^4}{2\pi^2}\int_0^\infty dx\, x^2\log\left(1\pm e^{-\sqrt{x^2+m^2(\phi)/T^2}}\right)\right). \tag{1}$$

The factor N is an integer that counts the particle's spin and multiplicity due to internal degrees of freedom. N is set positive for bosons and negative for fermions. The sign on $\pm$ also depends on the spin. It is $+$ for fermions and $-$ for bosons.

The Higgs effective potential is a real function of real variables, the coordinate independent classical fields. The effective potential minima determine the Higgs fields v.e.v.s. However the effective potential also depends on temperature and hence Higgs fields vacuum expectation values are temperature dependent. The temperature dependence for the effective potential can be separated in temperature independent and dependent terms, the first and second terms of (1) respectively. Temperature independent terms correspond to the zero temperature field theory Higgs effective potential. The temperature dependent terms vanish for $T = 0$.

Temperature corrections to the potential are such that even if the Higgs field has some non zero v.e.v.s at zero temperature there is a certain temperature that gives all the Higgs v.e.v.s equal to zero. At this temperature the Higgs mechanism is ruined and no particles acquire mass through it. Such temperature is called the critical temperature, T_c.

If a system with dynamics like we have previously described goes from $T > T_c$ to $T < T_c$ then it could undergo a phase transition. An order parameter in this phase transition is a Higgs vacuum expectation value that is nonzero for temperatures below the critical value. If at the phase transition the change in the order parameter is continuous then the phase transition in second order, if the change is discontinuous then the phase transition is first order and the system is outside thermal equilibrium.

This type of phase transition is called electroweak phase transition since the Higgs mechanism breaks the $SU(2)_L \times U(1)_Y$ gauge symmetry and gives rise to the electroweak interactions.

During the electroweak phase transition, EWPT, two different phases of matter coexist, one where the Higgs mechanism is active, called broken phase, and one where it is not active, unbroken phase.

As the universe expands after the Big Bang its temperature decreases. The temperature right after the Big Bang is expected to be much higher than the critical temperature for a model which describes the dynamics of the universe so we think the universe underwent an electroweak phase transition.

3. Baryogenesis in the Standard Model and extensions

Four conditions are necessary to dynamically generate a baryonic asymmetry. They were stated by A.D. Sakharov in 1967 [3]. These conditions are that particle interactions have:

(i) Baryon and lepton number non conservation.

(ii) C violation.

(iii) CP violation.

(iv) Departure from thermal equilibrium.

The Standard Model of particle physics could have all these features.

Because of an anomaly called the Adler-Bell-Jackiw anomaly there is baryon and lepton number non conservation in the Standard Model. In fact the $B + L$ current suffers the anomaly while $B - L$ is still conserved [4]. However the rate for baryon and lepton number non conservation in the Standard Model at this temperature is well below current experimental sensibilities.

Due to its chiral nature the Standard Model has explicit C no conservation.

CP violating interactions have been observed and the SM provides an explanation to this measurements through the Cabbibo-Kobayashi-Maskawa matrix [5].

If the Standard Model parameters are such that it has a first order electroweak phase transition then the universe is out of thermal equilibrium during the electroweak phase transition.

The Standard Model could fulfill all the Sakharov conditions, however this conditions are necessary but not sufficient to dynamically generate the universe baryonic asymmetry.

Now let us look at how the baryonic asymmetry is thought to be generated during the electroweak transition. The Higgs field effective potential for the Standard Model depends on a single Higgs related variable, ϕ. The universe reaches the critical temperature when the Higgs effective potential has two degenerate minima for ϕ. Near the temperature T_c regions in space where the Higgs mechanism is in action begin to form. This regions are called broken phase bubbles and they expand filling all the universe. The formation of this bubbles is somewhat similar to the formation of bubbles of steam on a liquid water container whose temperature is elevated until boiling temperature is reached.

Figure 1. Typical first order phase transition for an effective potential. Curves in red correspond to temperatures above T_c. The green curve corresponds to $T = T_c$. The blue curve corresponds to $T < T_c$.

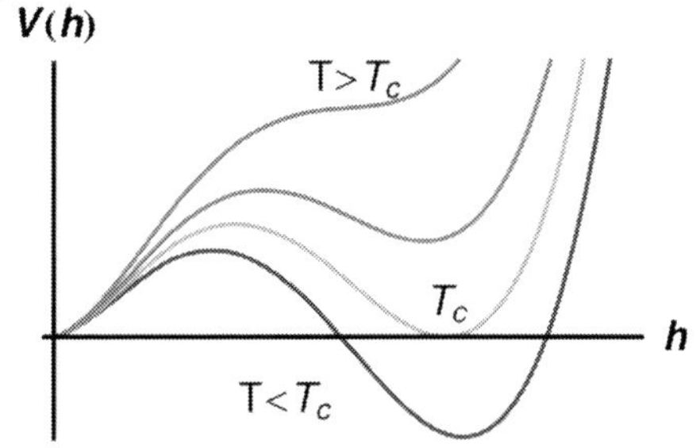

However particles diffuse between bubbles and also interact with bubble walls. CP violating interactions with the bubble walls give different diffusion properties for left and right quarks and leptons. This increases left quark and lepton densities on the unbroken phase.

Excess of left quarks and leptons on the unbroken phase enhances baryon and lepton number non conservation through processes called sphalerons. This processes are also supposed to have much bigger rates at higher temperatures, like that of the electroweak phase transition, than at the present temperature for the universe. Baryon and lepton number violation occurs on the unbroken phase where a baryonic and leptonic asymmetry is generated.

It is not sufficient to generate the correct baryonic asymmetry with this mechanism. It is also necessary that an asymmetry is not erased after the EWPT so that it persists to present time. For this to happen we demand that B and L violating processes are very suppressed inside the broken phase. This processes are such that

$$\frac{\phi_{\min}(T_c)}{T_c} \geq 1 \tag{2}$$

then the asymmetry is conserved inside the broken phase thus allowing it to survive up to now. This condition corresponds to demanding a large jump, bigger than the temperature, on the

Higgs v.e.v. at the phase transition. If there is such a jump we can also conclude that the phase transition is first order.

To generate a jump in the minimum of the effective potential it is necessary to have T dependent odd powers of ϕ. Otherwise such jump can not be generated. If an expansion is performed on the temperature dependent contributions in (1) it is found that only bosons have such contributions [2]. Hence integer spin fields with significant couplings to the Higgs sector are needed to generate a first order phase transition.

The Standard Model of particles only satisfies this condition for currently ruled out Higgs mass values. For the currently experimentally allowed Higgs mass values [6]. The current lower experimental bound for the Higgs mass is at about 115 GeV. However for the electroweak phase transition within the Standard Model dynamics to fulfill condition (2) the Higgs mass has to be below 80 GeV.

One possible solution to this problem is to use an extension to the Standard Model. The minimal supersymmetric extension to the Standard Model is a viable candidate.

In supersymmetric versions of the Standard Model the Higgs sector is extended. Instead of one Higgs doublet there are two Higgs doubles, H_1 and H_2, each with a component which acquires a non zero v.e.v.. After the electroweak symmetry breaking the MSSM has five Higgs particles. Two charged particles and three neutral, one of these three is CP odd and the other two are CP even. In this work we assume that only one linear combination of two CP even particles is light and all the other Higgs particles have much higher mass. The linear combination that remains light is:

$$h = \mathrm{Re}(H_1^0)\cos\beta + \mathrm{Re}(H_2^0)\sin\beta. \qquad (3)$$

Where $\tan\beta = v_2/v_1$ and $v_i/\sqrt{2}$ is the v.e.v. for the neutral component of H_i.

In this work we use the four generation Minimal Supersymmetric Standard Model. The Higgs and gauge sectors are the same for this model and the conventional MSSM. Only the lepton and quark sectors are extended. The fourth generation includes one left handed lepton doublet; a right handed charged lepton; a left handed quark doublet and two right handed quarks copies of the up and down right quarks.

Not all particles in the four generation MSSM contribute to the effective potential.

(i) From the Higgs sector only the scalar linear combination h is relevant.

(ii) The gauge sector only contributes with the $W^{\pm}$ and Z boson.

(iii) Quarks and leptons from the first and second generation play an unimportant role due to their small couplings with the Higgs sector compared to the top quark and the quarks and leptons from the fourth generation. From the third generation only the top quark has an important coupling. Quarks and leptons belonging to the fourth generation all have couplings similar to that of the top quark and hence have to be included. For the quark and lepton sector both fermions and their supersymmetric partners have important contributions.

We calculate the effective potential and find values for some of the 4gMSSM free parameters that are relevant to the phase transition that satisfy condition (2).

4. CP violation and the bubble wall

In case the phase transition in strong enough diffusion equations between the bubble walls for quarks and leptons still have to be solved to see if left quark and lepton densities are such that the correct baryonic and leptonic asymmetry is generated by $B + L$ violating processes in the unbroken phase.

Thermal Quantum Field Theory is used to compute transport equations between the two sides of the bubble wall.

Journal of Physics: Conference Series **315** (2011) 012022 doi:10.1088/1742-6596/315/1/012022

For a generic scalar, $\varphi(x)$, we have the current:

$$\begin{aligned}\langle J^\mu_\varphi(x)\rangle_\beta &= \mathrm{i}\langle\varphi^\dagger(x)(\partial^\mu_x\varphi(x)) - (\partial^\mu_x\varphi^\dagger(x))\varphi(x)\rangle_\beta \\ &= -(\partial^\mu_x - \partial^\mu_y)G^<_\varphi(x,y)\big|_{y=x}\end{aligned} \tag{4}$$

Taking $\langle J^\mu_\varphi(x)\rangle_\beta$ for the up type s-quarks and expanding the Green functions and the self energy in terms of the free Green functions, G_0, and the Lagrangian density, $\mathcal{L}_{\mathrm{MSSM}}$ we can write:

$$\begin{aligned}\langle\partial_\mu J^\mu_A(x)\rangle_\beta = \sum_{i,k=1}^{n_g}\sum_{B=L,R}\int_{\mathcal{R}^4} d^4y\{(g(x)^{AB}_{ik}\ G^>_0(x,y)^B_k \\ \cdot\ g(y)^{BA}_{ki}\ G^<_0(y,x)^A_i - g(x)^{AB}_{ik}\ G^<_0(x,y)^B_k\ g(y)^{BA}_{ki}\ G^>_0(y,x)^A_i \\ +\ g(y)^{AB}_{ik}\ G^>_0(y,x)^B_k\ g(x)^{BA}_{ki}\ G^<_0(x,y)^A_i - g(y)^{AB}_{ik}\ G^<_0(y,x)^B_k \\ \cdot\ g(x)^{BA}_{ki}\ G^>_0(x,y)^A_i)\theta(x^0,y^0)\}.\end{aligned} \tag{5}$$

The g^{AB}_{ij} come from $\mathcal{L}_{\mathrm{bilinear}\tilde{U}}$, the part of $\mathcal{L}_{\mathrm{MSSM}}$ which is bilinear in the up s-quarks. Upper indices, A, B, in g^{AB}_{ij} distinguish L and R s-quarks. Lower indices, i, j, are generation indices.

$$\mathcal{L}_{\mathrm{BL}\tilde{Q}} = \tilde{U}_{L,i}\, g^{LL}_{ij}\ \tilde{U}_{L,j} + \tilde{U}_{R,i}\, g^{RR}_{ij}\ \tilde{U}_{R,j}, +\tilde{U}_{L,i}\, g^{LR}_{ij}\ \tilde{U}_{R,j} + \tilde{U}_{R,i}\, g^{RL}_{ij}\ \tilde{U}_{L,j}. \tag{6}$$

More specifically:

$$\begin{aligned}g(x)^{LL}_{ij} &= [M^2_{\tilde{Q}}]_{ij}\ +\ v^2_2(x)[h^*_u h^T_u]_{ij}\ +\ \delta_{ij}\, D^2_L(x) \\ g(x)^{RR}_{ij} &= [M^2_{\tilde{U}_R}]_{ij}\ +\ v^2_2(x)[h^T_u h^*_u]_{ij}\ +\ \delta_{ij}\, D^2_R(x) \\ g(x)^{LR}_{ij} &= -v_1(x)\mu\ [h^*_u]_{ij}\ +\ v_2(x)[(Y_u)^*]_{ij} \\ g(x)^{RL}_{ij} &= (g(x)^{RL}_{ji})^*\end{aligned} \tag{7}$$

- $[M^2_{\tilde{Q}}]_{ij}$, $[M^2_{\tilde{U}_R}]_{ij}$ are Soft breaking terms.
- $[h^*_u h^T_u]_{ij}$ are scalar Yukawa mass terms.
- $[(Y_u)^*]_{ij}$ is a soft trilinear term.
- μ is the Higgs bilinear term coefficient.

Higgs v.e.v.'s v_1, v_2 have a coordinate dependence since there are two regions in space, the broken ($v_1 \neq 0$, $v_2 \neq 0$) and unbroken ($v_1 = 0$, $v_2 = 0$) phases.

Equation (5) is a g^2 order expression for the up, both left and right, s-quark asymmetry. It has contributions from diagrams of the following kind: Equation (5) is the Quantum Boltzmann equation for the up s-quark particle density asymmetry. It is an integral equation. It comes from the g^2 approximation. Eq. (5) and (7) can be used to get:

$$\begin{aligned}\langle\partial_\mu J^\mu_R(x)\rangle_\beta = 4\int_{\mathcal{R}^4} d^4y\{\theta(x^0,y^0)\ (v_1(x)v_2(y) - v_2(x)v_1(y)) \\ \Big(\mathrm{Im}[G^<_0(x,y)^L_k\,(G^<_0(y,x)^R_i)^*]\ \mathrm{Im}[\,\mu\,(Y_u)^T_{ik}\,(h_u)^*_{ki}]\Big)\}\end{aligned} \tag{8}$$

The form of free propagators G_0 is known from Thermal Quantum Field Theory [7].

The x coordinate will be considered as the coordinate in the rest frame for the bubble wall.

Higgs v.e.v. v_1 and v_2 have an x coordinate dependence since there it has a different value depending on whether it is inside or outside the bubble.

The bubble wall will be considered flat in the calculation of diffusion equations.

So far this work has already been performed for three generations in the light s-top scenario [8]. We intend to repeat the calculation for the four generation case.

5. Conclusions

It is important to find a convenient basis for the Yukawa matrices which allows us to clearly visualize the CP violation sources. Down s-quarks could also play a role, Quantum Boltzmann equations for theme also should be included.

References

[1] Discussion in the case of zero temperature Quantum Field Theory. S. Coleman and E. Weinberg, "Radiative Corrections as the Origin of Spontaneous Symmetry Breaking," Phys.Rev. D **7**, 1888-1910 (1973).

[2] Discussion for the case of Thermal Quantum Field Theory. L. Dolan and R. Jackiw, "Symmetry Behavior at Finite Temperature," Phys.Rev. D **9**, 3320-3341 (1974).

[3] A.D. Sakharov, "Violation of CP Invariance, C Asymmetry, and Baryon Asymmetry of the Universe," Pisma Zh.Eksp.Teor.Fiz. **5**, 23-35 (1967). This article has been reprinted several times: JETP Lett. **5**, 24-27 (1967). Sov.Phys.Usp. **34** 392-393 (1991).

[4] G. 't Hooft, "Symmetry Breaking through Bell-Jackiw Anomalies," Phys.Rev.Lett. **37**, 8-11 (1976). This article discusses how the Standard Model has an anomaly that violates baryon and lepton number.

[5] M. Kobayashi and T. Maskawa, "CP Violation in the Renormalizable Theory of Weak Interaction," Prog.Theor.Phys. **49**, 652-657 (1973). This article discusses CP violation in the Standard Model.

[6] K. Rummukainen, M. Tsypin, K. Kajantie, M. Laine and M. E. Shaposhnikov, "The Universality Class of the Electroweak Theory," Nucl.Phys. B **532**, 283-314 (1998). [arXiv:hep-lat/9805013].

[7] N.P. Landsmann and Ch.G. van Weert, "Real- and Imaginary-Time Field Theory at Finite Temperature and Density," Phys. Rep. **145**, 141 (1987).

[8] D. Delepine *et al.*, "A light stop and electroweak baryogenesis," Nucl.Phys.B **657**, 355-377 (2003).

[9] M.S. Carena and M. Quiros and C.E.M. Wagner, "Electroweak baryogenesis and Higgs and stop searches at LEP and the Tevatron," Nucl.Phys.B **524**, 3-22 (1998).

Measuring Lepton Flavour Violation at LHC with Long-Lived Slepton in the Coannihilation Region

Satoru Kaneko[1], Joe Sato[2], Takashi Shimomura[3], Oscar Vives[3] and Masato Yamanaka[2]

[1]Instituto Física Corpuscular - C.S.I.C/Universitat de València Campus de Paterna, Apt 22085, E46071, València, Spain

[2] Department of Physics, Saitama University, Shimo-Okubo, Sakura-ku, Saitama, 338-8570, Japan

[3]Departament de Física Teòrica and IFIC, Universitat de València - CSIC, E46100, Burjassot, València, Spain

E-mail: joe@phy.saitama-u.ac.jp

Abstract. When the mass difference between the lightest slepton, the NLSP, and the lightest neutralino, the LSP, is smaller than the tau mass, the lifetime of the lightest slepton is extraordinarily long and strongly dependent on lepton flavor violation. Therefore the long-lived slepton scenario offers an excellent opportunity to study lepton flavor violation at ATLAS and CMS detectors in the LHC. This is based on the work [1]

1. Introduction

Cosmological observations have confirmed the existence of the non-baryonic dark matter. The observed dark matter relic abundance is $\Omega_{DM}h^2 \simeq 0.11$ and points to the existence of a stable and weakly interacting particle with a mass $m = 100 - 1000$ GeV. In supersymmetric (SUSY) models with conserved R-parity, the Lightest Supersymmetric Particle (LSP), usually the lightest neutralino, is stable and it is a perfect candidate for the dark matter. In this case the mechanisms ,called coannihilation [2] is employed. In the coannihilation region, the Next to Lightest Supersymmetric Particle (NLSP) has a mass nearly degenerate to the LSP. The degeneracy required for the coannihilation to occur is generally $\delta m/m_{\rm LSP} <$ a few %, where $\delta m = m_{\rm NLSP} - m_{\rm LSP}$.

Furthermore, it was pointed out in [3, 4, 5] that if staus are the NLSP and its mass difference with the LSP is less than the tau mass, they destroy ^{7}Li/^{7}Be nucleus through the internal conversion. With this small mass difference, staus become long-lived, they survive until Big-Bang Neucleosynthesis (BBN) starts and form bound states with nucleus. The stau-nucleus bound states decay immediately by virtual exchange of the hadronic current. In this way, it was shown that the relic abundance of the light elements is lowered effectively and the discrepancy between the observed value of ^{7}Li/^{7}Be abundance [6, 7] and the predicted value from the standard BBN [8] with WMAP data [9, 10] can be solved. Thus, a scenario with stau NLSP and neutralino LSP

	mass (GeV)		mass (GeV)
$\tilde{\chi}_1^0$	290 – 320	$\tilde{\chi}_2^0$	540 – 600
$\tilde{\chi}_3^0$	750 – 820	$\tilde{\chi}_4^0$	770 – 830
$\tilde{\chi}_1^\pm$	540 – 600	$\tilde{\chi}_2^\pm$	770 – 830
$\tilde{g}$	1540 – 1680		
$\tilde{t}_1$	1150 – 1260	$\tilde{t}_2$	1330 – 1460
$\tilde{b}_1$	1300 – 1420	$\tilde{b}_2$	1340 – 1460
$\tilde{u}_R$	1370 – 1500	$\tilde{u}_L$	1430 – 1560
$\tilde{d}_R$	1370 – 1500	$\tilde{d}_L$	1430 – 1560
$\tilde{\tau}_1$	290 – 320	$\tilde{\tau}_2$	510 – 560
$\tilde{e}_R$	350 – 385	$\tilde{e}_L$	520 – 570
$\tilde{\nu}_1$	500 – 550	$\tilde{\nu}_3$	510 – 560
h	116 – 118	H	800 – 900
A	800 – 900	$H^\pm$	800 – 900

Table 1. Approximate mass ranges for points in the coannihilation region with $m_{\tilde{\tau}_1} - m_{\tilde{\chi}_1^0} \leq 1.77$ GeV and within the 2σ allowed region for a_μ for $\tan\beta = 30$, $A_0 = 600$ GeV. These points correspond to $m_0 \in (230, 250)$ GeV and $M_{1/2} \in (700, 800)$ GeV. $\tilde{u}$, $\tilde{d}$ and $\tilde{e}$, $\tilde{\nu}_1$ denote the first two generation up-type, down-type squarks, and sleptons and sneutrinos.

with $\delta m \leq m_\tau$ could explain the relic abundance of the light elements as well as the abundance of the dark matter. As shown in [3], if the mass difference, δm, between neutralino and stau is less than the tau mass, the two-body decay is forbidden and the lifetime of stau increases by more than 10 orders of magnitude.

Strictly speaking the previous discussion is correct only in the framework of the MSSM without intergenerational mixing in the slepton sector. In general, we expect that the lightest slepton, $\tilde{l}_1$, is not a pure stau, but there is some mixing with smuon and selectron. In the presence of this small intergenerational mixing, even with $\delta m \leq m_\tau$, other two-body decay channels like $\tilde{l}_1 \to \tilde{\chi}_1^0 + e(\mu)$, are still open [3]. In this case, given that the flavor conserving two-body decay channel is closed, the slepton lifetime will have a very good sensitivity to small Lepton Flavor Violation (LFV) parameters.

Long-lived charged-particles are very interesting since they provide a clear experimental signature at the LHC [11, 12, 13]. In Ref. [14] they conclude that even if the lifetime of the decaying particle is much longer than the size of the detector, some decays always take place inside the detector and it is possible to measure lifetimes as long as $10^{-5} - 10^{-3}$ seconds in a particular gauge mediation scenario.

2. mass spectrum and long-lived stau in MSSM

We found mass range which were obtained in the CMSSM parameter space by requiring the mass difference, δm, between the lightest neutralino mass, $m_{\tilde{\chi}_1^0}$, and the lighter stau mass, $m_{\tilde{\tau}_1}$, be smaller than tau mass, m_τ. Here we also imposed the constraints from the relic density of the dark matter abundance , the anomalous magnetic moment of the muon, the bound on $\mathrm{Br}(b \to s\gamma)$.

In Table 1, we show the SUSY mass spectrum for the interesting region at $\tan\beta = 30$ and $A_0 = 600$ GeV. As can be seen in this table, masses of the lightest neutralino and the lightest

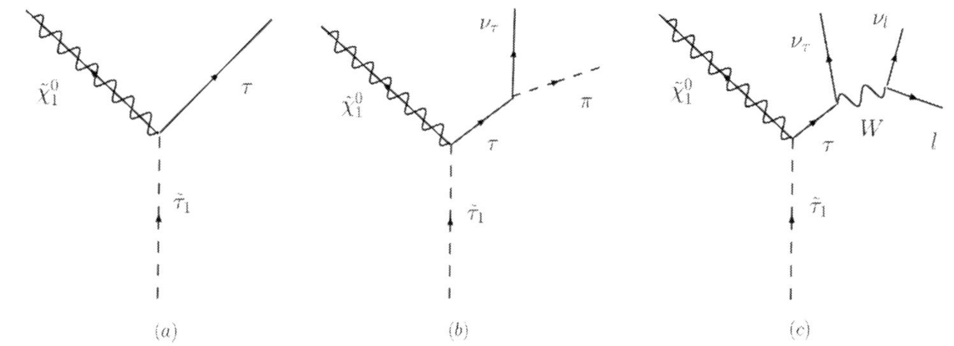

Figure 1. Feynman diagrams of stau decay: (a) two-body decay, (b) three-body decay, (c) four-body decay. l denotes the charged leptons.

slepton are around 300 GeV. The lightest neutralino is mainly bino and the second lightest neutralino is wino with a small admixture of higgsino. The second lightest neutralino and the lightest chargino have masses ~ 550 GeV. The right-handed sleptons are in the range 300–390 GeV and the left-handed sleptons have masses of 510–570 GeV. The squarks are in the range 1150–1560 GeV while gluinos are heavier than the squarks and about 1540 – 1680 GeV.

The decay chain of the NLSP stau into two, three and four bodies are shown in Fig. 1. The exact decay rates are given in Appendix of [1]. Notice that when $\delta m \gtrsim m_\tau$ the two-body decay is open and the lifetime of the stau, $\tau_{\tilde{\tau}_1}$, is 10^{-22} sec. However, for $\delta m m_\tau$ this decay is closed and the three or four-body decays are suppressed at least by an additional $(\delta m)^4 G_F^2 (f_\pi/m_\tau)^2 1/(30(2\pi)^2) \simeq 10^{-13}$ with $\delta m \sim 2$ GeV. Therefore the stau becomes long-lived and the phenomenology of the MSSM changes dramatically.

3. LFV and long-lived slepton

In the previous section we have seen that $\delta m \leq m_\tau$ is indeed possible in the framework of a CMSSM without LFV couplings. The lifetime of the NLSP stau in this scenario is increased by many orders of magnitude. However, once a new source of LFVs is introduced, the NLSP two-body decay channels into electron and/or muon can open again. In this case, the lifetime is inversely proportional to the square of the mixing of selectron and smuon with stau and therefore the measurement of the lifetime shows a strong sensitivity to LFV parameters.

To understand the dependence of lifetimes on LFV parameters, it is convenient to introduce the so-called Mass Insertions (MI), $(\delta^e_{RR/LL})_{\alpha\beta}$, defined

$$(\delta^e_{RR/LL})_{\alpha\beta} = \frac{\Delta M^{e\,2}_{RR/LL_{\alpha\beta}}}{M^e_{R/L\alpha} M^e_{R/L\beta}}, \tag{1}$$

where α, β denote the lepton flavors. $M^e_{R/L\alpha}$ and $M^e_{R/L\beta}$ are diagonal elements of the slepton mass matrix, and $\Delta M^e_{RR/LL\alpha\beta}$ are off-diagonal elements we introduced. In terms of these mass insertions, the two-body decay rate is approximately given by

$$\Gamma_{2-\text{body}} = \frac{g_2^2}{2\pi m_{\tilde{\tau}_1}} (\delta m)^2 (|g^L_{1\alpha 1}|^2 + |g^R_{1\alpha 1}|^2), \tag{2}$$

VI International Workshop on the Dark side of the Universe (DSU 2010) IOP Publishing
Journal of Physics: Conference Series **315** (2011) 012023 doi:10.1088/1742-6596/315/1/012023

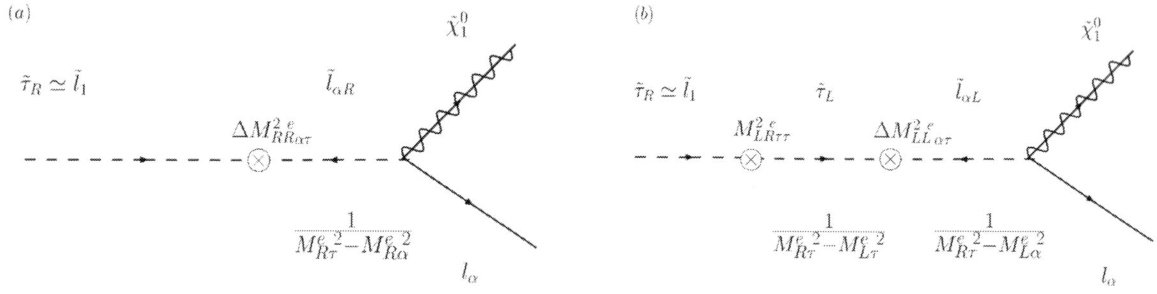

Figure 2. MI Feynman diagrams of two-body slepton decay: The diagram on the left side depicts two-body decay in the presence of δ^e_{RR} and the diagram on the right side in the presence of δ^e_{LL}. The circle-crosses represent mass insertions for flavor and left-right chirality changes. Propagators for intermediate states are shown below corresponding scalar lines.

No.	δm (GeV)	$m_{\tilde{\chi}^0_1}$ (GeV)	$m_{\tilde{l}_1}$ (GeV)	$\Omega_{\tilde{\chi}^0_1}h^2$	a_μ ($\times 10^{-10}$)
A	2.227	323.1549	325.3817	0.110	10.32
B	1.650	325.5601	326.2147	0.102	10.25
C	0.407	327.6294	328.0365	0.085	10.09
D	0.092	328.4060	328.4981	0.081	10.06

Table 2. Table of the mass difference and the lightest slepton, neutralino masses. m_0, A_0 and $\tan\beta$ are fixed to 260 GeV, 600 GeV and 30, respectively. The values of neutralino abundance and a_μ are shown for the reference.

where $\alpha = e,\ \mu$. $g^{L,R}_{1\alpha 1}$ can be approximated in the mass insertion as shown in Fig. 2. In the case of right-handed slepton mixing, we have,

$$g^L_{1\alpha 1} \simeq 0, \quad g^R_{1\alpha 1} \simeq \tan\theta_W \frac{M^e_{R\tau} M^e_{R\alpha}}{M^{e\,2}_{R\tau} - M^{e\,2}_{R\alpha}} (\delta^e_{RR})_{\alpha\tau}, \tag{3}$$

while in the left-handed slepton mixing case, these couplings are given

$$g^L_{1\alpha 1} \simeq \frac{1}{2} \tan\theta_W \frac{m_\tau (A_0 - \mu\tan\beta)}{M^{e\,2}_{R\tau} - M^{e\,2}_{L\tau}} \frac{M^e_{L\alpha} M^e_{R\tau}}{M^{e\,2}_{R\tau} - M^{e\,2}_{L\alpha}} (\delta^e_{LL})_{\alpha\tau}, \quad g^R_{1\alpha 1} \simeq 0. \tag{4}$$

To analyze the effects of the presence of a non-vanishing leptonic mass insertion on the NLSP lifetime we choose four points with different mass differences as shown in Table 2. In Fig. 3 (a), we show the lifetime of the lightest slepton, $\tau_{\tilde{l}_1}$, as a function of $(\delta^e_{RR})_{e\tau}$, encoding the right-handed selectron-stau mixing, that we vary from 10^{-10} to 10^{-2} [15, 16]. We can see that the lifetime for $\delta m > m_\tau$ (case A in Table. 2) does not change, because the decay of slepton to tau and neutralino is always the dominant decay mode and the lifetime is insensitive to δ^e_{RR} ($\leq 10^{-2}$). On the other hand, for $\delta m < m_\tau$, the lifetime grows more than 13 orders of magnitude in the limit $(\delta^e_{RR})_{e\tau} \to 0$, where the three- or four-body decay processes are dominant. Then, the two-body decay into τ and $\tilde{\chi}^0_1$ is forbidden but those into e (or μ for $(\delta^e_{RR})_{\mu\tau} \neq 0$) and $\tilde{\chi}^0_1$ are allowed through LFV couplings. The lifetime decreases proportionally to $|(\delta^e_{RR})_{e\tau}|^{-2}$ when the two-body decay dominates total decay width.

Figs. 4 (a) and (b) show the lifetime of the lightest slepton as a function of $(\delta^e_{LL})_{e\tau}$ and $(\delta^e_{LL})_{\mu\tau}$, respectively. The different curves correspond to the same mass differences used in

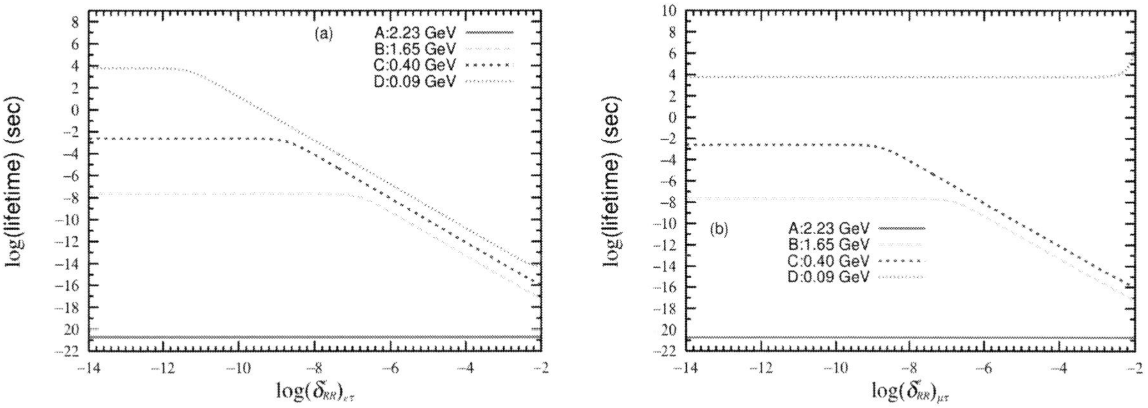

Figure 3. The lifetime of the lightest slepton as a function of δ^e_{RR}. The left panel,(a), is the lifetime of the lightest slepton with the right-handed selectron and stau mixing, and the right panel, (b), is the one with the right-handed smuon and stau mixing. In both panel, the solid (red), dashed (green), dotted (blue), dotted-dashed (pink) line correspond to $\delta m = 2.23$ (A), 1.65 (B), $0.41 (C)$, 0.09 (D) GeV, respectively. These lifetimes are calculated with the exact formulas in the Appendix of [1].

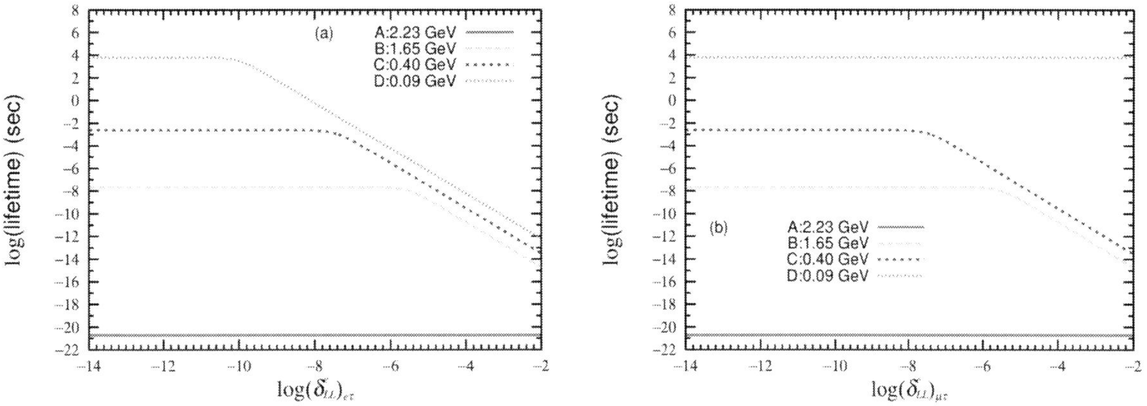

Figure 4. The lifetime of the lightest slepton as a function of δ^e_{LL}. The lines and the parameters are the same as Fig.3.

Fig. 3. The dependence on these mass insertions in these figures is completely analogous to the behavior observed in Fig. 3. In fact, both figures would be identical if we replace (δ^e_{LL}) by $10 \times (\delta^e_{RR})$. This is due to the fact that in this case, the lightest slepton decays into the electron or the muon through the left-handed stau component. The mixing of right and left-handed staus, Eq. (4), is proportional to $m_\tau(A_0 - \mu \tan\beta)/(M^{e\,2}_{R\tau} - M^{e\,2}_{L\tau})$. In the region of parameter space we are considering this factor is approximately 0.1. Thus, we can see that, also in the case of (δ^e_{LL}), the lifetime is sensitive to the presence of very small lepton flavor violating couplings and therefore to the presence of mass insertions orders of magnitude smaller than the present bounds.

	5 cm	50 cm	3.1 m	5.8 m	25.0 m
10^{-5} sec.	0.04	0.36	2.2	4.1	17.8
10^{-6} sec.	0.36	3.6	22.1	41.3	175.1
10^{-7} sec.	3.6	35.6	216.0	395.3	1461.9
10^{-8} sec.	35.6	343.0	1731.0	2658.3	4223.5
10^{-9} sec.	343.0	2425.6	4265.5	4289.7	4290.0
10^{-10} sec.	2425.6	4289.0	4290.0	4290.0	4290.0
10^{-11} sec.	4289.0	4290.0	4290.0	4290.0	4290.0
10^{-12} sec.	4290.0	4290.0	4290.0	4290.0	4290.0

Table 3. The expected number of slepton decay in the ATLAS detector. The lifetime is varied from 10^{-12} to 10^{-5} seconds. The number of sleptons produced at LHC assuming the integrated luminosity, $\mathcal{L}_{\rm int} = 30$ fb^{-1} is 4290.

4. LHC phenomenology

In this section, we discuss the expected phenomenology at LHC experiments, focusing mainly on the ATLAS detector [17, 18], of the long-lived slepton scenario. The lightest slepton is the NLSP, and therefore a large number of sleptons is expected to be produced via cascade decays of heavier SUSY particles, say,

$$\tilde{q}_L \to \tilde{\chi}_1^{\pm} + q, \qquad \tilde{\chi}_1^{\pm} \to \tilde{l}_1 + \nu_\tau \tag{5}$$

where q, $\tilde{\chi}_1^{\pm}$ and $\tilde{\chi}_2^0$ denote the SM quarks, the lighter charginos and the second lightest neutralinos and ν_τ, W and $\tilde{\nu}_\tau$ denote tau neutrino, weak boson and tau sneutrino, respectively. The 3rd generation squarks have many different decay chains.

When the long-lived sleptons have lifetimes between 10^{-5} and 10^{-12} sec., we will have a chance to observe decays of the slepton inside the detector.

The total cross section of SUSY pair production in our scenario is given in [19],

$$\sigma_{\rm SUSY} = 130 \text{ fb} \tag{6}$$

and we assume the integrated luminosity, $\mathcal{L}_{\rm int} = 30$ fb^{-1}. Then the number of lightest sleptons produced, $N_{\tilde{l}_1}$, is estimated as 4290. Therefore we would have 4290 long-lived sleptons produced in ATLAS that could decay inside the detector depending on the lifetime.

In Table 3, we show the expected number of slepton decays in each detector, assuming the integrated luminosity, $\mathcal{L}_{\rm int} = 30$ fb^{-1} and $\beta\gamma = 2$. As can be seen, when the lifetime, $\tau_{\tilde{l}_1}$, is below 10^{-9} sec., most of the sleptons decay inside the pixel detector. When $\tau_{\tilde{l}_1} \sim 10^{-8}$ sec., almost half of them decay inside the inner detector and nearly all of them decay within the detector. If $\tau_{\tilde{l}_1}$ is between 10^{-8} and 10^{-6} sec., several hundreds of slepton decays occur inside the ATLAS detector. Almost all of them escape from the detector when $\tau_{\tilde{l}_1} > 10^{-5}$ sec., although we expect of the order of 10 decays inside the detector for $\tau_{\tilde{l}_1} \simeq 10^{-5}$ sec.

If we can fix the mass difference between the lightest slepton and neutralino, we can determine the value of the mass insertion parameters from Figs. 3 and 4. In the case of right-handed slepton mixing case, lifetimes between 10^{-10} and 10^{-8} sec. would correspond to $(\delta^e_{RR})_{e\tau}$ between 10^{-7} and 10^{-4} with the mass difference, $m_e < \delta m < m_\tau$, and $(\delta^e_{RR})_{\mu\tau}$ between 10^{-7} and 10^{-5} with $m_\mu < \delta m < m_\tau$. Similarly, in the case of left-handed slepton mixing, the same lifetimes would correspond to $(\delta^e_{LL})_{e\tau}$ between 4×10^{-6} and 10^{-3} with $m_e < \delta m < m_\tau$, and $(\delta^e_{LL})_{\mu\tau}$ between 4×10^{-6} and 10^{-4} with $m_\mu < \delta m < m_\tau$. For $\tau_{\tilde{l}_1} \sim 10^{-8}$ sec., about half of sleptons would decay inside the inner detector and the rest would decay inside the calorimeters and/or muon spectrometer.

5. summary

In this work, we have studied lepton flavor violation in a long-lived slepton scenario where the mass difference between the NLSP, the lightest slepton, and the LSP, the lightest neutralino, is smaller than the tau mass. This small δm is possible even in the framework of the CMSSM.

Then, we have analyzed the dependence of the lightest slepton lifetimes on different mass insertions, $(\delta^e_{RR/LL})_{e\tau,\ \mu\tau}$ for values of δm from 2.23 to 0.09 GeV. We found that the lifetimes are proportional to $|\delta^e|^{-2}$ until three- or four-body decays become comparable. There is a difference of approximately a factor of 10 in the sensitivity of the lifetime on δ^e_{RR} and δ^e_{LL}. This is due to the different proportion of left-handed and right-handed staus in the lightest slepton. We can see that, in this scenario, the lifetimes are sensitive to much smaller values of these δ^es.

Finally, we have discussed the expected phenomenology at LHC experiments, mainly concentrating on the ATLAS detector. We have estimated the number of slepton decays in the different detectors, assuming an integrated luminosity $\mathcal{L}_{\rm int} = 30\ {\rm fb}^{-1}$ and $\beta\gamma = 2$ (Table. 3). We have seen that the ATLAS detector can observe lifetimes in the range of 10^{-11} to 10^{-6} sec., and these lifetime would correspond to $(\delta^e_{RR})_{e\tau,\mu\tau}$ between 10^{-7} and 10^{-3} and $(\delta^e_{LL})_{e\tau,\mu\tau}$ between 4×10^{-6} and 10^{-3}. Therefore we have shown that in the long-lived slepton scenario, the LHC offers a very good opportunity to study lepton flavor violation.

[1] Kaneko S, Sato J, Shimomura T, Vives O, Yamanaka M *Phys. Rev.* D **78** 116013 (2008) [arXiv:0811.0703[hep-ph]].
[2] Griest K and Seckel D, *Phys. Rev.* D **43** (1991) 3191.
[3] Jittoh T, Sato J, Shimomura T and Yamanaka M, *Phys. Rev.* D **73**, 055009 (2006) [arXiv:hep-ph/0512197].
[4] Jittoh T, Kohri K, Koike M, Sato J, Shimomura T and Yamanaka M, *Phys. Rev.* D **76**, 125023 (2007) [arXiv:0704.2914 [hep-ph]].
[5] Jittoh T, Kohri K, Koike M, Sato J, Shimomura T and Yamanaka M, *Phys. Rev.* D **78**, 055007 (2008) [arXiv:0805.3389 [hep-ph]].
[6] Ryan S G, Beers T C, Olive K A, Fields B D and Norris J E, *Astrophys. J.* **530** (2000) L57.
[7] Cyburt R H, Fields B D and Olive K A, *Phys. Lett.* B **567**, 227 (2003) [arXiv:astro-ph/0302431].
[8] Coc A, Vangioni-Flam E, Descouvemont P, Adahchour A and Angulo C, *Astrophys. J.* **600**, 544 (2004) [arXiv:astro-ph/0309480].
[9] Jarosik N*et al.* [WMAP Collaboration], *Astrophys. J. Suppl.* **170**, 263 (2007) [arXiv:astro-ph/0603452].
[10] Dunkley J*et al.* [WMAP Collaboration], arXiv:0803.0586 [astro-ph].
Raidal M *et al.*, arXiv:0801.1826 [hep-ph].
[11] Hamaguchi K, Nojiri M M and de Roeck A, *JHEP* **0703**, 046 (2007) [arXiv:hep-ph/0612060].
[12] Tarem S *et al.* [ATLAS Collaboration] ATL-SN-ATLAS-2008-071
[13] Tarem S *et al.* ATL-PHYS-PUB-2005-02
[14] Ishiwata K, Ito T and Moroi T, arXiv:0807.0975 [hep-ph].
[15] Calibbi L, Jones-Perez J and Vives O, arXiv:0804.4620 [hep-ph].
[16] Masina I and Savoy C A, *Nucl. Phys.* B **661**, 365 (2003) [arXiv:hep-ph/0211283].
[17] Bentvelsen S*et al.* [ATLAS Collaboration], JINST **3** (2008) S08003
[18] For more information on the ATLAS Exepriment, visit website http://atlas.ch/
[19] Skands P Z, *Eur. Phys. J.* C **23**, 173 (2002) [arXiv:hep-ph/0110137].

Journal of Physics: Conference Series **315** (2011) 012024 doi:10.1088/1742-6596/315/1/012024

Higgs decay in Higgs portal dark matter models

Mayumi Aoki[1], **Shinya Kanemura**[2] **and Osamu Seto**[3,4]

[1] Institute for Theoretical Physics, Kanazawa University, Kanazawa 920-1192, Japan
[2] Department of Physics, University of Toyama, Toyama 930-8555, Japan
[3] Department of Architecture and Building Engineering, Hokkai-Gakuen University, Sapporo 062-8605, Japan
[4] speaker

E-mail: seto@physics.umn.edu

Abstract. In a so-called Higgs portal dark matter model, Higgs exchange processes are essential for both dark matter annihilation in the early Universe and direct search experiments. We study a scalar dark matter model with two Higgs doublets, and find that the possible maximal value for the branching ratio of the invisible decay of the Higgs boson can be significantly greater than that in the Higgs portal model with one Higgs doublet. Therefore, the search for the invisible decay of the Higgs boson at the CERN Large Hadron Collider and future collider experiments would provide useful information not only for the nature of dark matter but also for the structure of the Higgs sector even without directly detecting any extra scalar boson.

1. Introduction

Various astrophysical and cosmological observations provide evidence of the existence of dark matter (DM) [1]. The most interesting and promising candidate for DM is weakly interacting massive particles (WIMPs) whose relic abundance can be estimated in the standard thermal freeze out scenario without specifying the thermal history of the very early Universe. Among WIMP DM candidates, a class of models is categorized as "Higgs portal dark matter", in which a DM particle interacts with Standard Model (SM) particles through only Higgs exchange processes. The minimal model was constructed by adding only one new Z_2 parity-odd real scalar field to the SM [2, 3]. A variety of such models has been proposed [4, 5, 6, 7, 8, 9]. Some of them are motivated in the context of radiative seesaw models [5, 7].

The Higgs sector is the last unknown piece of the SM of particle physics. Although the Weinberg-Salam model contains only one Higgs doublet, the Higgs sector might consist of multi-Higgs doublet fields. In fact, such multi-Higgs models are realized in various new physics models including supersymmetry. In Higgs portal DM models, the interactions between DM and SM particle is very sensitive not only to the size of DM-DM-Higgs boson coupling constants but also to the structure and nature of Higgs sector.

There are several approaches to explore the property of a WIMP DM candidate. One is WIMP direct detection through the elastic scattering of WIMP with target nuclei through nuclear recoil. Since these experiments with improved sensitivity have been carried out by many collaborations, the null results reported by EDELWEISS II [10], ZEPLIN II [11] and XENON [12, 13] give the stringent constraints on the scattering cross section between WIMP and a nucleon, while two suspicious events in CDMS-II [14] and an excess at low energy events in CoGeNT [15] were also reported. Higgs-boson-exchange processes mainly lead to scalar (in other words, spin-independent (SI)) couplings between nuclei and WIMP. Hence, the structure of a Higgs sector and its couplings with DM, namely again DM-DM-Higgs boson couplings, are crucial for direct DM searches.

Another approach to explore DM may be collider experiments, especially those at the Large Hadron Collider (LHC), where the signal of WIMP is expected to appear as a missing transverse momentum. Moreover, another signal is possible in Higgs portal DM models. That is the invisible decay of the Higgs boson due to DM-DM-Higgs couplings for the case that the DM mass is smaller than one half of the Higgs boson mass [16, 17]. In the minimal Higgs portal DM model, the upper bound is obtained for the branching ratio of the invisible decay from the CDMS II final results [18, 19, 20, 21].

We consider a Z_2-odd scalar Higgs portal DM scenario in the framework of multi-Higgs doublet models [22]. Such a scenario can appear, for example, in the effective theory of the three-loop-induced neutrino mass model [7], in which not only tiny neutrino masses but also DM as well as baryon asymmetry may be explained simultaneously by the TeV scale physics. The upper bound on the branching ratio of the Higgs boson invisible decay is evaluated in the model with two Higgs doublets and a real Z_2-odd singlet scalar field. In the analysis, a specific Yukawa interaction (the Type-X Yukawa interaction [8, 23, 24, 25]) is employed, which is used in Ref. [7] and is defined under the other (softly broken) discrete symmetry ($\tilde{Z}_2$) for avoiding flavor changing neutral current (FCNC). We also give a comment on the results assuming the other types of Yukawa interaction, such as so-called Type-II. We then discuss the difference of the upper bound from that in the minimal model with one Higgs doublet. We show exclusive features of scalar Higgs portal DM with multi-Higgs doublets.

2. Model

We consider the model with two Higgs doublet fields Φ_1 and Φ_2 and one real singlet scalar field η. A discrete Z_2 parity is introduced in the model, and the odd charge is assigned for η to guarantee the stability as a candidate of DM. The scalar potential is given by

$$V = \frac{1}{2}\mu_\eta^2\eta^2 + \lambda_\eta\eta^4 + \sum_{i=1,2}\sigma_i|\Phi_i|^2\eta^2 + V(\Phi_1, \Phi_2), \tag{1}$$

where μ_η^2 is the invariant squared mass of η, and $V(\Phi_1, \Phi_2)$ is the potential of the two Higgs doublet model. We neglect the CP violating phase, so that all the coupling constants are real. After electroweak symmetry breaking, neutral components in the Higgs doublets are parametrized as

$$\phi_i^0 = \frac{1}{\sqrt{2}}(v_i + h_i + iz_i), \quad (i = 1, 2), \tag{2}$$

where v_i are the vacuum expectation values (VEVs) that satisfy $v_1^2 + v_2^2 = v^2 \simeq (246\,\text{GeV})^2$ and $\tan\beta = v_2/v_1$. The mass matrix for h_1 and h_2 is diagonalized by introducing the mixing angle α, and two CP-even states h and H are the mass eigenstates of the CP-even bosons. The CP-odd scalar bosons z_1 and z_2 mix with each other, and become the CP-odd Higgs A and the longitudinal mode of the Z boson. In total, from Φ_1 and Φ_2 five physical states appear; i.e., two CP-even (h, H), one CP-odd (A), and charged $(H^\pm)$ scalar bosons.

In the limit of $\sin(\beta - \alpha) = 1$, h is the SM-like Higgs boson; i.e, all the coupling constants with SM fields coincide with those of the SM Higgs boson at the tree level [26]. On the other hand, H does not receive the VEV and does not couple to the weak gauge bosons in this limit. Throughout this work, for simplicity, we always take this limit (the SM-like limit) where the mass of the SM-like Higgs boson h is bounded from below ($m_h > 114$ GeV) from the LEP

	ξ_h^u	ξ_h^d	ξ_h^ℓ	ξ_H^u	ξ_H^d	ξ_H^ℓ	ξ_A^u	ξ_A^d	ξ_A^ℓ
Type-I	c_α/s_β	c_α/s_β	c_α/s_β	s_α/s_β	s_α/s_β	s_α/s_β	$\cot\beta$	$-\cot\beta$	$-\cot\beta$
Type-II	c_α/s_β	$-s_\alpha/c_\beta$	$-s_\alpha/c_\beta$	s_α/s_β	c_α/c_β	c_α/c_β	$\cot\beta$	$\tan\beta$	$\tan\beta$
Type-X	c_α/s_β	c_α/s_β	$-s_\alpha/c_\beta$	s_α/s_β	s_α/s_β	c_α/c_β	$\cot\beta$	$-\cot\beta$	$\tan\beta$
Type-Y	c_α/s_β	$-s_\alpha/c_\beta$	c_α/s_β	s_α/s_β	c_α/c_β	s_α/s_β	$\cot\beta$	$\tan\beta$	$-\cot\beta$

Table 1. The mixing factors in Yukawa interactions in Eq. 3

experiment while that of H can be lower than 100 GeV because of no coupling to the weak gauge bosons.

Multi-Higgs doublet models in general suffer from dangerous FCNC. To avoid FCNC, we impose a softly broken discrete symmetry $\tilde{Z}_2$ under the transformation $\Phi_1 \to \Phi_1$ and $\Phi_2 \to -\Phi_2$. The Yukawa interactions are expressed in terms of mass eigenstates of the Higgs bosons as

$$\begin{aligned} \mathcal{L}_{\text{yukawa}}^{\text{THDM}} &= -\sum_{f=u,d,\ell} \left(\frac{m_f}{v}\xi_h^f \overline{f} f h + \frac{m_f}{v}\xi_H^f \overline{f} f H - i\frac{m_f}{v}\xi_A^f \overline{f}\gamma_5 f A \right) \\ &\quad - \left\{ \frac{\sqrt{2}V_{ud}}{v}\overline{u}\left(m_u \xi_A^u \mathrm{P}_L + m_d \xi_A^d \mathrm{P}_R \right) d H^+ + \frac{\sqrt{2} m_\ell \xi_A^\ell}{v} \overline{\nu_L} \ell_R H^+ + \text{H.c.} \right\}, \end{aligned} \tag{3}$$

where $\mathrm{P}_{L/R}$ are projection operators for left-/right-handed fermions, and the factors ξ_φ^f are listed in TABLE 1. There are four ways of charge assignment under this $\tilde{Z}_2$ parity, thus correspondingly four independent types of Yukawa interaction are possible [27, 28]. The typical example of so-called Type-II Yukawa interactions is that of the minimal supersymmetric standard model. The Type-X Yukawa interaction [8, 23, 24, 25], where one of the Higgs doublet couples to only quarks and the other does to only leptons, is adopted in the model for radiative generation of tiny neutrino masses with including the scalar DM proposed in Ref. [7], whose Higgs sector contains two Higgs doublets and a DM candidate Z_2-odd singlet scalar field as well as some heavier particles. Therefore, our present model given in Eq. (1) can be regarded as the effective theory of the model in Ref. [7]. We mainly study the model with the Type-X Yukawa interaction, and then give a short comment on the cases of the other types for Yukawa interactions.

Even in the SM-like limit, the total decay width of the SM-like Higgs boson h in our model can drastically change from the SM value when $m_\eta < m_h/2$ because of the additional invisible $h \to \eta\eta$ decay. The total width of h is given by

$$\Gamma_{\text{tot}} = \Gamma_{\text{vis}} + \Gamma_{\text{inv}}, \tag{4}$$

where Γ_{vis} denotes the width for Higgs boson decays into SM particle contents. In the SM-like limit, Γ_{vis} in our model coincides with that in the SM at the lowest order. The invisible decay width Γ_{inv} of the SM-like Higgs boson is computed as

$$\Gamma_{\text{inv}}(h \to \eta\eta) = \frac{v^2}{32\pi m_h}\sqrt{1 - \frac{4m_\eta^2}{m_h^2}}\, |-\sigma_1 \sin\alpha\cos\beta + \sigma_2 \cos\alpha \sin\beta|^2 . \tag{5}$$

The corresponding formula in the minimal Higgs portal DM model with a scalar doublet Φ and a real scalar field η is obtained from Eq. (5) by replacing $(-\sigma_1 \sin\alpha\cos\beta + \sigma_2\cos\alpha\sin\beta)$ by $2\sigma_{\text{m}}$ when the DM-Higgs coupling is given by

$$\mathcal{L}_{\text{int}} = \cdots - \sigma_{\text{m}}\eta^2|\Phi|^2 + \cdots. \tag{6}$$

The branching ratio for the invisible decay is given by

$$B_{\text{inv}}(h \to \eta\eta) \equiv \frac{\Gamma_{\text{inv}}}{\Gamma_{\text{tot}}}. \tag{7}$$

3. The invisible decay branching ratio

3.1. Astrophysical and cosmological constraints

Before we consider the invisible decay branching ratio of the SM-like Higgs boson h, we summarize the constraints on the $h\eta\eta$ coupling σ_is.

One is the cosmological DM abundance determined by thermal freeze out. The relic mass density is evaluated as

$$\Omega_\eta h^2 = 1.1\times 10^9 \frac{m_\eta/T_d}{\sqrt{g_*}M_P\langle\sigma v\rangle}\mathrm{GeV}^{-1}, \tag{8}$$

with the Planck mass M_P, the total number of relativistic degrees of freedom in the thermal bath g_*, and the decoupling temperature T_d. For the Type-X Yukawa interaction, the processes of $\eta\eta\to b\bar{b}$ and $\eta\eta\to\tau^+\tau^-$ are dominant when $m_\eta < m_W$, and the thermal averaged product of annihilation cross section and relative velocity is evaluated as [7]

$$\begin{aligned}\langle\sigma v\rangle \;\simeq\; & \frac{s}{16\pi m_\eta^2}\left[3m_b^2\left|\frac{-\sigma_1\sin\alpha\cos\beta+\sigma_2\cos\alpha\sin\beta}{s-m_h^2+im_h\Gamma_{\mathrm{tot}}^h}\left(\frac{\cos\alpha}{\sin\beta}\right)\right.\right.\\ &\left.+\frac{\sigma_1\cos\alpha\cos\beta+\sigma_2\sin\alpha\sin\beta}{s-m_H^2+im_H\Gamma_{\mathrm{tot}}^H}\left(\frac{\sin\alpha}{\sin\beta}\right)\right|^2\\ &+m_\tau^2\left|\frac{-\sigma_1\sin\alpha\cos\beta+\sigma_2\cos\alpha\sin\beta}{s-m_h^2+im_h\Gamma_{\mathrm{tot}}^h}\left(\frac{-\sin\alpha}{\cos\beta}\right)\right.\\ &\left.\left.+\frac{\sigma_1\cos\alpha\cos\beta+\sigma_2\sin\alpha\sin\beta}{s-m_H^2+im_H\Gamma_{\mathrm{tot}}^H}\left(\frac{\cos\alpha}{\cos\beta}\right)\right|^2\right]\Bigg|_{s=4m_\eta^2},\end{aligned} \tag{9}$$

where Γ_{tot}^H is the total width of H. In the minimal Higgs portal DM model, it is given as $\langle\sigma v\rangle\sim 3s/(4\pi m_\eta^2)\,|\sigma_{\mathrm{m}}/(s-m_h^2+im_h\Gamma_{\mathrm{tot}}^h)|^2$ with $s\simeq 4m_\eta^2$. Too large (small) coupling constants σ_i correspond to the over-annihilation (over-abundance) of DM. We evaluate the consistent parameter region of the $h\eta\eta$ coupling and DM mass.

The other is from the direct DM search, because the DM proton scattering is induced by only the t-channel Higgs bosons exchange in the Higgs portal models and its strength is proportional to the square of the $h\eta\eta$ coupling σ_is. The DM SI cross section for a proton is given by

$$\sigma_p^{\mathrm{SI}} = \frac{m_p^2}{\pi(m_\eta+m_p)^2}f_p^2, \tag{10}$$

with

$$\frac{f_p}{m_p} = \left(\sum_{q=u,d,s}f_{Tq}^{(p)}+\frac{2}{27}\sum_{q=c,b,t}f_{TGq}^{(p)}\right)\frac{f_q}{m_q}, \tag{11}$$

where m_p is the proton mass. f_p is the effective coupling with proton and $f_q^{(p)}$ is the hadronic matrix elements. The effective coupling of DM particle with a quark f_q depends on models. In the model in Eq. (1) with the Type-X Yukawa coupling, this is calculated at the tree level as

$$\begin{aligned}\frac{f_q}{m_q} \;=\; & \frac{(-\sigma_1\sin\alpha\cos\beta+\sigma_2\cos\alpha\sin\beta)}{2m_h^2}\frac{\cos\alpha}{\sin\beta}\\ &+\frac{(\sigma_1\cos\alpha\cos\beta+\sigma_2\sin\alpha\sin\beta)}{2m_H^2}\frac{\sin\alpha}{\sin\beta}.\end{aligned} \tag{12}$$

In the minimal Higgs portal DM model, it is given by $f_q/m_q = \sigma_{\mathrm{m}}/m_h^2$.

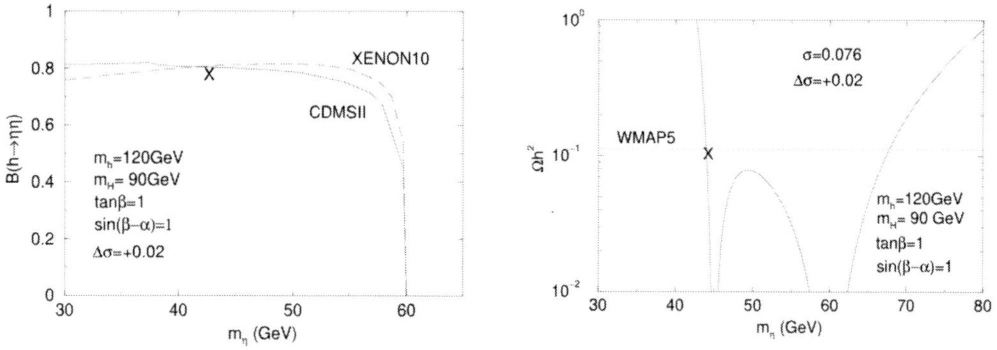

Figure 1. [Left] The constraint on the decay branching ratio $B_{\rm inv}(h \to \eta\eta)$ for the invisible decay of the SM-like Higgs boson into a DM pair from the CDMS II results and the XENON 10 results. [Right] The thermal abundance $\Omega_\eta h^2$ of DM as a function of the DM mass with $\sigma = 0.076$.

3.2. The invisible decay branching ratio

Now we consider the invisible decay branching ratio of the SM-like Higgs boson h. We examine the upper bound on $B_{\rm inv}(h \to \eta\eta)$ from the CDMS II and the XENON 10 for some parameter sets in the case of the Type-X interaction. As stated, we work in the SM-like limit $\sin(\beta-\alpha) = 1$. We use the average of the coupling constants $\sigma \equiv (\sigma_1 + \sigma_2)/2$ to show the typical scale of couplings and the difference $\Delta\sigma \equiv \sigma_1 - \sigma_2$ to see the effect of the difference instead of σ_1 and σ_2. The mass of the SM-like Higgs boson h is set to be $m_h = 120$ GeV. The other input parameters are commonly taken as $m_H = 90$ GeV and $\Delta\sigma = 0.02$. These parameter sets are not excluded by the current data. We here only show the results of the case $\tan\beta = 1$. One may find a case of another choice of $\tan\beta$ in Ref. [22].

The left panel of Fig. 1 shows the constraint on the Higgs invisible decay branching ratio from direct DM searches in the model with the Type-X Yukawa interaction and $\sin(\beta - \alpha) = 1$ for $(m_h, m_H, \Delta\sigma, \tan\beta) = (120\ {\rm GeV}, 90\ {\rm GeV}, 0.02, 1)$. The upper bound on $B_{\rm inv}(h \to \eta\eta)$ does not depend on the DM mass much and about $B_{\rm inv}(h \to \eta\eta) \sim 0.8$ is allowed for $m_\eta \lesssim 50$ GeV, while the bound becomes stringent for $m_\eta > 55$ GeV. Around $m_\eta \simeq 43$ GeV (near the H-resonance), we can obtain the maximal value of $B_{\rm inv}(h \to \eta\eta) \simeq 0.8$ which corresponds to $\sigma \simeq 0.076$ through Eqs. (5) and (7). The right figure shows the point $(m_\eta, \sigma) \simeq (43\ {\rm GeV}, 0.076)$ with the same other parameters indeed satisfies the WMAP constraint $\Omega_\eta h^2 \simeq 0.1$.

We have found that various parameter sets satisfy $\Omega_\eta h^2 \simeq 0.1$ and the bound from the direct DM searches. For instance, with the above parameter set,

$$B_{\rm inv}(h \to \eta\eta) = 0.8, \tag{13}$$

can be realized for $m_h = 120$ GeV and $m_\eta \simeq 43$ GeV around the edge of the resonance of H. On the other hand, in the model with the minimal Higgs portal DM model, we obtain

$$B_{\rm inv}(h \to \eta\eta) \lesssim 0.63, \tag{14}$$

for the same value of m_h but for $m_\eta \sim 55$ GeV (near the h resonance) in the same calculation manner. Therefore, if $B_{\rm inv}(h \to \eta\eta) \gg 0.63$ will be measured at the LHC, it will indicate a non-minimal Higgs sector in the Higgs portal DM scenario even when no extra Higgs boson will be found there yet. Here, we note that our result of the upper bound of $B_{\rm inv}(h \to \eta\eta)$ in the minimal Higgs portal DM model is somewhat smaller than those reported in Refs. [18] and [19].

The difference between ours and theirs mainly comes from the different choice for the values of the hadronic matrix elements. We have consistently used the values in Ref. [29].

The invisible decay of the Higgs boson can be detected at the LHC if $B_{\rm inv}(h \to \eta\eta) > 0.25$ [30]. At the International Linear Collider (ILC), invisible decays of the SM-like Higgs boson h can be tested when $B_{\rm inv}(h \to \eta\eta) >$ only a few % [31]. Therefore, we can distinguish the maximal value of the branching ratio evaluated in our model from the upper bound in the minimal one doublet model with a Z_2-odd singlet scalar boson.

As we have seen, the extra scalar boson H has to be lighter than the SM-like Higgs boson h in order to obtain $B_{\rm inv}(h \to \eta\eta) \gtrsim 0.63$. Phenomenology of extra Higgs bosons in the Type-X two Higgs doublet model has been studied in Ref. [8, 23, 24, 25, 32]. At the LHC such a light H ($m_H \sim 90$ GeV) can be produced via gluon fusion processes $gg \to H$ and $gg \to gH$, and also the Drell-Yan type processes $q\bar{q}' \to HH^+$ and $q\bar{q} \to AH$. The decay pattern of H largely depends on $\tan\beta$. For $\tan\beta \sim 1$, it decays mainly into bb, while for $\tan\beta \sim 10$ the leptonic decay modes into $\tau^+\tau^-$ and $\mu^+\mu^-$ are dominant. If such a light H is identified and large $B_{\rm inv}(h \to \eta\eta) \gtrsim 0.63$ is confirmed at the LHC, then the two Higgs portal DM scenario can be tested. On the other hand, if only $B_{\rm inv}(h \to \eta\eta) \gg 0.63$ is measured at the LHC without detecting H, then we could obtain indirect information on the extended Higgs sector in the Higgs portal DM scenario before direct detection of the extra Higgs bosons.

We have studied the two Higgs doublet model with Type-X Yukawa coupling. However, relaxation of the upper bound on the invisible decay branching ratio seems to be generic for other multi-Higgs doublet models as well. It is easily understood from Eq. (3) that for $\sin(\beta - \alpha) = 1$ and $\tan\beta = 1$ the visible width of the extra Higgs boson H is independent of the types of Yukawa interaction [23], and the abundance of η is also calculated to be almost common when the mass of η to be near the H and h resonances. Therefore, our result of $B_{\rm inv}(h \to \eta\eta) \gg 0.63$ for the above parameter set is essentially independent of the type of Yukawa interaction. The essence of this enhancement comes from the fact that the relevant interaction of DM for direct DM search experiments is both h-mediation and H-mediation, while only the coupling with h is relevant to the invisible decay of the SM-like Higgs boson.

4. Summary

We studied the branching ratio of the Higgs invisible decay in the model with two Higgs doublets and one scalar singlet DM field, mainly assuming the Type-X Yukawa interaction. We could rewrite the latest CDMS II and XENON 10 excluded region into an upper bound of Higgs invisible decay for a given parameter set.

As compared to the case of the minimal Higgs portal DM model, in the two Higgs doublet portal DM model it is still allowed to have a larger value of the invisible decay branching ratio such as 0.8 or even larger. Although the Type-X Yukawa interaction has been used in our analysis of the invisible decay branching ratio, this main results and its physics behind is independent of the type of Yukawa interaction. Thus, we conclude that precise determination of the invisible decay branching ratio at the LHC or future collider experiments would give useful information not only for the nature of dark matter but also for the structure of the Higgs sector even without detecting any extra scalar boson directly.

Acknowledgments

The speaker (OS) would like to thank the organizers for their financial support to him. The work was supported, in part, by Grant-in-Aid for scientific research (C), Japan Society for the Promotion of Science (JSPS), No. 19540277 (SK) and by the scientific research grants from Hokkai-Gakuen (OS).

References

[1] e.g., Spergel D N *et al.* [WMAP Collaboration] 2007 *Astrophys. J. Suppl.* **170** 377
[2] McDonald J 1994 *Phys. Rev.* D **50** 3637
[3] Burgess C P, Pospelov M and ter Veldhuis T 2001 *Nucl. Phys.* B **619** 709
[4] Deshpande N G and Ma E 1978 *Phys. Rev.* D **18** 2574
[5] Ma E 2006 *Phys. Rev.* D **73** 077301
[6] Barbieri R, Hall L J and Rychkov V S 2006 *Phys. Rev.* D **74** 015007
[7] Aoki M, Kanemura S and Seto O 2009 *Phys. Rev. Lett.* **102** 051805 and *Phys. Rev.* D **80** 033007
[8] Goh H S, Hall L J and Kumar P 2009 *J. High Energy Phys.* JHEP05(2009)097
[9] Okada N and Seto O 2010 *Phys. Rev.* D **82** 023507
[10] Broniatowski A *et al.* 2009 *Phys. Lett.* B **681** 305
[11] Lebedenko V N *et al.* 2009 *Phys. Rev.* D **80** 052010
[12] Angle J *et al.* [XENON10 Collaboration] 2008 *Phys. Rev. Lett.* **101** 091301
[13] Aprile E *et al.* [XENON100 Collaboration] 2010 *Phys. Rev. Lett.* **105** 131302
[14] Ahmed Z *et al.* [The CDMS Collaboration] 2010 *Science* **327** 1619
[15] Aalseth C E *et al.* [CoGeNT collaboration] 2011 *Phys. Rev. Lett.* **106** 131301
[16] Bento M C, Bertolami O, Rosenfeld R and Teodoro L 2000 *Phys. Rev.* D **62** 041302
[17] Bento M C, Bertolami O and Rosenfeld R 2001 *Phys. Lett.* B **518** 276
[18] He X G, Li T, Li X Q, Tandean J and Tsai H C 2010 *Phys. Lett.* B **688** 332
[19] Farina M, Pappadopulo D and Strumia A 2010 *Phys. Lett.* B **688** 329
[20] Kadastik M, Kannike K, Racioppi A and Raidal M 2010 *Phys. Lett.* B **694** 242
[21] Cheung K and Yuan T C 2010 *Phys. Lett.* B **685** 182
[22] Aoki M, Kanemura S and Seto O 2010 *Phys. Lett.* B **685** 313
[23] Aoki M, Kanemura S, Tsumura K and Yagyu K 2009 *Phys. Rev.* D **80** 015017
[24] Su S and Thomas B 2009 *Phys. Rev.* D **79** 095014
[25] Logan H E and MacLennan D 2009 *Phys. Rev.* D **79** 115022
[26] Gunion J F and Haber H E 2003 *Phys. Rev.* D **67** 075019
[27] Barger V D, Hewett J L and Phillips R J N 1990 *Phys. Rev.* D **41** 3421
[28] Grossman Y 1994 *Nucl. Phys.* B **426** 355
[29] Ellis J R, Olive K A and Savage C 2008 *Phys. Rev.* D **77** 065026
[30] Warsinsky M [ATLAS Collaoration] 2008 *J. Phys.: Conf. Series* **110** 072046
[31] Schumacher M 2003 *Report No.* LC-PHSM-2003-096.
[32] Belyaev A, Guedes R, Moretti S and Santos R 2010 *J. High Energy Phys.* JHEP07(2010)051

Scalar Field Dark Matter from Two Concentric Spherical Branes Universe

Tonatiuh Matos[1]

Departamento de Física, Centro de Investigación y de Estudios Avanzados del IPN,
A.P. 14-740, 07000 México D.F., México.

Miguel A. García Aspeitia

Departamento de Física, Centro de Investigación y de Estudios Avanzados del IPN,
A.P. 14-740, 07000 México D.F., México.

Abstract. The Scalar Field Dark Matter (SFDM) model has provided to be a very successful one to explain the large scale structure of the Universe. Nevertheless, one of the main problems of the SFDM model is the ultralight mass of the scalar field $m_\Phi \sim 10^{-22}$eV. In this work we explore a five-dimensional Universe with a special topology with the aim to explain this mass using the Randal-Sumdrum prescription for the hierarchy problem using two branes. In this model we show two limits where the branes have mutual interaction and have no interaction. We analyze the Friedmann equations and the linear corrections introduced by this model.

1. INTRODUCTION

The Scalar Field Dark Matter (SFDM) model has provided to be very successful model for explaining the large scale structure of the Universe [1]. The hypothesis of this model is that the dark matter of the Universe is a scalar field with mass $m_\Phi \sim 10^{-22}$eV. Of course this mass is too small and provokes a strong hierarchy problem. Nevertheless, this model has very interesting features which makes the model worth to keep studying. For example, with this mass, that means, with only one free parameter, the scalar field contains the following important features:
The ultra-light scalar field mass ($m_\Phi \sim 10^{-22}$eV) fits:

(i) The evolution of the cosmological densities [2].
(ii) The rotation curves of galaxies [5] and the central density profile of LSB galaxies [4],
(iii) With this mass, the critical mass of collapse for a real scalar field is just $10^{12} M_\odot$, i.e., the one obeserved in galaxies halos [3].
(iv) The central density profile of the dark matter is flat [4].
(v) The scalar field has a natural cut off, thus the substructure in clusters of galaxies is avoided naturally. With a scalar field mass of $m_\Phi \sim 10^{-22}$eV the amount of substructure is compatible with the observed one [6].

[1] Part of the Instituto Avanzado de Cosmología (IAC) collaboration http://www.iac.edu.mx/

Journal of Physics: Conference Series **315** (2011) 012025 doi:10.1088/1742-6596/315/1/012025

(vi) SFDM forms galaxies earlier than the cold dark matter model, because they form Bose-Einstein Condensates at a critical temperature $T_c >>$TeV. So, if SFDM is right, we have to observe big galaxies at big redshifts.

In this work we present a model using two branes, where the mass of the scalar field is not too small, but the brane where we are living reads an ultralight scalar field mass, using the Randal Sundrum prescription. Recent works [8]-[15] assume that the Universe have more than the four dimensions and in fact we live in a membrane that lives in a five dimensional space time obtaining a new degree of freedom (the radion). The imposition of extra dimensions generate different behavior in the Newtonian law at long distances with the correction in the potential and solving the problem of the acceleration as a consecuence of the fifth dimension.

Particulary we porpose a model with two concentric spherical branes embedded in a five dimensional AdS-S space-time. Each brane is filled with different field with the aim of solve the problem of the dark matter. The shape of the action to model this physical structure is given by

$$S = \int dX^5 \sqrt{-g_{(5)}}\, m^3_{(5)} \left(R_{(5)} + \Lambda\right) - \sum_{\pm} \int_{\pm} dx^4 \sqrt{-g^{\pm}} \left(2m^3_{(5)} K^{\pm} + \pounds^{\pm}\right),$$

being $\pm$ the exterior or interior region of the brane respectively, $g_{(5)}$ is the determinant of the five-dimensional (5D) metric and g the determinant of the four-dimensional (4D) one, $m_{(5)}$ is the 5D Planck mass, $R_{(5)}$ is the 5D scalar curvature, K is the extrinsic curvature and Λ is the 5D cosmological constant, $\pounds^+$ is the Lagrangian of the fields content in the exterior brane (spin zero) and $\pounds^-$ is the Lagrangian of the fields content in the interior brane (spin one).
Following the next ideas it is possible to enumerate the next characteristics of the branes and the bulk

(i) The bulk is a five dimensional AdS-S with two branes embedded and with no energy momentum tensor in the bulk $T_{AB} = 0$ $i.e.$ no fields in the bulk.
(ii) The inner brane is our Universe and content the fields predicted by the standard model of particles
(iii) The outer brane content a particular scalar field that behave like a dark matter [15] and imprint the gravitational well potential in the evolution of our Universe (inner brane).

With the next features it is possible to analyze when the branes have no dynamical interaction between them and with interaction. The next section analyze the behavior of this two cases in the Friedmann equations and the fields that lives in the branes.

2. Non interaction between the branes.

In this case we propose two concentric branes embedded in a five dimensional bulk (FIG. 1). For simplicity the problem of each brane can be treaty in separate way in concordance with the superposition method due to the theory is linear in the same way as the electromagnetism. Generalizing the metric associated with the bulk it is possible to write

$$ds^2_{(5)} = -A(a)_{\pm} dt^2_{\pm} + \frac{1}{A(a)_{\pm}} da^2 + a^2 \left[d\chi^2 + \sin^2(\chi)\left(d\theta^2 + \sin^2\theta d\varphi^2\right)\right], \quad (1)$$

with the functions $A(a)_{\pm}$ directly solved by the five dimensional Einstein equations. From here, we use natural units ($\hbar = c = 1$).

Following the idea, it is possible to find the Friedmann equation in each brane with the junction conditions in the next way [15]

$$\left(\frac{\dot{a}_1}{a_1}\right)^2 + \frac{1}{a_1^2} = \frac{k^4_{(5)}\rho_1^2}{36} + \frac{\Lambda_1 + \Lambda_2}{12} + \frac{M_1}{m^3_{(5)} a_1^4} + \frac{(\Lambda_2 - \Lambda_1)^2}{16\rho_1^2 k^4_{(5)}} + \frac{9M_1^2}{m^6_{(5)} k^4_{(5)} \rho_1^2 a_1^8} + \frac{3M_1(\Lambda_2 - \Lambda_1)}{2m^3_{(5)} a_1^4 k^4_{(5)} \rho_1^2}, \quad (2)$$

Journal of Physics: Conference Series **315** (2011) 012025 doi:10.1088/1742-6596/315/1/012025

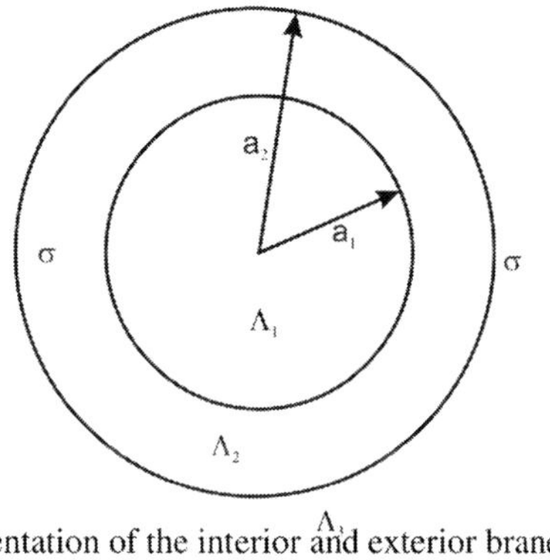

Figure 1. Schematic representation of the interior and exterior branes and the different values of the 5D cosmological constant Λ_i $i = 1, 2, 3$ and the brane tensions σ.

and

$$\left(\frac{\dot{a}_2}{a_2}\right)^2 + \frac{1}{a_2^2} = \frac{k_{(5)}^4 \rho_2^2}{36} + \frac{\Lambda_3 + \Lambda_2}{12} + \frac{2M_1 + M_2}{m_{(5)}^3 a_2^4} + \frac{(\Lambda_3 - \Lambda_2)^2}{16\rho_2^2 k_{(5)}^4} + \frac{9M_2^2}{m_{(5)}^6 k_{(5)}^1 \rho_2^2 a_2^8} + \frac{3M_2(\Lambda_3 - \Lambda_2)}{2m_{(5)}^3 a_2^1 k_{(5)}^4 \rho_2^2}. \tag{3}$$

with ρ_1, ρ_2 the density matter, M_1, M_2 is the mass of each branes respectively. The constants Λ_1, Λ_2 and Λ_3 is associated with the vacumm energy in the five dimensional space time and $k_{(5)}$ is the five dimensional gravitational constant. $k_{(5)}$ can be associated with the five dimensional Planck mass in the next way

$$\kappa_{(5)}^2 = 8\pi G_{(5)} = \frac{8\pi}{M_{(5)}^3}, \tag{4}$$

where $G_{(5)}$ is the 5-dimensional Newtonian constant and $M_{(5)}$ is the 5-dimensional Planck mass.

It is possible to observe both regimes in the Friedmann equations (2), (3) with an apropiate limit between the brane tension and the energy density of the brane.

2.1. High Energy Limit

In this scenario the Universe is very young and the corrections in the Friedmann equations (2), (3) appear when the energy density is much bigger than the brane tension $\rho >> \sigma$ in the next way

$$\left(\frac{\dot{a}_1}{a_1}\right)^2 + \frac{1}{a_1^2} = \frac{\kappa_{(4)}^2}{3}\frac{\rho^2}{2\lambda} + \frac{\Lambda_1}{3} + \frac{M}{m_{(5)}^3 a_1^4}, \tag{5}$$

and

$$\left(\frac{\dot{a}_2}{a_2}\right)^2 + \frac{1}{a_2^2} = \frac{\kappa_{(4)}^2}{3}\frac{\rho^2}{2\lambda} + \frac{\Lambda_2}{3} - \frac{M}{m_{(5)}^3 a_2^4}, \tag{6}$$

with

$$\Lambda_1 = -\frac{\lambda_{(5)}}{2}, \quad \Lambda_2 = -\frac{\lambda_{(5)}}{2}, \tag{7}$$

where $m_{(5)}$ is the five dimensional Planck mass and λ is associated with the tension in the brane. This is because in the high energy limit it follows that

$$\rho^2 \sim 2\sigma\rho\left(1 + \frac{\rho}{2\sigma}\right) + \sigma^2,$$

for simplicity we assume the ansatz $M_1 = -M_2 = -M$ with non physical interpretation, where we impose $\kappa_{(4)}^2\sigma \approx \lambda_{(5)}$.

2.2. *Low Energy Limit*

In this limit we choose that the energy density is much less than the brane tension on the brane $\rho << \sigma$, obtaining the classical Friedmann equation with corrections provided by the brane theory.

$$\left(\frac{\dot{a}_1}{a_1}\right)^2 + \frac{1}{a_1^2} = \frac{\kappa_{(4)}^2}{3}\rho + \frac{\Lambda_1}{3} + \frac{3M^2}{2m_{(5)}^5 \lambda a_1^8}, \tag{8}$$

and

$$\left(\frac{\dot{a}_2}{a_2}\right)^2 + \frac{1}{a_2^2} = \frac{\kappa_{(4)}^2}{3}\rho + \frac{\Lambda_2}{3} + \frac{3M^2}{2m_{(5)}^5 \lambda a_2^8}, \tag{9}$$

with

$$\Lambda_{1,2} = \Lambda_{(5)}, \tag{10}$$

where $m_{(5)}$ is the five dimensional Planck mass and λ is associated with the tension in the brane. In the same way as before, this is because $\rho^2 = 2\sigma\rho\left(1+\frac{\rho}{2\sigma}\right)+\sigma^2$,. where we impose the condition $\kappa_{(4)}^2\sigma \approx \lambda_{(5)}$.

3. Interaction between the branes.

In this scenario, the branes have topological interactions between them, obtaining restrictions in the matter contents of each brane. This scenario assume flat five dimensional bulk embedded in the next way

$$ds^2 = -n^2(t,y)dt^2 + a^2(t,y)\delta_{ij}dx^i dx^j + b^2(t,y)dy^2. \tag{11}$$

with two energy momentum tensor for each brane fixed in the orbifold at $y = 0$ and $y = 1/2$ as

$$T_{AB} = \frac{\delta(y)}{b} diag(-\rho, \vec{p}, 0), \tag{12}$$

$$T_{AB*} = \frac{\delta(y-1/2)}{b} diag(-\rho_*, \vec{p}_*, 0), \tag{13}$$

If we assume that the five dimensional Einstein equations can be written as $G_{AB} = \kappa_{(5)}^2(T_{AB}|_{bulk} + \widetilde{T}_{AB}|_{branes})$, where $A, B = 0, 1, 2, 3, 4$ and the energy-momentum tensor $T_{AB}|_{bulk}$ is for the bulk and $\widetilde{T}_{AB}|_{brane} = T_{AB} + T_{AB*}$ is for the visible brane and the hidden brane respectively.
Then we obtain the following constraint equations in the next way [11]

$$a(t,y)_0\rho = -a(t,y)_{1/2}\rho_*. \tag{14}$$

$$(3p + 2\rho)n(t,y)_0 = -(3p_* + 2\rho_*)n(t,y)_{1/2}. \tag{15}$$

where ρ, p and ρ_*, p_* it is the energy density and pressure of our Universe and Hidden brane respectively. Assuming that $p = \omega\rho$ and $p_* = \omega_*\rho_*$. If we assume linear solution to the functions $a(t,y)$, $n(t,y)$ and $b(t,y)$ in the next way [11]

$$a(t,y) = a_0(t)\left(1 + \lambda|y|\right), \tag{16}$$

$$n(t,y) = n_0(t)\left(1 + \mu|y|\right), \tag{17}$$

$$b(t,y)=b_0, \tag{18}$$

the constraint equations can be reduced as

$$\omega_* = \frac{1}{3}\left((2+3\omega)\frac{2-b_0H}{2+b_0H(2+3\omega)}-2\right), \tag{19}$$

$$\rho_* = -\rho(1-b_0H)^{-1}, \tag{20}$$

where H is the Hubble parameter of our Universe. If $H=0$ we demonstrate that $\omega_*=\omega$ and $\rho_*=-\rho$, where the sign is caused by the imposition of Z_2-symmetry. In similar way exist corrections in the Friedmann equations caused by the constraints.

3.1. *The Cosmology*

For our Universe, the Friedmann equation can be written in the next way [12]

$$\left(\frac{\dot a_1}{a_1}\right)^2 = \frac{\kappa_{(5)}^2}{36}\sum_i \rho_i^2, \tag{21}$$

where i is for baryons, neutrinos, radiation, cosmological constant etc.. The equation of energy conservation in the visible brane can be written as

$$\dot\rho + 3\frac{\dot a_1^2}{a_1^2}(1+\omega)\rho = 0. \tag{22}$$

For the hidden brane, the Friedmann equation can be written as [10]

$$\left(\frac{\dot a_2}{a_2}\right)^2 = \frac{\kappa_{(5)}^2}{36}\sum_i H_i^2\frac{(2+b_0H_i(2+3\omega_i))^2}{(2-b_0H_i)^2}, \tag{23}$$

the equivalent energy conservation equation for the hidden brane

$$\dot\rho_* + 3\frac{\dot a_2^2}{a_2^2}(1+\omega_*)\rho_* = 0. \tag{24}$$

3.2. *The Poisson Equation For The Hidden Brane.*

It is possible to study the Poisson equation for the hidden branen in the case when the Newtonian potential it is principally influenced by the scalar field as dark matter. In a region around of a cluster of matter, is better work the Klein-Gordon equation in a spherical coordinates in the next way

$$\Phi'' + \frac{2}{r}\Phi' - m_\Phi^2\Phi = \ddot\Phi, \tag{25}$$

where $\prime$ indicates differentiation of r. It is well known that the solution of the above equation (25) can be written as

$$\Phi(t,r) = \frac{e^{\pm ikr}}{r}e^{\pm i\omega t}, \tag{26}$$

and the dispersion relation $k^2=\omega^2-m_\Phi^2$ [7]. To study this case, we choose the particular solution of the equation (26) used by Bernal *et al* [7] in the next way

$$\Phi(t,x) = \Phi_0 \frac{sin(x)}{x} cos(\omega t), \tag{27}$$

where Φ_0 is an a constant and $x = kr$ and ρ_Φ can be written in the next way

$$\rho_\Phi = \frac{\Phi_0^2}{2}\left(\omega^2 - k^2 cos^2(\omega t)\right)\left(\frac{sin(x)}{x}\right)^2. \tag{28}$$

Then the Poisson equation can be written as

$$\nabla^2 \Psi_* = -2\pi G \Phi_0^2 \left(\omega^2 - k^2 cos^2(\omega t)\right)\left(\frac{sin(x)}{x}\right)^2 (1 - b_0 H)^{-1}, \tag{29}$$

in spherical coordinates

$$\frac{k^2}{x^2}\frac{\partial}{\partial x}\left(x^2 \frac{\partial \Psi_*}{\partial x}\right) + \frac{k^2}{x^2 sin(\theta)}\frac{\partial}{\partial \theta}\left(sin(\theta)\frac{\partial \Psi_*}{\partial \theta}\right) + \frac{k^2}{x^2 sin^2(\theta)}\frac{\partial^2 \Psi_*}{\partial \varphi^2} = -2\pi G \Phi_0^2 \left(\omega^2 - k^2 cos^2(\omega t)\right)\left(\frac{sin(x)}{x}\right)^2 \times (1 - b_0 H)^{-1}, \tag{30}$$

where we use the next change of variable $x = rk$.

Now, it is possible to solve the last differential equation (30) when the Hubble parameter depend of $\vec{r}$ and t and H only depends of t.

(i) *Non dependence of the spatial variables in* H. In this case, we assume that H only depends of t for simplicity, in general H depends of t and the spatial variables $\vec{r}$.
To solve the last equation we assume the next separation of variables $\Psi(x,\theta,\varphi)_* = R(x)Y(\theta,\varphi)$. Then we obtain the next two differential equations

$$\frac{1}{sin(\theta)}\frac{d}{d\theta}\left(sin(\theta)\frac{dY(\theta,\varphi)}{d\theta}\right) + \frac{1}{sin^2(\theta)}\frac{d^2Y(\theta,\varphi)}{d\varphi^2} = -l(l+1)Y(\theta,\varphi), \tag{31}$$

and

$$\frac{d}{dx}\left(x^2\frac{dR(x)}{dx}\right) - \left(l(l+1) - \frac{\gamma(t)}{k^2} sin^2 x\right) R(x) = 0, \tag{32}$$

where $\gamma(t) = 2\pi G \Phi_0^2 \left(\omega^2 - k^2 cos^2(\omega t)\right)(1 - b_0 H(t))^{-1}$ only depends of t. From the equation (32) it is evident that the solution is the spherical harmonics

$$Y_l^m(\theta,\varphi) = \sqrt{\frac{(2l+1)(l-m)!}{4\pi(l+m)!}} P_l^m(cos\theta) e^{im\varphi}, \tag{33}$$

where $P_l^m(cos\theta)$ are the Legendre Polynomial.
For the equation 31 it is neccesary obtain a numerical solution.

(ii) *Dependence of the spatial variables in* H. In this general case, H depends of the spatial variables and the time, this implies that ρ has the same characteristics because $H \sim \rho$. Then we obtain the same differential equation for the angular variables (33), but the equation (31) can be generalized in the next way

$$\frac{d}{dx}\left(x^2\frac{dR(x)}{dx}\right) - \left(l(l+1) - \frac{\beta(t) x^2 sin^2 x}{x^2 - \alpha(t) sin^2 x}\right) R(x) = 0, \tag{34}$$

where $\beta(t) = 2\pi G \Phi_0^2 \left(\omega^2 - k^2 cos^2(\omega t)\right)$ and $\alpha(t) = b_0 \frac{\kappa_{(5)}^2 \Phi_0^2}{12}\left(\omega^2 - k^2 cos^2(\omega t)\right)$ in similar way the last equation (34) must be integrated numerically.

4. CONCLUSIONS

Braneworld theory is an exciting possibility to understand the real topology of the Universe and the fields inside of the universe we observe. As it is possible to show, the dynamic equations change with high energy corrections in the energy density, predicting new observations in the very early Universe and posible evidence in the CMB imprints in the next few year with the Planck mission. Similarly, the topological constraints caused by the imposition of higher dimensions, cause a different behavior in the fields contents in the branes. As we observe in the section III, the dynamic of the fields in the hidden brane is correlated with the evolution parameter of our Universe and the fields contents giving new dynamics. In similar way, the equations in the hidden brane change drastically when we impose interaction between them and it is related with the evolution of our universe and the field contents.

Recently, new experiments (LHC, Planck satellite, etc...) could obtain evidence of the existence of higher dimensions and giving us the real behavior of the gravitation, the real topology and the dynamics of the fields that lives in the Universe.

ACKNOWLEDGEMENT

This work was partially supported by CONACyT México, under grant 49865-F, 216536/219673. Instituto Avanzado de Cosmologia (IAC) collaboration.

References

[1] Matos Tonatiuh and Siddhartha Guzmán F. *et al.*, Class. Quant. Grav. **17**, L9-116 (2000), arXiv: gr-qc /9810028
[2] Matos Tonatiuh, Vazquez Alberto and Magaña J.A. *et al* MNRAS., **389**, 13957 (2009) arXiv:0806.0683
[3] Alcubierre Miguel, Siddhartha Guzmán F., Matos Tonatiuh, Núñez Darío, Ureña Luis A. and Wiederhold Petra. Class. Quant. Grav. **19** 5017-5024 (2002), gr-qc /0110102.
[4] Bernal Argelia, Matos Tonatiuh and Núñez Darío., Rev. Mex. A.A. **44**, 149-160 (2008), arXiv:astro-ph/0303455
[5] C. G. Boehmer and T. Harko., *et al.*, JCAP **0706** 025 (2007), arXiv:astro-ph/0303455.
[6] Matos Tonatiuh and Ureña Luis A., *et al.*, Phys Rev. D **63**, 063506 (2001), arXiv:astro-ph/ 0006024
[7] Bernal A., Matos T., Nuñez D., *et al* Rev. Mex. A. A., **44** 149-160 (2008), arXiv:astro-ph/0303455v3 Flat Central Density Profiles from Scalar Field Dark Matter.
[8] Randall Lisa and Sundrum Raman., *et al.*, Phys. Rev. Lett. **83** 3370-3373 (1999), arXiv:hep-ph/9905221v1 A Large Mass Hierarchy from a Small Extra Dimension
[9] Dvali G., Gabadadze G., Porrati M. *et al.*, Phys. Lett. B **485** 208-214, (2000), arXiv:hep-th/0005016v2 4D Gravity on a Brane in 5D Minkowski Space
[10] Khoury, Ovrut, Steinhardt, Turok., *et al.*, Phys. Rev. D **64**:123522 (2001)
[11] Binétruy P., Deffayet C., Lanoglois D., *et al.*, Nucl. Phys. **B** 565: 269-287 (2000), arXiv:hep-th/9905012v2. Non-conventional cosmology from a brane-universe
[12] García Aspeitia M. A. and Matos T. To be Published. Φ^2 Scalar Field Dark Matter in a Braneworld
[13] Räsänen S., Ph.D. thesis, (2002). arXiv:astro-ph/0208282v2 A primer on the ekpyrotic scenario
[14] Steinhardt, Turok., arXiv:hep-th/0111098
[15] García Aspeitia M. A. and Matos T., arXiv:gr-qc/0906.3278 (2009). To be Published. The Universe Dynamics from Topological Considerations

Probing the Majorana neutrino CP phases and masses in neutrino-antineutrino conversion

D. Delepine[1], V. González Macías[1], S. Khalil[2] and G. López Castro[3]

[1] División de Ciencias e Ingenierías de la Universidad de Guanajuato, C.P. 37150, León, Guanajuato, México.

[2] Centre for Theoretical Physics, The British University in Egypt, El Sherouk City, Postal No, 11837, P.O. Box 43, Egypt

[3] Departamento de Fisica, Cinvestav, Apartado Postal 14-740, 07000 Mexico D.F., Mexico

E-mail: delepine@fisica.ugto.mx

E-mail: vanniagm@fisica.ugto.mx

E-mail: skhalil@bue.edu.eg

E-mail: glopez@fis.cinvestav.mx

Abstract. We propose a new strategy for detecting the CP-violating phases and the effective mass of muon Majorana neutrinos by measuring observables associated with neutrino-antineutrino oscillations in $\pi^\pm$ decays. Within the generic framework of quantum field theory, we compute the non-factorizable probability for producing a pair of same-charged muons in $\pi^\pm$ decays as a distinctive signature of $\nu_\mu - \bar{\nu_\mu}$ oscillations. We show that an intense neutrino beam through a long baseline experiment is favored for probing the Majorana phases. Using the neutrino-antineutrino oscillation probability reported by MINOS collaboration, a new stringent bound on the effective muon-neutrino mass is derived.

1. Introduction.

As is well known, the Majorana nature of neutrinos can be established via the observation of $\Delta L = 2$ processes [1]. The parameter characterizing the rate of such transitions, the effective neutrino mass $\langle m_{ll} \rangle \equiv \sum_i U_{li}^2 m_{\nu_i}$, involves a combination of neutrino masses, mixings and phases. Although it is clear that a combination of neutrino oscillation and nuclear beta decay experiments currently provide strong constraints on neutrino masses, having direct constraints on effective Majorana masses would be useful to have a more complete picture of Majorana neutrinos. Actually, it turns out that the only way to access the values of Majorana phases is through observables associated to $\Delta L = 2$ transitions [2].

Measurements of the effective electron-neutrino mass in the neutrinoless double beta decay $(0\nu\beta\beta)$ experiments can not restrict the two Majorana CP violating phases present in the PMNS mixing matrix. This may be expected since in $(0\nu\beta\beta)$ one measures the lifetime of the decay of two neutrons in a nucleus into two protons and two electrons, which is a CP conserving quantity [3]. On the other hand, direct bounds on other effective neutrino mass parameters $\langle m_{ll} \rangle$ from present experimental data are very poor. Currently, the strongest bound for the muon-neutrino case from the $K^+ \to \pi^- \mu^+ \mu^+$ branching fraction [4] is only $|\langle m_{\mu\mu} \rangle| \leq 0.04$ TeV [5], which leads to a negligible constraint on the neutrino masses and CP violating phases.

Here we propose a mechanism, based on neutrino-antineutrino oscillation [6, 7, 8, 9], to derive a strong bound on the effective mass of the muon-neutrino, using the results reported by the MINOS collaboration [10]. In addition, it provides a method for detecting the Majorana neutrino CP violating phases through measuring the CP asymmetry of the $\pi^{\pm}$ decay. It is worth noting that the probability of a process associated to neutrino oscillation is usually assumed to be factorized into three independent parts: the production process, the oscillation probability and the detection cross section. Here, we adopt the S-matrix amplitude method described in [11], in order to avoid the usual factorization scheme.

2. Neutrino-Antineutrino Oscillation

Let us start by considering a positive charged pion which decays into a virtual neutrino at the space-time location (x, t) together with a positive charged muon. After propagating, the neutrino can be converted into an antineutrino which produces a positive charged muon at the point (x', t') when it interacts with a target, as shown in Fig. 1. The non-observation of neutrino-antineutrino conversion can be used to set bounds on the effective Majorana mass of the muon neutrino.

For definiteness, we illustrate this process with the production of the neutrino in π^+ decay and its later detection via its weak interaction with a target nucleon N

$$\pi^+(p_1) \to \quad \mu^+(p_2) \quad +\nu_\mu^s(p) \hookrightarrow \overline{\nu_l}^d(p) + N(p_N) \to N'(p_{N'}) + \mu^+(p_l)$$

where the superscript $s(d)$ refers to the virtual neutrino (antineutrino) at the source (detection) vertex. This is a $|\Delta L| = 2$ process.

If one ignores other flavors, the time evolution of the $\nu_\mu - \bar{\nu}_\mu$ system would be analogous to that of the $K^0 - \bar{K}^0$ or $B^0 - \bar{B}^0$ systems. Instead, we prefer to use the formalism developed in Ref.[11], where the whole reaction includes the production and detection processes of neutrinos. The decay amplitude becomes:

$$\begin{aligned} T_{\nu_\mu - \bar{\nu}_\mu}(t) &= (2\pi)^4 \delta^4(p_l - p_N + p_{N'} + p_2 - p_1)(G_F V_{ud})^2 (J_{NN'})_\mu f_\pi \\ &\times \sum_i \bar{v}_\mu(p_l) \gamma^\mu (1+\gamma_5) \not{p}_1 v(p_2) \times U_{\mu i} U_{\mu i} (m_{\nu_i}) \frac{e^{-itE_{\nu_i}}}{2E_{\nu_i}} , \end{aligned} \tag{1}$$

where the relation $\nu_k = \sum U_{k\alpha} \nu_\alpha$ between flavor k and mass α neutrino eigenstates has been used, $f_\pi = 130.4$ MeV is the $\pi^{\pm}$ decay constant, and $J_{NN'}$ parametrizes the interaction with the nucleon. For simplicity, one assumes that

$$(J_{NN'})_\mu = \overline{u}_{N'}(p_{N'}) \gamma_\mu [g_V(q^2) + g_A(q^2)\gamma_5] u_N(p_N) \tag{2}$$

where we keep only the contributions of leading vector $g_V(q^2)$ and axial-vector $g_A(q^2)$ form factors, with $q = p_{N'} - p_N$.

If we neglect terms of $O(m_\mu / m_{N,N'})$, one obtains

$$\begin{aligned} \left|T_{\nu_\mu - \bar{\nu}_\mu}(t)\right|^2 &= (2\pi)^4 \delta^4(p_l - p_N + p_{N'} + p_2 - p_1)(G_F V_{ud})^4 \left|f_\pi\right|^2 64(g_A - 1)^2 m_N m_\mu^2 \\ &\times \sum_{i,j} U_{\mu i} U_{\mu j}^* U_{\mu i} U_{\mu j}^* e^{-it\triangle E_{\nu_{ij}}} \frac{m_{\nu_i} m_{\nu_j}}{4E_{\nu_i} E_{\nu_j}} (E_2 - E_p) \left(\left[1 - \frac{m_N}{(E_2 - E_p)} G(g_A)\right] p_l \cdot p_2 \right. \\ &\left. -2m_N F(g_A) \left[E_2 - E_l \left(1 + \frac{m_\pi^2}{m_\mu^2}\right)\right] - \frac{1}{2}(m_\mu^2 - m_\pi^2) \right) \end{aligned} \tag{3}$$

where $E_2(E_l), E_p$ are, respectively, the initial (final) muon and the pion energies and $\triangle E_{\nu_{ij}} = E_{\nu_i} - E_{\nu_j}$. The functions $F(g_A)$ and $G(g_A)$ are given by: $F(g_A) = \frac{g_A^2 + 1}{(g_A - 1)^2}$, $G(g_A) = \frac{g_A + 1}{g_A - 1}$. One can easily check that Eq. (3) is not factorizable into (production)×(propagation)×(detection) subprocesses due to the terms proportional to $p_l \cdot p_2 = E_l E_2 - |\vec{p_l}||\vec{p_2}| \cos\alpha$, where α is the angle between the directions of μ^+ particles. This is an important difference with respect to the case of neutrino-neutrino ($\Delta L = 0$) oscillations where it was shown in Ref.[11] that the S-matrix formalism reproduces the hypothesis of factorization of the probabilities.

We shall neglect the q^2-dependence of the nucleon form factors (namely, we take $g_V = g_V(q^2 = 0) = 1$ and $g_A = g_A(q^2 = 0) \approx -1.27$ [2]). As is well known [12], the cross section of charged current neutrino-nucleon quasielastic scattering is sensitive to the q^2-dependence of these form factors. However, as long as we confine to the CP rate asymmetry for neutrino↔antineutrino oscillations (see below) we expect that the effects of the momentum-transfer dependence of $g_{V,A}$ will partially cancel in the ratio of oscillation rates. Thus, after integration over kinematical variables, it is possible to write the rate of the complete process as

$$\Gamma_{\nu_\mu - \bar{\nu}_\mu} = \left| \sum_i U_{\mu i}^2 \frac{m_{\nu_i}}{2E_{\nu_i}} e^{itE_{\nu_i}} \right|^2 \times F(M, \phi), \tag{4}$$

where $F(M, \phi)$ denotes the kinematical function

There are two interesting limits for this process. At very short times, (as in short-baseline neutrino experiments), $\Gamma_{\nu_\mu - \bar{\nu}_\mu} \simeq \frac{|\langle m_{\mu\mu} \rangle|^2}{E_\nu^2} \times F(M, \phi)$, where $\langle m_{\mu\mu} \rangle$ is the effective Majorana mass for the muon neutrino. However in the long time limit (long-baseline neutrino experiments) the oscillation terms cannot be neglected. In the limit of $\theta_{13} = 0$, the Majorana phases $\alpha_{1,2}$ are the only sources of CP violation and hence

$$a_{CP} \simeq \tan\left[2(\alpha_2 - \alpha_3)\right] \sin\gamma \qquad \text{where } \gamma = \frac{\Delta m_{23}^2 L(km)}{2E_\nu(GeV)} \tag{5}$$

Thus, in the case of LBL neutrino experiment like MINOS where the distance L is given by $L = 735$ km and the energy E_ν is typically around $2 - 3$ GeV, one finds that $\sin\gamma \sim \mathcal{O}(1)$ [13],[14]. Thus, measuring CP asymmetry will be unavoidable indication for large CP violating Majorana phases.

3. Application to MINOS results on neutrino-antineutrino oscillations

Recently, MINOS [14] has measured the spectrum of ν_μ events which are missing after travelling 735 km. It is these missing events which are the potential source of $\bar{\nu}_\mu$ appearance. In their preliminary analysis, they were able to put a limit on the fraction of muon neutrinos transition to muon anti-neutrinos [10]: $P(\nu_\mu \to \bar{\nu}_\mu) < 0.026$ (90% c.l.). Assuming CPT, this limit can be written as $\frac{\Gamma_{\nu_\mu - \bar{\nu}_\mu}}{\Gamma_{\nu_\mu - \nu_\mu}} < 0.026$.

In the limit of ultrarelativistic neutrinos, $E_{\nu_i} \simeq E_\nu(1 + m_{\nu_i}^2/2E_\nu)$ and keeping the leading terms in the m_{ν_i}/E_ν terms, we get

$$\left| \sum_i U_{\mu i}^2 m_{\nu_i} e^{it\frac{m_{\nu_i}^2}{2E_\nu}} \right|^2 \lesssim 0.001 \times E_\nu^2 \tag{6}$$

To illustrate the usefulness of this relation, let us consider the general case of 3 generations. In this case, one finds

$$\begin{aligned} 0.001 E_\nu^2 &\gtrsim |\langle m_{\mu\mu} \rangle|^2 - 4 \sum_{i>j} \mathrm{Re}\left(U_{\mu i}^2 U_{\mu j}^{*2}\right) m_{\nu_i} m_{\nu_j} \sin^2 \frac{\Delta m_{ij} L}{4E_\nu} \\ &- 2 \sum_{i>j} \mathrm{Im}\left(U_{\mu i}^2 U_{\mu j}^{*2}\right) m_{\nu_i} m_{\nu_j} \sin \frac{\Delta m_{ij} L}{2E_\nu} \end{aligned} \tag{7}$$

Journal of Physics: Conference Series **315** (2011) 012026 doi:10.1088/1742-6596/315/1/012026

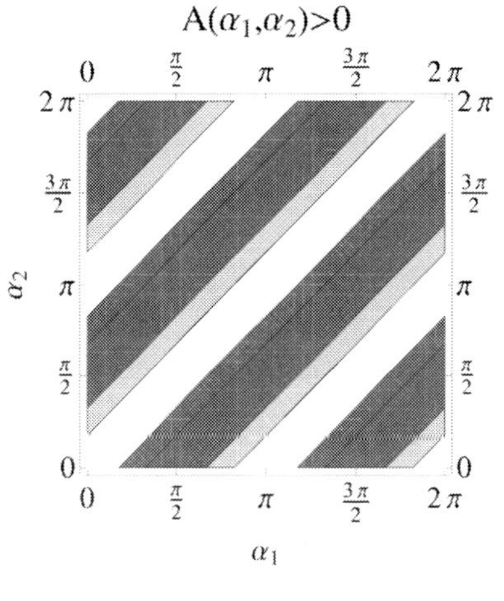

Figure 1. Region of the parametric space (α,β) for which the $A(\alpha,\beta)$ coefficient is positive. The light and dark gray zones correspond to an inverted and normal hierarchy schemes respectively.

Assuming that the only phases that appear in the neutrino mixing matrix are the Majorana phases, it is possible to get a bound on the effective muon-neutrino Majorana mass, only depending on the values of the Majorana phases as the oscillation terms cannot be neglected. In such a case, Eq.(7) can be written as:

$$\begin{aligned} 0.001E_\nu^2 &\gtrsim |\langle m_{\mu\mu}\rangle|^2 \pm 4\,\sin\left(\frac{\gamma_{32}}{2}\right) m_{\nu_2}m_{\nu_3}|U_{\mu 2}^2 U_{\mu 3}^{*2}|\sin\left(2\alpha_2 - 2\alpha_3 \pm \frac{\gamma_{32}}{2}\right) \\ &- 4\sin\left(\frac{\gamma_{21}}{2}\right) m_{\nu_2}m_{\nu_1}|U_{\mu 1}^2||U_{\mu 2}^{*2}|\sin\left(\alpha_1 + \frac{\gamma_{21}}{2}\right) \\ &\pm 4\sin\left(\frac{\gamma_{31}}{2}\right) m_{\nu_1}m_{\nu_3}|U_{\mu 1}^2||U_{\mu 3}^{*2}|\sin\left(\alpha_2 \pm \frac{\gamma_{31}}{2}\right) \end{aligned} \quad (8)$$

where $\gamma_{ij} = \frac{\Delta m_{ij} L(km)}{2E_\nu(GeV)}$, and the positive and negative signs refer to normal and inverted hierarchies, respectively. We can further neglect $\sin\gamma_{21} \approx 0$ such that $m_{\nu_1} \approx m_{\nu_2}$ at first order in $\mathcal{O}(\frac{\Delta m_{12}^2}{4E_\nu})$ for the fixed experimental parameters in MINOS, then Eq.(8) can be written as

$$0.001 \times E_\nu^2 \gtrsim |\langle m_{\mu\mu}\rangle|^2 + A(\alpha_1,\alpha_2) m_{\nu_2} m_{\nu_3} \gtrsim |\langle m_{\mu\mu}\rangle|^2 \quad (9)$$

where the coefficient $A(\alpha_1,\alpha_2)$ is a function of the Majorana phases. In Fig. (1) we show the regions in which $A(\alpha_1,\alpha_2) > 0$ is satisfied for both cases of normal or inverted hierarchies, and therefore we can find a stringent bound on the effective Majorana mass.

Thus, using $E_\nu \approx 2$ GeV, one gets the following bound on $|\langle m_{\mu\hat{\mu}}\rangle|$64 MeV

Over the excluded regions it is not possible to get a conservative bound on $|\langle m_{\mu\mu}\rangle|$ without making extra assumptions on the neutrino mass matrix, however Eq.(7) can be used to bound Majorana parameters (masses and phases) which appear in $|\langle m_{\mu\mu}\rangle|$. As an example, if $0 \leq 2(\alpha_2 - \alpha_3) \leq \pi - \gamma/2$ and assuming that the effective muon-neutrino Majorana mass is dominated by m_{ν_2} and m_{ν_3} (the two flavour limit case) , it is still possible to get the following conservative bound: $|\langle m_{\mu\hat{\mu}}\rangle|$109 MeV

A bound on the effective Majorana mass of the muon neutrino, which is independent of the mass hierarchies and Majorana phases, can be obtained using the fluxes of ν_μ and $\bar{\nu}_\mu$ measured with the near detector of the MINOS experiment [15]. Since the near detector is located L=1.04 km away from the target and for neutrino energies above 1 GeV, all oscillatory terms in Eq. (5) are equal to 1. Under the assumption that the excess of $\bar{\nu}_\mu$ events arises from $\nu_\mu \to \bar{\nu}_\mu$ transitions we get (note the muon-neutrino and muon-antineutrino total cross sections induced by charged currents are flat for neutrino energies above 2 GeV [2]):

$$\frac{|\langle m_{\mu\mu}\rangle|^2 \int \frac{dE_\nu}{E_\nu^2}}{\int dE_\nu} \leq \frac{\int(\Phi^{Obs}_{\bar{\nu}_\mu}(E_\nu) - \Phi^{MC}_{\bar{\nu}_\mu}(E_\nu))dE_\nu}{\int \Phi^{Obs}_{\nu_\mu}(E_\nu)dE_\nu} . \tag{10}$$

where $\Phi^{Obs(MC)}_{\nu,\bar{\nu}}$ denote the observed(expected) fluxes. Using the expected and measured integrated fluxes by the MINOS collaboration for the energy region $5 \leq E_\nu \leq 50$ GeV, we get the following bound: $|\langle m_{\mu\mu}\rangle| \leq 2.7$ GeV , which looks rather poor compared to the value reported above.

4. Conclusions

The production of leptons with same charges at the production and detection vertices of neutrinos will be a clear manifestation of $|\Delta L| = 2$ processes. One interesting result is that the time evolution probability of the whole process is not factorizable into production, oscillation and detection probabilities, as is the case in neutrino oscillations [11]. We find that, for very short times of propagation of neutrinos, the observation of $\mu^+\mu^+$ events would lead to a direct bound on the effective mass of muon Majorana neutrinos. In the case of long-baseline neutrino experiments, the CP rate asymmetry for production of $\mu^+\mu^+/\mu^-\mu^-$ events would lead to direct bounds on the difference of CP-violating Majorana phases. Finally, using the current bound on muon neutrino-antineutrino oscillations reported by the MINOS Collaboration we are able to set the bound $\langle m_{\mu\bar{\mu}}\rangle$64 MeV, which is several orders of magnitude below current bounds reported in the literature.

Future results from MINOS are expected from the analysis of twice the data set used to get the bound reported so far [10] and quoted in Eq. (12) above. Since current uncertainties in the observed and expected number of $\bar{\nu}_\mu$ events are dominated by statistical errors [10], we could expect only a slight improvement by a factor $1/\sqrt{2}$ on the effective Majorana mass of the muon neutrino. Neutrino factories may improve this bound by more than one order of magnitude.

As a consequence of these results, neutrino experiments aiming to measure neutrino-antineutrino oscillations with different short- and long-baseline setups can be useful to get direct and complementary constraints on the masses and phases of Majorana neutrinos.

References

[1] M. Doi, T. Kotani, H. Nishiura, K. Okuda and E. Takasugi, Phys. Lett. B **102**, 323 (1981); J. Schechter and J.W.F. Valle, Phys. Rev. D 22, 2227 (1980); S. Bilenky, J. Hosek, and S. Petcov, Phys. Lett. 94B, ¿ 495 (1980).
[2] C. Amsler *et al.* [Particle Data Group], Phys. Lett. B **667**, 1 (2008).
[3] A. de Gouvea, B. Kayser and R. N. Mohapatra, Phys. Rev. D **67**, 053004 (2003).
[4] R. Appel et al, Phys. Rev. Lett. 85, 2877 (2000).
[5] M. Flanz, W. Rodejohann, and K. Zuber, Phys. Lett. B473, 324 (2000); 480, 418(E) (2000); W. Rodejohann, J. Phys. G28, 1477 (2002).
[6] J. N. Bahcall and H. Primakoff, Phys. Rev. D **18**, 3463 (1978).
[7] J. Schechter and J. W. F. Valle, Phys. Rev. D **23**, 1666 (1981).
[8] P. Langacker and J. Wang, Phys. Rev. D **58**, 093004 (1998).
[9] H. Minakata, H. Nunokawa and S. J. Parke, Phys. Rev. D **66**, 093012 (2002); J. Burguet-Castell, M. B. Gavela, J. J. Gomez-Cadenas, P. Hernandez and O. Mena, Nucl. Phys. B **646**, 301 (2002); R. N. Mohapatra *et al.*, Rept. Prog. Phys. **70**, 1757 (2007).

[10] A. I. Himmel [MINOS Collaboration], AIP Conf. Proc. **1182**, 56 (2009); I. Danko [MINOS Collaboration], arXiv:0910.3439 [hep-ex].
[11] D. Delepine, V. G. Macias, S. Khalil and G. Lopez Castro, Phys. Rev. D **79**, 093003 (2009).
[12] M. Fukugita and T. Yanagida, *Physics of Neutrinos and Applications to Astrophysics*, Springer-Verlag, Berlin (2003).
[13] K. E. Arms [MINOS and NOνA Collaborations], Nucl. Phys. Proc. Suppl. **189**, 271 (2009).
[14] A. Blake [MINOS Collaboration], J. Phys. Conf. Ser. **120**, 052041 (2008).
[15] P. Adamson et al. [MINOS Collaboration], Phys. Rev. **D81**, 072002 (2010).

Low-energy Theorems for Strongly Interacting W's and Z's in $SU(2)_L \times SU(2)_R \times U(1)_{B-L}$

C. A. Vaquera-Araujo and M. Napsuciale

Departamento de Física, Universidad de Guanajuato, Campus Leon, C.P. 37150, León, Guanajuato, México.

E-mail: vaquera@fisica.ugto.mx

Abstract. In this work, low energy theorems for gauge boson (GB) longitudinal modes of the $SU(2)_L \times SU(2)_R \times U(1)_{B-L}$ electroweak gauge theory are derived. These theorems are valid for energies in the domain of applicability of the Equivalence Theorem, and hold for all left-right symmetric models assuming that there are no extra contributions from light scalars to the scattering amplitudes.

1. Introduction

Even though the standard model (SM) is extremely successful, the actual electroweak symmetry breaking (EWSB) mechanism is still unknown. Perhaps the most important goal of the LHC is to elucidate this phenomenon, and an essential piece of the puzzle is to identify the nature of GB scattering reactions $VV \to VV$. If we do observe these reactions, it means that the quanta of the symmetry breaking sector are of strong magnitude; otherwise, EWSB is accomplished by weakly coupled quanta below $\mathcal{O}(1)$ TeV.

L-R symmetric extensions of the SM based on $SU(2)_L \times SU(2)_R \times U(1)_{B-L}$ have several attractive features[1]. Plenty of L-R symmetric models have been proposed with different contents of Higgs multiplets, but there also exist the possibility of strong dynamics taking place in this kind of models. In this work we find LETs for the scattering of longitudinally polarized GB's in a general L-R context, without the contribution of any explicit Higgs. Our results hold for all $SU(2)_L \times SU(2)_R \times U(1)_{B-L}$ models provided they contain no other fundamental bosons with masses below the symmetry breaking scale M_{SB}.

2. Effective Lagrangian

The most general $\mathcal{O}(p^2)$ electroweak chiral Lagrangian density invariant under local $SU(2)_L \times SU(2)_R \times U(1)_{B-L}$ transformations [2] can be written as

$$\begin{aligned}
\mathcal{L}_{\rm eff} &= -\frac{f_L^2}{4}\operatorname{Tr}\left\{X_{L\mu}X_L^{\mu}\right\} - \frac{f_R^2}{4}\operatorname{Tr}\left\{X_{R\mu}X_R^{\mu}\right\} + \frac{\beta_L f_L^2}{8}\operatorname{Tr}\left\{\tau^3 X_L^{\mu}\right\}^2 + \frac{\beta_R f_R^2}{8}\operatorname{Tr}\left\{\tau^3 X_R^{\mu}\right\}^2 \\
&+ \frac{\kappa f_L f_R}{2}\operatorname{Tr}\left\{X_{L\mu}X_R^{\mu}\right\} + \frac{\beta f_L f_R}{4}\operatorname{Tr}\left\{\tau^3 X_L^{\mu}\right\}\operatorname{Tr}\left\{\tau^3 X_R^{\mu}\right\} \\
&- \frac{1}{2}\operatorname{Tr}\left\{A'_{L\mu\nu}A'^{\mu\nu}_L\right\} - \frac{1}{2}\operatorname{Tr}\left\{A'_{R\mu\nu}A'^{\mu\nu}_R\right\} - \frac{1}{2}\operatorname{Tr}\left\{B'_{\mu\nu}B'^{\mu\nu}\right\}. \qquad (1)
\end{aligned}$$

Here $A_A^{j\mu}$ and B^μ are the electroweak gauge fields, with indices $j = 1, 2, 3$, $A = L, R$, and $X_A^\mu \equiv U_A^\dagger D^\mu U_A$ with U_A as the unimodular matrices that parameterize the corresponding Nambu-Goldstone bosons (NGB). Using the definitions $A'^\mu_A = T^j A_A^{j\mu}$, $B' = T^3 B^\mu$, $U_A = \exp(i\tau^j \pi_A^j / f_A)$, and $T^j = \tau^j/2$, the covariant derivative of U_A is $D^\mu U_A = \partial^\mu U_A - i g_A A'^\mu_A U_A + i g_1 U_A B'^\mu$, where g_A and g_1 are the gauge couplings of $SU(2)_A$ and $U(1)_{B-L}$, respectively. On the other hand, f_L and f_R stand for the scales of EWSB and parity symmetry breaking (PSB), while κ is a L-R mixing parameter and the coefficients β_L, β_R and β are responsible for L-R mixing caused by custodial symmetry violating interactions at $\mathcal{O}(p^2)$.

Writing $\mathcal{L}_{\text{eff}}$ in unitary gauge $U_L = U_R = 1$ and defining $W_\mu^\pm = (A_\mu^1 \mp i A_\mu^2)/\sqrt{2}$, the physical basis is given by the following orthogonal transformations:

$$\begin{bmatrix} W_L^\pm \\ W_R^\pm \end{bmatrix} = \begin{bmatrix} c_\xi & s_\xi \\ -s_\xi & c_\xi \end{bmatrix} \begin{bmatrix} W_1^\pm \\ W_2^\pm \end{bmatrix} \tag{2}$$

$$\begin{bmatrix} A_L^3 \\ A_R^3 \\ B \end{bmatrix} = \begin{bmatrix} c_\theta c_\phi & c_\theta s_\phi & s_\theta \\ -s_\theta s_\omega c - c_\omega s_\phi & -s_\theta s_\omega s + c_\omega c_\phi & c_\theta s_\omega \\ -s_\theta c_\omega c_\phi + s_\omega s_\phi & -s_\theta c_\omega s_\phi - s_\omega c_\phi & c_\theta c_\omega \end{bmatrix} \begin{bmatrix} Z_1 \\ Z_2 \\ A \end{bmatrix} \tag{3}$$

with $g_L = g_1 c_\omega c_\theta / s_\theta$, $g_R = g_1 c_\omega / s_\omega$ and θ as the Weinberg angle for L-R symmetric models (notice that $\cos\chi = c_\chi$, $\sin\chi = s_\chi$). If one assumes L-R manifest symmetry ($g_L = g_R$) then $\sin\omega = \tan\theta$. In an arbitrary R_ξ gauge, GB-NGB mixing terms are removed using the following gauge fixing Lagrangian:

$$\mathcal{L}_{\text{fix}} = -\frac{1}{2}\left[F_\gamma^2 + \sum_{j=1}^{2}\left(F_{Z_j}^2 + 2F_{W_j^+}F_{W_j^-}\right)\right],$$

$$F_{Z_j} = \frac{\partial^\mu Z_{j\mu} + M_{Z_j}\xi^{Z_j} z_j}{\sqrt{\xi^{Z_j}}}, \quad F_{W_j^\pm} = \frac{\partial^\mu W_{j\mu}^\pm + M_{W_j}\xi^{W_j} w_j^\pm}{\sqrt{\xi^{W_j}}}.$$

Where $w_j^\pm$, z_j are the linear combinations of NGB's that become the longitudinal degrees of freedom of each physical GB.

3. Low Energy Theorems

Although NGB's are not part of the physical spectrum, one can compute S-matrix elements with them [3, 4]. Fixing the gauge with $\mathcal{L}_{\text{fix}}$ implies that the one-particle state $w_1^+ |0\rangle$ has some overlap with the one-particle state $\partial_\mu W_1^{+\mu} |0\rangle$. Equivalence theorem (ET) [5] codifies the fact that two overlapping fields can be traded for each other in S-matrix calculations at high energy $E \gg M$:

$$\mathcal{M}\left[W_1(p_1), \ldots, W_1(p_n)\right]_U = \mathcal{M}\left[w_1(p_1), \ldots, w_1(p_n)\right]_R + \mathcal{O}\left(M/E\right). \tag{4}$$

Low-energy theorems (LET's) can be obtained calculating the amplitudes of NG boson scattering which follow from (1) assuming the global symmetry breaking pattern $SU(2) \times SU(2) \times U(1)/U(1)$. Written in terms of $\pi_A^\pm = \left(\pi_A^1 \mp i\pi_A^2\right)/\sqrt{2}$, the kinetic part of Lagrangian (1) must be diagonalized and properly normalized to satisfy the definitions of $w_j^\pm$ and z_j. Now, parameters f_A, β_A, κ, and β can be eliminated in favor of the masses and mixing angles of the GB's. Definitive expressions for two-body scattering of the lighter NGB's in the energy range where both $\mathcal{L}_{\rm fix}$ and ET are valid, $M^2 \ll s \ll \Lambda_{SB}^2$, are:

$$\begin{aligned}\mathcal{M}(w_1^+w_1^- \to w_1^+w_1^-) &= \frac{e^2u}{4M_{W_1}^4 O_{13}^2 O_{23}^2}\{-4M_{W_1}^2(O_{23}^2c_\xi^4 + O_{13}^2s_\xi^4) \\ &+3M_{Z_1}^2(O_{11}O_{23}c_\xi^2 + O_{21}O_{13}s_\xi^2)^2 \\ &+3M_{Z_2}^2(O_{12}O_{23}c_\xi^2 + O_{22}O_{13}s_\xi^2)^2\} \end{aligned} \tag{5}$$

$$\begin{aligned}\mathcal{M}(w_1^+w_1^- \to z_1z_1) &= \frac{e^2s}{4M_{W_1}^4 M_{W_2}^2 M_{Z_1}^2 O_{13}^2 O_{23}^2}\Big\{M_{W_2}^2 M_{Z_1}^4\left(O_{11}O_{23}c_\xi^2 + O_{21}O_{13}s_\xi^2\right)^2 \\ &+s_\xi^2c_\xi^2 M_{W_1}^2\left[4M_{W_2}^2\left(M_{W_1}^2 - M_{W_2}^2\right) + (M_{Z_1}^2 + M_{W_2}^2 - M_{W_1}^2)^2\right]O_{32}^2\Big\} \end{aligned} \tag{6}$$

$$\mathcal{M}(z_1z_1 \to z_1z_1) = 0. \tag{7}$$

where O_{ij} are the elements of the neutral GB mixing matrix (3).

4. Conclusions

We have found LETs for the longitudinal modes of the lighter $SU(2)_L \times SU(2)_R \times U(1)_{B-L}$ GB's. These theorems are valid for energies in the domain $M^2 \ll s \ll \Lambda_{SB}^2$ and hold for all L-R symmetric models assuming that there are no extra contributions from light zero-spin particles to the scattering amplitudes.

Acknowledgments

Support by CONACyT grant No. 50471-F is gratefully acknowledged.

References

[1] R. N. Mohapatra, Unification and Supersymmetry. The Frontiers of Quark-lepton Physics, *Berlin, Germany: Springer (1986) 309 P. (Contemporary Physics)*
[2] Y. Zhang, S. Z. Wang, F. J. Ge and Q. Wang, Phys. Lett. B **653**, 259 (2007) [arXiv:0704.2172 [hep-ph]].
[3] M. S. Chanowitz, M. Golden and H. Georgi, Phys. Rev. Lett. **57**, 2344 (1986). M. S. Chanowitz, M. Golden and H. Georgi, Phys. Rev. D **36**, 1490 (1987).
[4] G. J. Gounaris, R. Kogerler and H. Neufeld, Phys. Rev. D **34**, 3257 (1986). J. Bagger and C. Schmidt, Phys. Rev. D **41**, 264 (1990).
[5] A. Dobado and J. R. Pelaez, Phys. Lett. B **329**, 469 (1994) [Addendum-ibid. B **335**, 554 (1994)] [arXiv:hep-ph/9404239]. A. Dobado and J. R. Pelaez, Nucl. Phys. B **425**, 110 (1994) [Erratum-ibid. B **434**, 475 (1995)] [arXiv:hep-ph/9401202]. A. Dobado, J. R. Pelaez and M. T. Urdiales, Phys. Rev. D **56**, 7133 (1997) [arXiv:hep-ph/9702206].

Direct dark matter detection: The diurnal variation in directional experiments

J.D. Vergados[1] **and Ch.C. Moustakidis**[2]

[1]Theoretical Physics Division, University of Ioannina, Ioannina, Gr 451 10, Greece
[2]Department of Theoretical Physics, Aristotle University of Thessaloniki, 54124 Thessaloniki, Greece

E-mail: vergados@uoi.gr; moustaki@auth.gr

Abstract. We present some theoretical elements relevant to the direct dark matter detection experiments, paying particular attention to directional experiments, i.e. experiments in which, not only the energy but the direction of the recoiling nucleus is observed. Since the direction of observation is fixed with respect the earth, while the Earth is rotating around its axis, in a directional experiment the angle between the direction of observation and the Sun's direction of motion will change during the day. So, since the event rates sensitively depend on this angle, the observed signal in such experiments will exhibit very interesting and characteristic periodic diurnal variation.

1. Introduction

The combined MAXIMA-1 [1], BOOMERANG [2], DASI [3] and COBE/DMR Cosmic Microwave Background (CMB) observations [4] imply that the Universe is flat [5] and that most of the matter in the Universe is Dark [6], i.e. exotic. These results have been confirmed and improved by the recent WMAP data [7]. Combining the the data of these quite precise experiments one finds:

$$\Omega_b = 0.0456 \pm 0.0015, \quad \Omega_{CDM} = 0.228 \pm 0.013, \quad \Omega_\Lambda = 0.726 \pm 0.015.$$

Since any "invisible" non exotic component cannot possibly exceed 40% of the above Ω_{CDM} [8], exotic (non baryonic) matter is required and there is room for cold dark matter candidates or WIMPs (Weakly Interacting Massive Particles).

Since the WIMP's are expected to be very massive and extremely non relativistic, they are not likely to excite the nucleus. So they can be directly detected mainly via the recoiling of a nucleus (A,Z) in elastic scattering. In the standard nuclear recoil experiments, first proposed more than 30 years ago [9], one has to face the problem that the reaction of interest does not have a characteristic feature to distinguish it from the background. So for the expected low counting rates the background is a formidable problem. Some special features of the WIMP-nuclear interaction can be exploited to reduce the background problems. Such are: i) the modulation effect: this yields a periodic signal due to the motion of the earth around the sun. ii) backward-forward asymmetry expected in directional experiments, i.e. experiments in which the direction of the recoiling nucleus is also observed. iii) transitions to excited states: in this case one need not measure nuclear recoils, but the de-excitation γ rays. (iv) detection of electrons produced during the WIMP-nucleus collision. two very interesting patterns i) its magnitude in certain

In our previous work [10, 11, 12] we have found that the observed directional rate is characterized by a strong dependence on the angle between the direction of nuclear recoil and the direction of the sun's motion, with respect to the center of the galaxy. The apparatus will, of course, be oriented in a direction defined in the local frame, i.e. the line of observation will point to a point in the sky specified, in the equatorial system, by right ascension $\tilde{\alpha}$ and inclination $\tilde{\delta}$. Thus due to the rotation of the earth in a directional experiment the angle between the direction of observation and the Sun's direction of motion will change during the day. So, since the event rates sensitively depend on this angle, as we have recently suggested[13] one may be able to observe a diurnal variation, even if one does not change the orientation of the apparatus in the laboratory. In the present paper we will explore this novel feature of directional experiments, namely the characteristic diurnal variation, i.e. the variation of the data with a period of 24 hours. This variation is independent of the Earth's rotational velocity. Its amplitude depends, in addition to some of the usual parameters specifying the standard WIMP event rate, on the inclination of observation, i.e. the angle between the direction of observation and the axis of the Earth's rotation. Those features cannot be masked by any known background.

2. Standard (non Directional) Rates

The event rate for a given WIMP velocity υ is given by

$$\frac{\mathrm{d}R}{\mathrm{d}u}=\frac{\rho_{\chi}}{m_{\chi}}\frac{m_t}{Am_p}\sqrt{<\upsilon^2>}A^2\sigma_n\left(\frac{\mu_r}{\mu_p}\right)^2\frac{\mathrm{d}t(u,\upsilon)}{\mathrm{d}u}, \tag{1}$$

where m_t/Am_p is the number of nuclei in a target of mass m_t and also

$$\frac{\mathrm{d}t(u,\upsilon)}{\mathrm{d}u}=\frac{\upsilon}{\sqrt{<\upsilon^2>}}F^2(u)\frac{1}{2(\mu_r b\upsilon)^2}. \tag{2}$$

We must now fold $\mathrm{d}t(u,\upsilon)/\mathrm{d}u$ with the velocity distribution $f(y,\xi)$ in the local frame. We will assume in this work that the velocity distribution is Maxwell-Boltzmann in the galactic frame, namely

$$f(\mathbf{y}')=\frac{1}{\pi\sqrt{\pi}}e^{-(\mathbf{y}')^2},\qquad \mathbf{y}'=\frac{\upsilon'}{\upsilon_0}, \tag{3}$$

where υ' is the WIMP velocity in the galactic frame and υ_0 is the sun's velocity with respect to the center of the galaxy. To simplify the notation we have chosen to exhibit the relative variable $\mathbf{y}'$, but we have suitably normalized the expressions for the event rates. Obviously one has to transform this distribution to the local frame taking into account the motion of the Sun and Earth:

$$\mathbf{y}'=\mathbf{y}+\hat{z}+\delta\left(\sin\alpha\hat{x}-\cos\alpha\cos\gamma\hat{y}+\cos\alpha\sin\gamma\hat{z}\right), \tag{4}$$

where $\delta=\upsilon_1/\upsilon_0=0.136$ with υ_1 the velocity of the earth around the sun, α the phase of the earth (α=0 on June 3rd) and $\gamma\approx\pi/6$.

In the case where $\delta=0$, we get for the differential (with respect to the energy transfer) rate

$$\frac{\mathrm{d}R}{\mathrm{d}u}=\frac{\rho_{\chi}}{m_{\chi}}\frac{m_t}{Am_p}\sigma_n\left(\frac{\mu_r}{\mu_p}\right)^2\sqrt{<\upsilon^2>}A^2\frac{dt}{du},\qquad \frac{\mathrm{d}t}{\mathrm{d}u}=\sqrt{\frac{2}{3}}a^2F^2(u)\Psi_0(a\sqrt{u}), \tag{5}$$

$$\Psi_0(a\sqrt{u})=\int_{a\sqrt{u}}^{y_{esc}}y\mathrm{d}y\;2\pi\int_{-1}^{1}\mathrm{d}\xi f(y,\xi), \tag{6}$$

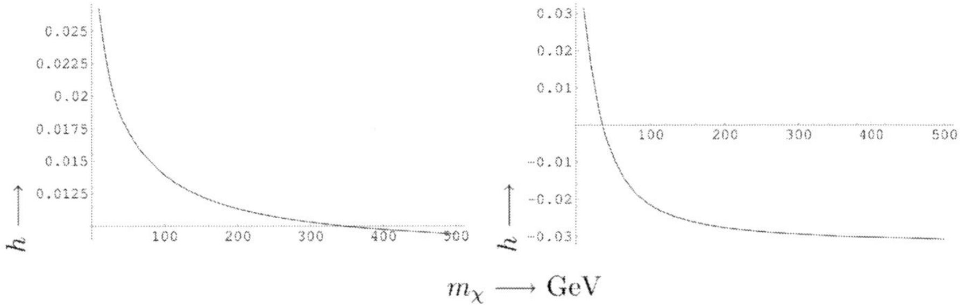

Figure 1. The quantity h, entering in the coherent mode discussed in this work, is shown as a function of the wimp mass in GeV for $Q_{min} = 0$ for a light system, ^{32}S (left) and ^{127}I or ^{131}Xe (right).

where

$$a = (\sqrt{2}\mu_r b v_0)^{-1}. \tag{7}$$

We note that the function $\Psi_0(x)$ depends only on the velocity distribution and is independent of nuclear physics.

An additional suppression as the energy transfer increases comes, of course, from the nuclear form factor $F(u)$. Note that the nuclear dependence of the differential rate comes not only from the form factor, but via the parameter a as well. Integrating the differential event rate from $u_{min} = E_{th}/Q_0$, which depends on the energy threshold, to $u_{max} = y_{esc}^2/a^2$ (where $y_{esc} = 2.84$ since $v_{esc} = 2.84 v_0$) we obtain the total rate

$$R = \frac{\rho_\chi}{m_\chi}\frac{m_t}{Am_p}\sigma_n\left(\frac{\mu_r}{\mu_p}\right)^2\sqrt{<v^2>}A^2 t, \qquad t = \int_{u_{min}}^{u_{max}} \mathrm{d}u \frac{\mathrm{d}t}{\mathrm{d}u}. \tag{8}$$

In the case where $\delta \neq 0$ we proceed as above and finally get for the differential rate an expression similar to (5) where now one has to replace:

$$\Psi_0(x) \longrightarrow \Psi_0(x) + \Psi_1(x,\gamma,\delta)\cos\alpha, \tag{9}$$

Additionally, one has to replace:

$$\frac{\mathrm{d}t}{\mathrm{d}u} \longrightarrow \frac{\mathrm{d}r}{\mathrm{d}u} = \frac{\mathrm{d}t}{\mathrm{d}u} + \frac{\mathrm{d}\tilde{h}}{\mathrm{d}u}\cos\alpha, \tag{10}$$

$$t \longrightarrow t(1 + h\cos\alpha), \quad h = \frac{1}{t}\int_{u_{min}}^{u_{max}} \mathrm{d}u \frac{\mathrm{d}\tilde{h}}{\mathrm{d}u}. \tag{11}$$

Thus we finally get

$$R = \frac{\rho_\chi}{m_\chi}\frac{m_t}{Am_p}\sigma_n\left(\frac{\mu_r}{\mu_p}\right)^2\sqrt{<v^2>}A^2\, t\,(1 + h\cos\alpha)\,. \tag{12}$$

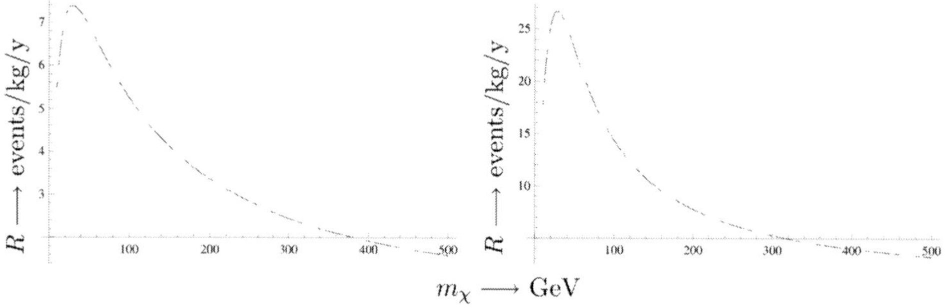

Figure 2. The coherent total event rate, as a function of the WIMP mass in GeV, is exhibited assuming a nucleon cross section $\sigma_n = 10^{-43}$ cm^2=10^{-7}pb for $Q_{min} = 0$ for a light system, ^{32}S (left) and a heavy system, ^{131}Xe (right).

3. Directional Rates

In this instance the experiments will attempt to measure not only the energy, but the direction of the recoiling nucleus as well [12], [16]. Let us indicate the direction of observation by:

$$\hat{e} = (e_x, e_y, e_z) = (\sin\Theta\cos\Phi, \sin\Theta\sin\Phi, \cos\Theta). \tag{13}$$

Now (2) must be notified since the variables u and υ are no longer independent. Furthermore no integration over the azimuthal angle specifying the direction of the outgoing nucleus is performed. Eqs. (1) and (2) now become

$$\left(\frac{dR}{du}\right)_{dir} = \frac{\rho_\chi}{m_\chi}\frac{m_t}{Am_p}\sqrt{<\upsilon^2>}A^2\sigma_n\left(\frac{\mu_r}{\mu_p}\right)^2\frac{1}{2\pi}\left(\frac{dt(u,\upsilon)}{du}\right)_{dir}, \tag{14}$$

$$\left(\frac{dt(u,\upsilon)}{du}\right)_{dir} = \frac{\upsilon}{\sqrt{<\upsilon^2>}}F^2(u)\frac{1}{2(\mu_r b\upsilon)^2}\delta\left(\hat{\upsilon}.\hat{e} - \frac{a\sqrt{u}}{\upsilon/\upsilon_0}\right). \tag{15}$$

The factor $1/(2\pi)$ enters since we are going to make use of the same nuclear cross section as in the non directional case.
We find it convenient to use a more convenient system (X, Y, Z), in which the polar axes $\hat{Z}$ is in the direction of the recoiling nucleus. The needed transformation is and thus the three components of the velocity are given by:

$$\begin{pmatrix} \hat{\upsilon}_X \\ \hat{\upsilon}_Y \\ \hat{\upsilon}_Z \end{pmatrix} = \begin{pmatrix} \cos(\Theta)\cos(\Phi) & \cos(\Theta)\sin(\Phi) & -\sin(\Theta) \\ -\sin(\Phi) & \cos(\Phi) & 0 \\ \cos(\Phi)\sin(\Theta) & \sin(\Theta)\sin(\Phi) & \cos(\Theta) \end{pmatrix}\begin{pmatrix} \delta\sin(\alpha) \\ -\delta\cos(\alpha)\cos(\gamma) \\ \delta\cos(\alpha)\sin(\gamma)+1 \end{pmatrix}. \tag{16}$$

In the directional case without modulation ($\delta = 0$) the corresponding function $\Psi_0^{dir}(a\sqrt{u}, \Theta)$ can only be obtained numerically. Once this function is known the computation of the event rate proceeds in a fashion analogous to the standard (non directional) case.

The total event rate is obtained after integrating the above expression over the energy transfer. It can be cast in the form:

$$R_{dir} = \frac{\rho_\chi}{m_\chi}\frac{m_t}{Am_p}\sigma_n\left(\frac{\mu_r}{\mu_p}\right)^2\sqrt{<\upsilon^2>}A^2\frac{1}{2\pi}t_{dir}(\Theta,\Phi)\left(1 + h_c(\Theta,\Phi)\cos\alpha + h_s(\Theta,\Phi)\sin\alpha\right). \tag{17}$$

We found it convenient to use a new relative parameter κ[1] as well as the parameters h_m and α_0 given by:

$$\kappa(\Theta,\Phi)=\frac{t_{dir}}{t},\qquad h_m(\Theta,\Phi)\cos(\alpha+\alpha_0)=h_c(\Theta,\Phi)\cos\alpha+h_s(\Theta,\Phi)\sin\alpha. \tag{18}$$

These parameters depend, of course, on the reduced mass μ_r, the WIMP velocity distribution and, to some extent, on the nuclear physics via the nuclear form factor. The parameter κ gives the retardation factor of the directional rate, over and above the factor of $1/(2\pi)$, compared to the standard rate. Since, however, the unmodulated amplitude is independent of Φ one can integrate over Φ so that the suppression of $1/(2\pi)$ drops out for this term. Because of the existence of both $\cos\alpha$ and $\sin\alpha$ terms the time dependence will be of the form $\cos(\alpha+\alpha_0)$, with the phase α_0 being direction dependent. Thus the time of the maximum and minimum will depend on the direction. In other words the seasonal dependence will depend on the direction of observation. So it cannot be masked by irrelevant seasonal effects.

Before concluding this section we like to consider the case connected with the partly directional experiments, i.e. experiments which can determine the line along which the nucleus is recoiling, but not the sense of direction on it. The results in this case can be obtained by summing up the events in both directions. i.e. those specified by (Θ,Φ) as well as $(\pi-\Theta,\Phi+\pi)$. A given line of observation is now specified by Θ, Φ in the range:

$$0\le\Theta\le\pi/2,\qquad 0\le\Phi\le\pi.$$

The parameter h entering the non directional case is shown in Fig. 1. Using the above parameters t and employing Eq. (12) we obtain the time independent total event rate. The obtained results are shown in Fig. 2. The total modulation, since it is defined relative to the time independent part, is still given by Fig. 1.

4. Diurnal Variation

Up to now we have considered the event rate in a directional experiment in fixed direction with respect to the galaxy in the galactic system discussed above. The apparatus, of course, will be oriented in a direction specified in the local frame, e.g. by a point in the sky specified, in the equatorial system, by right ascension $\tilde{\alpha}$ and inclination $\tilde{\delta}$[2]. This will lead to a diurnal variation of the event rate [13].

The galactic frame, in the so called J2000 system, is defined by the galactic pole with ascension $\tilde{\alpha}_1=12^h\ 51^m\ 26.282^s$ and inclination $\tilde{\delta}_1=+27^0\ 7'\ 42.01''$ and the galactic center at $\tilde{\alpha}_2=17^h\ 45^m\ 37.224^s$, $\tilde{\delta}_2=-(28^0\ 56'\ 10.23'')$. Thus the galactic unit vector $\hat{y}$, specified by $(\tilde{\alpha}_1,\tilde{\delta}_1)$, and the unit vector $\hat{s}$, specified by $(\tilde{\alpha}_2,\tilde{\delta}_2)$, can be expressed in terms of the celestial unit vectors $\hat{i}$ (beginning of measuring the right ascension), $\hat{k}$ (the axis of the Earth's rotation) and $\hat{j}=\hat{k}\times\hat{i}$. One finds

$$\begin{aligned}\hat{y}&=&-0.868\hat{i}-0.198\hat{j}+0.456\hat{k}\ \text{(galactic axis)}\ ,\\ \hat{x}&=&-\hat{s}=0.055\hat{i}+0.873\hat{j}+0.483\hat{k}\ \text{(radially out towards the sun)}\ ,\\ \hat{z}&=&\hat{x}\times\hat{y}=0.494\hat{i}-0.445\hat{j}+0.747\hat{k}\ \text{(the sun's direction of motion)}.\end{aligned} \tag{19}$$

[1] To avoid or minimize, e.g., the dependence on the parameters of the particle model and, in particular, the unknown WIMP-nucleon cross section. Furthermore the directional experiments can also obtain the rate in all directions. Thus κ maybe of experimental interest.

[2] We have chosen to adopt the notation $\tilde{\alpha}$ and $\tilde{\delta}$ instead of the standard notation α and δ employed by the astronomers to avoid possible confusion stemming from the fact that α has already been used to designate the phase of the Earth and δ has been used for the ratio of the rotational velocity of the Earth around the Sun by the velocity of the sun around the center of the galaxy

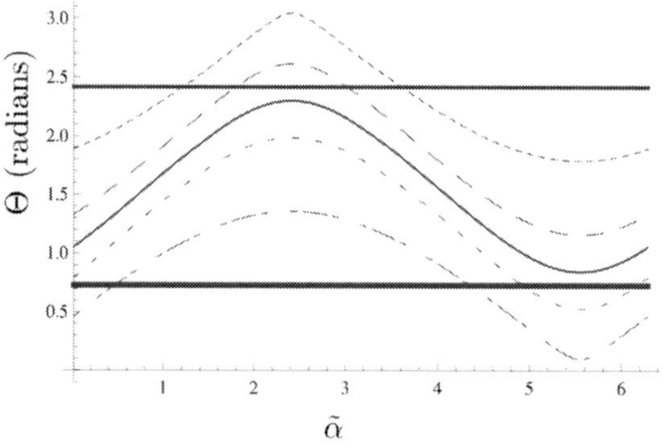

Figure 3. Due to the diurnal motion of the Earth different angles Θ in galactic coordinates are sampled as the earth rotates. The angle Θ scanned by the direction of observation is shown for various inclinations $\tilde{\delta}$. The intermediate thickness, the short dash, the long dash, the fine line, the long-short dash, the short-long-short dash and the thick line correspond to inclination $\tilde{\delta} = -\pi/2, -3\pi/10, -\pi/10, 0, \pi/10, 3\pi/10$ and $\pi/2$ respectively. We see that, for negative inclinations, the angle Θ can take values near π, i.e. opposite to the direction of the sun's velocity, where the rate attains its maximum.

Note in our system the x-axis is opposite to the s-axis used by the astronomers. Thus a vector oriented by $(\tilde{\alpha}, \tilde{\delta})$ in the laboratory is given in the galactic frame by a unit vector with components:

$$\begin{pmatrix} y \\ x \\ z \end{pmatrix} = \begin{pmatrix} -0.868\cos\tilde{\alpha}\cos\tilde{\delta} - 0.198\sin\tilde{\alpha}\cos\tilde{\delta} + 0.456\sin\tilde{\delta} \\ 0.055\cos\tilde{\alpha}\cos\tilde{\delta} + 0.873\sin\tilde{\alpha}\cos\tilde{\delta} + 0.4831\sin\tilde{\delta} \\ 0.494\cos\tilde{\alpha}\cos\tilde{\delta} - 0.445\sin\tilde{\alpha}\cos\tilde{\delta} + 0.747\sin\tilde{\delta} \end{pmatrix}, \quad (20)$$

where $\tilde{\alpha}_0 = 282.25^0$ is the right ascension of the equinox, $\gamma \approx 33^0$ was given above and $\theta_P = 62.6^0$ is the angle the Earth's north pole forms with the axis of the galaxy.
Due to the Earth's rotation the unit vector (x, y, z), with a suitable choice of the initial time, $\tilde{\alpha} - \tilde{\alpha}_0 = 2\pi(t/T)$, is changing as a function of time

The angle Θ scanned by the direction of observation is shown, for various inclinations $\tilde{\delta}$, in Fig. 3. We see that for negative inclinations, the angle Θ can take values near π, i.e. opposite to the direction of the sun's velocity, where the rate attains its maximum.

The equipment scans different parts of the galactic sky, i.e. observes different angles Θ. So the rate will change with time depending on whether the sense of the recoiling nucleus can be determined along the line of recoil. The results depend, of course, on the WIMP mass and the target employed. We will consider a light (the time dependence of κ is exhibited in Fig. 4) and an intermediate-heavy target (the time dependence of κ is exhibited in Fig. 5).

5. Discussion

In directional experiments one measures not only the recoil energy but the direction of the nuclear recoil as well. Some of the requirements that should be met by such detectors have recently been discussed [15, 14]. To fully exploit the advantages of such detectors one should be able to distinguish between recoils with momenta **p** and −**p** (sense of direction), which now appears to be feasible. Some of the predicted interesting features of directional event rates persist, even if it turns out that the sense of motion of recoils along their line of motion cannot be measured. Such experiments given a sufficient number of events provide an excellent signature to discriminate

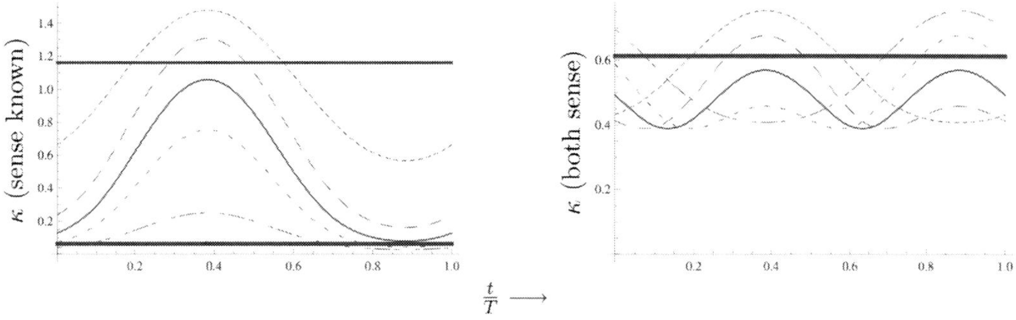

Figure 4. The time dependence(in units of the Earth's rotation period) of the parameter κ for various inclinations $\tilde{\delta}$ in the case of the target CS_2. The intermediate thickness, the short dash, the long dash, the fine line, the long-short dash, the short-long-short dash and the thick line correspond to inclination $\tilde{\delta} = -\pi/2, -3\pi/10, -\pi/10, 0, \pi/10, 3\pi/10$ and $\pi/2$ respectively. Due to the diurnal motion of the Earth different angles Θ are sampled as the earth rotates.

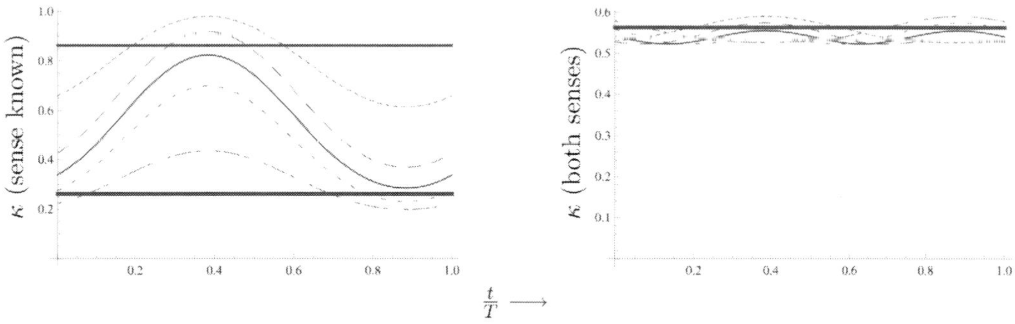

Figure 5. The same as in Fig. 4 in the case of the Iodine target. The intermediate thickness, the short dash, the long dash, the fine line, the long-short dash, the short-long-short dash and the thick line correspond to inclination $\tilde{\delta} = -\pi/2, -3\pi/10, -\pi/10, 0, \pi/10, 3\pi/10$ and $\pi/2$ respectively.

against background. One, of course, gets a smaller rate by observing in a given direction. In the most favored direction, opposite to the sun's direction of motion, the event rate is $\approx \frac{\kappa}{2\pi}$ down from that of the standard non directional experiments, if a specific angle Φ is chosen. Since, however, κ is independent of the angle Φ, one can integrate over all azimuthal angles and thus the retardation is only of order κ ($\kappa \approx 1$ in the most favored direction). Finally we have seen that in directional experiments the relative event rate for detecting WIMPs within our galaxy, as given by the parameter κ, will show a periodic diurnal variation due to the rotation of the Earth. The parameter κ is essentially independent of any particle model parameters other than the WIMP mass. It does depend, however, on the assumed velocity distribution and to the nuclear form factor. The time variation is larger in the case of light WIMP and/or light target. So from this perspective the lighter target is preferred. The time variation arising after the inclusion of the modulation parameters (h_m, α) or (h_c, h_s) is expected to be even more complicated, since these parameters depend on both Θ and Φ. One expects a diurnal variation on top of the annual variation characteristic of the usual modulation effect entering both directional and non directional experiments. Such effects will be discussed elsewhere.

Acknowledgments
The work of one of the authors (JDV) was partially supported by the grant MRTN-CT-2006-035863 (UniverseNet). He is indebted to Professors N. Spooner and J. Jochum for useful discussions on experimental aspects of this work.

References

[1] Hanary S *et al* 2000 *Astrophys. J.* **545** L5; Wu J H P *et al* 2001 *Phys. Rev. Lett.* **87** 251303; Santos M G 2002 *et al Phys. Rev. Lett.* **88** 241302.
[2] Mauskopf P D *et al* 2002 Astrophys. J. **536** L59; Mosi S *et al* 2002 Prog. Nuc.Part. Phys. **48** 243 ; Ruhl S B *al* (preprint astro-ph/0212229) and references therein.
[3] Halverson N W *et al* 2002 Astrophys. J. **568** 38; Sievers L S *et al* (preprint astro-ph/0205287) and references therein.
[4] Smoot G F and *et al (COBE Collaboration)* 1992 Astrophys. Journ. **396** L1.
[5] Jaffe A H Phys. Rev. Lett. 2001 **86** 3475.
[6] Spergel D N and *et al* 2003 Astrophys. J. Suppl. **148** 175.
[7] D. Spergel *et al* 2007 Astrophys. Journal. **170** 377; (Preprint arXiv:astro-ph/0603449v2).
[8] Bennett D P and *et al* Phys. Rev. Lett. 1995 **74** 2867.
[9] Goodman M W and Witten E 1985 Phys. Rev. **D 31** 3059.
[10] Vergados J D 2003 Phys. Rev. **D 57** 103003; (Preprint hep-ph/0303231).
[11] Vergados J D 2004 J. Phys. G: **30** 1127.
[12] Vergados J and Faessler A 2007 Phys. Rev. **D 75** 055007.
[13] Ahlen S 2010 The case for a directional dark matter detector and the status of current experimental efforts, cygnus2009Whitepaper, Edited by J. B. R. Battat, IJMPA 25 1; (Preprint arXiv:0911.0323 (astro-ph.CO)).
[14] Copi C J Krauss L M Simmons-Duffin D. and Stroiney S R 2007 Phys. Rev. **D 75** 023514; (Preprint arXiv:astro-ph/0508649).
[15] Green A M and Morgan B 2007 Astropart. Phys. **27** 142; (Preprint arXiv:0707.1488 (astro-ph)).
[16] Vergados J 2004 J. Phys. G: **30** 1127; (Preprint arXiv:hep-ph/0406134).

Journal of Physics: Conference Series **315** (2011) 012029 doi:10.1088/1742-6596/315/1/012029

The Photon Dispersion as an Indicator for New Physics ?

Wolfgang Bietenholz

Instituto de Ciencias Nucleares, Universidad Nacional Autónoma de México, A.P. 70-543, C.P. 04510 Distrito Federal, Mexico

Abstract. We first comment on the search for a deviation from the linear photon dispersion relation, in particular based on cosmic photons from Gamma Ray Bursts. Then we consider the non-commutative space as a theoretical concept that could lead to such a deviation, which would be a manifestation of Lorentz Invariance Violation. In particular we review a numerical study of pure $U(1)$ gauge theory in a 4d non-commutative space. Starting from a finite lattice, we explore the phase diagram and the extrapolation to the continuum and infinite volume. These simultaneous limits — taken at fixed non-commutativity — lead to a phase of broken Poincaré symmetry, where the photon appears to be IR stable, despite a negative IR divergence to one loop.

1. Lorentz Invariance

Lorentz Invariance is a central concept of relativity: it is a global symmetry in Special Relativity, and a local symmetry in General Relativity. Here we stay within the framework of particle physics as described by quantum field theory, so we refer to Special Relativity. Then this symmetry implies that some field Φ (scalar, spinor, vector or tensor) transforms globally in some representation D of the Lorentz group $SO(1,3)$,

$$\Phi(x) \to U(\Lambda)^{\dagger}\Phi(x)U(\Lambda) = D(\Lambda)\Phi(\Lambda^{-1}x) \qquad (\Lambda \in SO(1,3)\ ,\ U \text{ unitary})\ . \tag{1}$$

A number of high precision tests of Lorentz Invariance involve cosmic rays, see *e.g.* Refs. [1] for recent reviews. They provide perhaps the best chance to probe parameters not that far below the Planck scale (in specific cases even exceeding it). Cosmic rays attain by far the highest particle energies in the Universe (up to $\approx 10^{20}$ eV). In addition, the fact that they travel over tremendous distances may be conclusive for high precision properties even at moderate or low energies, because tiny effects could be accumulated over a very long trajectory. In view of the latter scenario, we discuss here the *photon dispersion relation*, as a direct kinematic test of Lorentz Invariance.

2. Cosmic γ-rays

Gamma Ray Bursts (GBRs) are emitted in powerful energy eruptions over short periods, typically a few seconds or minutes. Temporarily this causes the brightest spots in the sky. They were discovered from satellites since the 70s, but nowadays they can also be observed from ground. Their sources are small, and there are speculations that they could be generated by the merger of neutron stars, or of black holes. Here we are pragmatic about their origin: in any case GRBs exist, and we want to see what we can learn from the long journey of their photons through the Universe.

The distance to the source is usually well evaluated from the redshift (if we assume the Hubble parameter H_0 to be known). In particular, in the year 2005 a spectacular GRB was observed [2]: it occurred at a distance of 3.9 Gpc, *i.e.* it was emitted when the Universe was only $\approx 8 \cdot 10^8$ years old. Typically the photon energies are in the range of $10^4 \dots 10^8$ eV. Thus it is an obvious idea to use GRBs to probe if the speed of photons is really energy independent [3].

A systematic study of this question was performed in Ref. [4]. It probed the effective dispersion ansatz

$$\vec{p}^{\,2} = E^2\Big(1 + \frac{E}{M}\Big) , \quad v(E) = 1 - \frac{E}{M} + \mathcal{O}\Big(\frac{E^2}{M^2}\Big) , \tag{2}$$

where M is a very heavy mass parameter, which might emerge *somehow* from some kind of "quantum gravity foam". Here we set the speed of light at photon energy $E \ll |M|$ to $c = 1$. If a finite parameter M exists, deviations of the group velocity v from that speed should be manifest at very high energy E, *or* after a very long path. Ref. [4] analyzed data of 35 GRBs, detected by three satellites. In fact, photons of higher energy tend to arrive slightly later. This could be explained by a relative delay at the source, $\Delta_{\rm source}$, which would then be amplified by the redshift z to the observed delay $\Delta_{\rm obs} = \Delta_{\rm source}(1+z)$. So far this is standard physics. If one includes a finite parameter M, this formula is modified to

$$\frac{\Delta_{\rm obs}}{1+z} = \Delta_{\rm source} + a_{\rm LIV} K(z) , \quad a_{\rm LIV} = \frac{\Delta E}{M H_0} , \tag{3}$$

for photon energies that differ by ΔE. Here $a_{\rm LIV}$ is the Lorentz Invariance Violation (LIV) parameter, and the function $K(z)$ is specified in Ref. [4]. The question is now whether or not the data for $\Delta_{\rm obs}/(1+z)$ are compatible with a constant in z (or $K(z)$). The authors of Ref. [4] enhanced the errors by hand until the data became compatible with a linear fit. The resulting fit is in fact almost constant, and it suggests

$$|M| > 1.4 \cdot 10^{25} \text{ eV} \approx 0.001 \ M_{\rm Planck} \quad \text{(with 95 \% C.L.)} . \tag{4}$$

Some studies of single GRBs or blazar flares even conclude $|M| \gtrsim 0.01 \ M_{\rm Planck}$ [1]. All these results are *negative* regarding the hope to discover new physics, but it is impressive that phenomenological information about this energy regime is accessible at all.

Further hypothetical LIV effects, that are searched for in cosmic photons, are a decay $\gamma \to e^+ + e^-$, photon splitting, vacuum birefringence and an irregular threshold energy for long distance propagation through the cosmic background radiation [1].

3. $U(1)$ gauge theory in an non-commutative space

Here we address $U(1)$ gauge theory in an non-commutative (NC) space as one specific theoretical framework that could lead to a non-linear photon dispersion relation. Due to its discontinuous behavior in the IR limit, this approach does *not* belong to the large class of low energy effective actions — such effective LIV theories, including their predictions for photons, have also been reviewed in Refs. [1]. Conceptual issues related to causality and energy positivity are discussed in Ref. [5].

To describe our framework, let us start by considering a NC Euclidean plane. The (Hermitian) position operators obey the commutation relation

$$[\hat{x}_\mu, \hat{x}_\nu] = {\rm i}\Theta_{\mu\nu} = {\rm i}\theta\epsilon_{\mu\nu} \qquad (\mu,\nu = 1,2) , \tag{5}$$

where we assume the non-commutativity parameter θ to be constant. This implies a spatial uncertainty relation $\Delta x_1 \Delta x_2 = \mathcal{O}(\theta)$. In a Gedankenexperiment, Ref. [6] interpreted this relation (in $d = 4$) as the event horizon, when one tries to measure extremely short distances (of the order of the Planck length), which requires an enormous energy concentration in this range. Thus $\sqrt{\|\Theta\|}$ could be viewed as a "minimal length", *i.e.* an additional constant of Nature (similar to the parameter $1/M$ in Section 2).

A field theory on such a space is *non-local.* On the quantum level, the modification of the standard geometry at short distances implies not only UV, but also IR effects — in particular perturbation theory is in deep trouble due to the emergence of a new type of *IR divergences* ("UV/IR mixing") [7].

Here we consider a fully *non-perturbative approach*, which is even more motivated than in commutative quantum field theory. In analogy to that case, we impose a (fuzzy) lattice structure of spacing a by means of the operator identity

$$\exp\left(\mathrm{i}\frac{2\pi}{a}\hat{x}_\mu\right) = \hat{1} \ . \tag{6}$$

The momentum components p_μ commute, and they obey the usual periodicity over the Brillouin zone, which implies

$$e^{\mathrm{i}p_\mu \hat{x}_\mu} = e^{\mathrm{i}(p_\mu + 2\pi/a)\hat{x}_\mu} \ \Rightarrow \ \hat{1} = e^{\mathrm{i}(p_\mu + 2\pi/a)\hat{x}_\mu} e^{-\mathrm{i}p_\nu \hat{x}_\nu} = \dots = \hat{1}\exp\left(\frac{\mathrm{i}\pi}{a}\theta(p_2 - p_1)\right) .$$

At fixed θ and a, this means that the momenta are *discrete*, and the lattice is automatically periodic. This is of course in contrast to the commutative space. If the lattice is $N \times N$ periodic, the momentum components take the form

$$p_\mu = \frac{2\pi}{aN} n_\mu \quad (n_\mu \in \mathbb{Z}) \quad \Rightarrow \quad \theta = \frac{1}{\pi} N a^2 \ . \tag{7}$$

Therefore the *Double Scaling Limit*

$$\{a \to 0 \quad \text{and} \quad N \to \infty\} \quad \text{at} \quad Na^2 = \text{const.} \tag{8}$$

leads to a *continuous NC plane of an infinite extent.* We start from a finite lattice, where both the UV and the IR sector are regular, and we simultaneously remove both regularizations in a controlled manner. Other procedures could easily lead to $\theta \to 0$ or $\theta \to \infty$, which are different commutative limits. The requirement to link the UV and IR extrapolations is related to the notorious UV/IR mixing (for a review, see Ref. [8]).

A Fourier-type analysis shows that we can return to the use of ordinary coordinates x_μ, if all field multiplications are turned into *star products*, such as

$$\phi(x) \star \psi(x) := \phi(x) \exp\left(\frac{\mathrm{i}}{2} \overleftarrow{\partial}_\mu \, \theta \, \epsilon_{\mu\nu} \overrightarrow{\partial}_\nu\right) \psi(x) \ . \tag{9}$$

The $\star$-commutator $[x_\mu, x_\nu]_\star := x_\mu \star x_\nu - x_\nu \star x_\mu = \mathrm{i}\Theta_{\mu\nu}$ makes this formulation plausible.

Hence the action of pure $U(1)$ gauge theory on a NC plane can be written as

$$S[A] = \frac{1}{2}\int d^2x \, F_{\mu\nu} \star F_{\mu\nu} \ , \quad F_{\mu\nu} = \partial_\mu A_\nu - \partial_\nu A_\mu + \mathrm{i}g(A_\mu \star A_\nu - A_\nu \star A_\mu) \ . \tag{10}$$

This action is $\star$-gauge invariant (though this does not hold for $F_{\mu\nu}$). Note that non-commutativity brings in a Yang-Mills type self-interaction term.

At first sight, it looks hopeless to put this action on a lattice and simulated it: this seems to require $\star$-unitary link variables, $U^\dagger_{x,\mu} \star U_{x,\mu} = 1$ (no sum), which can only be constructed and updated with stringent constraints given by the complete configuration. There is a way out, however, namely the mapping onto a *matrix model*.

A long time ago, and in a different context, González-Arroyo and Okawa introduced the *twisted Eguchi-Kawai (TEK) model* [9]. It is formulated on a single point, with the action

$$S_{\mathrm{TEK}}[U] = N\beta \sum_{\mu\neq\nu} Z_{\mu\nu} \mathrm{Tr}[U_\nu^\dagger U_\mu^\dagger U_\nu U_\mu] \qquad (\beta = 1/g^2) \ , \tag{11}$$

where U_μ ($\mu = 1, 2$ in $d = 2$) are unitary $N \times N$ matrices (which emerged from link variables after dimensional reduction of $U(N)$ lattice gauge theory), and $Z_{21} = Z_{12}^* = \exp(2\pi i n/N)$ is the twist factor ($n \in \mathbb{Z}$, $n \bmod N \neq 0$). Much later it turned out that this model is equivalent to NC $U(1)$ gauge theory on a $N \times N$ lattice: in particular the algebras were shown to be identical (Morita equivalence) [10].

Referring to the historic interpretation of the matrices as contracted link variables, it is obvious to formulate (rectangular) Wilson loops in the TEK model as

$$W_{\mu\nu}(I \times J) = \frac{1}{N} Z_{\mu\nu}^{I \cdot J} \, \mathrm{Tr}[U_\nu^{\dagger\ J} U_\mu^{\dagger\ I} U_\nu^J U_\mu^I] \ . \tag{12}$$

Mapping this term back to the NC gauge theory leads to a sensible definition of a NC Wilson loop [11]. (There are no "closed loops" on a NC plane, but the essential criterion is $\star$-gauge invariance). A specific Wilson loop is complex in general, but the property $W_{\mu\nu} = W_{\nu\mu}^*$ guarantees that the action is real positive, since it involves a sum over both plaquette orientations. Therefore the TEK model formulation is suitable for Monte Carlo simulations, *i.e.* for a non-perturbative study of NC gauge theory.

4. The fate of the non-commutative photon

Let us now address the question how a possible non-commutativity could deform the photon dispersion relation. A 1-loop result of the form

$$E^2 = \vec{p}^{\,2} + \frac{Cg^2}{(p\Theta)^2} \qquad (C = \mathrm{const.}) \tag{13}$$

was first derived in Ref. [12]. Hence one could even be tempted to deduce a *lower* bound on $||\Theta||$ from the observed high-precision linear dispersion relation (*if* $\Theta \neq \mathbf{0}$) [13]. However, further calculations revealed that the constant C is actually *negative*, $C = -2/\pi^2$ [14]. This apparent IR instability is worrisome indeed: it seemed questionable if NC QED does have a ground state, *i.e.* if it actually represents a well-defined quantum theory. Hence people quickly proceeded to a supersymmetric version of this model, where the negative IR divergence is cancelled [14]. (It is not visible either, of course, when one treats the model classically [15].)

However, we were not happy with this combination of daring hypotheses, so we did not assume supersymmetry in addition. We wanted to verify if the NC photon as such is really ill-defined [16]. To this end, we considered a "minimally" NC Euclidean space-time, which involves a NC plane, $[\hat{x}_1, \hat{x}_2] = \mathrm{i}\theta = \mathrm{const.}$, plus a commutative plane (x_3, t) which includes the Euclidean time t. (A NC time coordinate would break reflection positivity, so in that case the transition to the Minkowski signature would no be on safe grounds [17]. Commutative time also alleviates the problems with causality.)

For the commutative plane we used the standard regularization of an $L \times L$ lattice, while the NC plane was treated with a TEK model as described in Section 3, *i.e.* we put a $N \times N$ matrix model on each of the L^2 lattice sites ($L \approx N$). The first challenge for the numerical measurements was the identification of a *scale*, in order to define the Double Scaling Limit

(8). A simple ansatz for the lattice spacing, $a \propto 1/\beta$, was successful. It corresponds to a Double Scaling Limit which keeps the ratio N/β^2 constant. Fig. 1 illustrates, for the example $N/\beta^2 = 20$, that the Wilson loops $\langle W_{\mu\nu} \rangle$ of different sizes, in various planes, are in fact stable as we vary N and β accordingly.

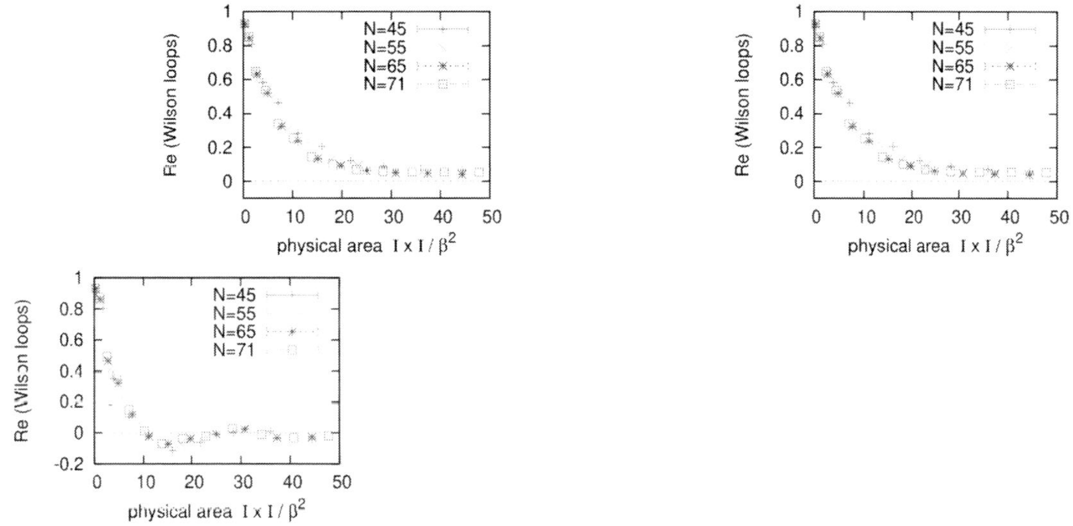

Figure 1. The real part of square-shaped Wilson loops at a constant ratio $N/\beta^2 = 20$. We show results in the commutative plane (left), a mixed plane (center) and the NC plane (right). In the former two planes the Wilson loops are real due to the symmetry in the signs of x_3 and t. In the NC plane $\langle W_{12} \rangle$ has a complex phase which rises linearly in the area for large loops. (This Aharonov-Bohm-type effect was also observed in 2d NC $U(1)$ gauge theory [18], though with some shape dependence of the Wilson loop expectation values [19].) In all cases we observe an excellent Double Scaling behavior.

Next we measured the open Wilson lines $P_\mu(n)$ in μ-direction of length n in lattice unit. In dimensional units this line follows the vector

$$\Theta_{\mu\nu} p_\nu = n a \hat{\mu} \qquad (\hat{\mu} \ : \ \text{unit vector in } \mu \text{ direction}) \ . \tag{14}$$

This observable is $\star$-gauge invariant, in agreement with the non-locality of the $\star$-gauge transformation [11]. Such an open Wilson line carries momentum p given in eq. (14). It is therefore a suitable order parameter to detect if translation invariance in the NC plane is broken. Fig. 2 shows a symmetric behavior at strong coupling ($\beta \lesssim 0.35$) and again at weak coupling, where the transition value of β depends on N. This agrees with the expectation based on strong and weak coupling expansions, but in between we discovered a *phase of broken symmetry*, which was not expected from any analytical consideration. For the weak coupling transition, Fig. 2 shows a marked hysteresis (the two curves for fixed N refer to increasing and decreasing coupling strength), which is characteristic for a first order phase transition.

Fig. 3 shows the corresponding phase diagram. For any N the strong coupling transition is at $\beta \approx 0.35$ (cf. Fig. 2), whereas the weak coupling transition occurs at a value of $\beta \propto N^2$. This means that the Double Scaling Limits — at fixed $\theta \propto N a^2 \propto N/\beta^2$ — all lead to the *broken phase*, and not to the weak coupling phase as one might have expected. Hence it is the broken phase which is really relevant for the question if observables in NC $U(1)$ gauge theory are stable or not. The 1-loop result (13) refers to the weak coupling phase; it is therefore not relevant for

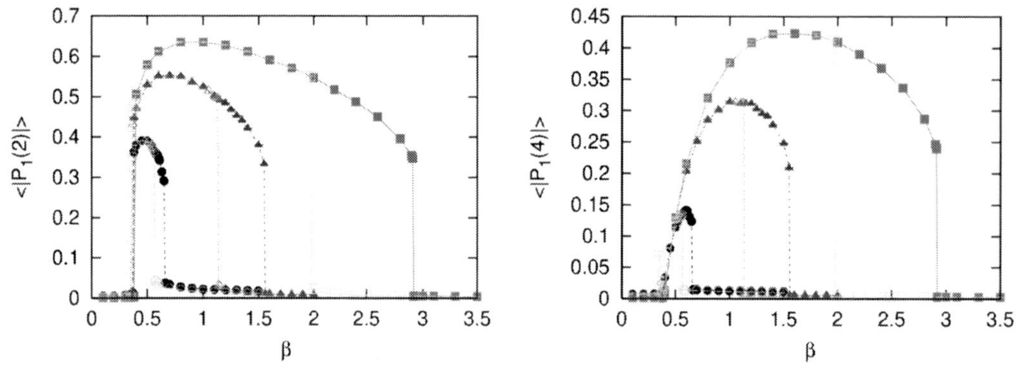

Figure 2. The measured vacuum expectation values for the open Wilson lines of length 2 and 4 lattice units, at $N = 15, 25, 35$ (from left to right), and β fixed by the Double Scaling condition. We observe symmetric phases at strong and at weak coupling, but at moderate coupling we discovered a phase of spontaneously broken translation symmetry. The weak coupling transition has a clear hysteresis behavior.

the NC continuum limit (the Double Scaling Limit). In the broken phase we found stability for all observables that we measured [16]. This holds in particular for our explicit results for the

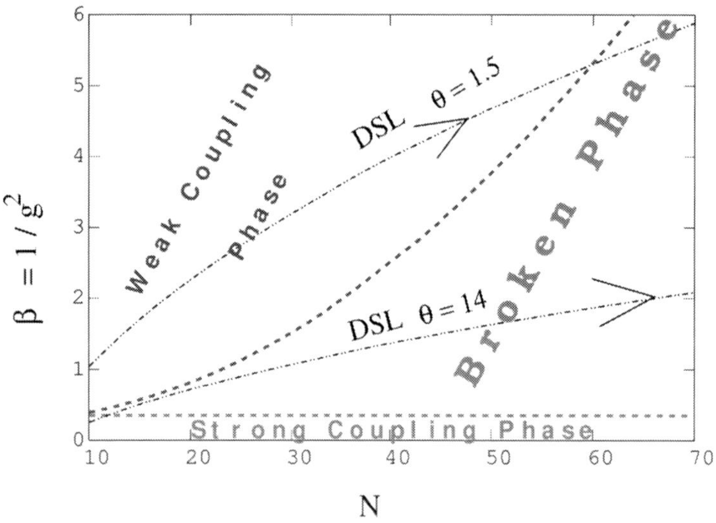

Figure 3. The phase diagram for pure $U(1)$ gauge theory in 4 dimensions, involving a (spatial) NC plane. At fixed size N, Poincaré symmetry is intact at strong and at weak coupling, but it is spontaneously broken in a moderate coupling phase. The Double Scaling Limits (DSL) — taken at fixed non-commutativity parameter θ — lead to the *broken* phase, which is therefore relevant for the NC continuum limit. In that phase, the (negative) IR singularity is not observed, and the system could be sensible.

dispersion relation of the NC photon. It was measured from the exponential decay of the Wilson line correlation function, separated in Euclidean time. We set the momentum components in the NC plane to zero, hence we measured the energy $E(p = p_3)|_{p_1=p_2=0}$.

Journal of Physics: Conference Series **315** (2011) 012029 doi:10.1088/1742-6596/315/1/012029

Fig. 4 (on the left) refers to the (symmetric) weak coupling phase. We see an energy dip at low momenta. This feature is fully consistent with the prediction (13) based on perturbation theory, which some people denote as "tachyonic behavior". (Of course we cannot measure negative energies based on the decay of 2-point functions, but the feature observed here matches exactly such an IR instability).

However, as we pointed out above, the phase which really matters for a continuum limit to a NC space of infinite extent is the phase of broken Poincaré symmetry. The result in that phase is illustrated on the right-hand-side of Fig. 4. It follows the standard linear dispersion relation. Hence we do not see any non-commutativity effect for the dispersion in the commutative plane, but we do see that the disastrous negative IR divergence disappears. Our attempts to measure the dispersion relation also at finite momentum components in the NC plane were not successful, since we did not obtain an exponential decay of the Wilson line correlation functions.

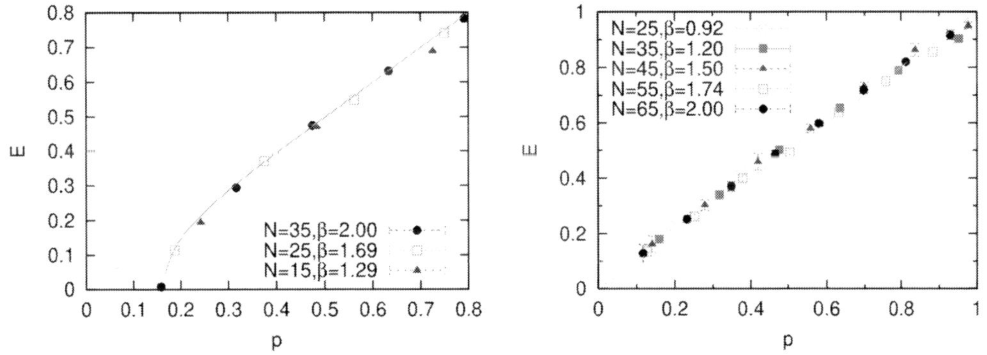

Figure 4. The dispersion relation of the NC photon in the symmetric weak coupling phase (on the left), and in the phase of broken Poincaré symmetry (on the right). We show the energy as a function of the commutative momentum component p_3, $E(p = p_3)|_{p_1=p_2=0}$. Our result in the symmetric phase is consistent with the IR instability, which was predicted perturbatively [14]. (In contrast, a positive IR divergence has been observed in the NC $\lambda\phi^4$ model, where translation symmetry breaks as well [20].) The physically relevant result, however, is the dispersion in the broken phase, where the Double Scaling Limit ends up, and there we observe an *IR stable* behavior.

5. Conclusions

Lorentz Invariance Violation (LIV) has been searched for intensively in recent years, but so far it has not been observed anywhere. We reviewed a failed attempt which checked the photon dispersion relation based on Gamma Rays Bursts.

As one of the theoretical frameworks which could give rise to LIV we discussed quantum field theory — in particular pure $U(1)$ gauge theory — in a non-commutative space. Our non-perturbative study [16] was based on Monte Carlo simulations in a space of Euclidean signature with one (spatial) non-commutative plane. The Double Scaling Limit extrapolates simultaneously to the continuum and to infinite volume at a fixed non-commutativity parameter. It leads to a phase of spontaneously broken Poincaré symmetry, which corresponds to a moderate coupling strength in finite volume.

In this limit we found evidence for convergent observables. This suggests that this model might be renormalizable, despite severe problems in its perturbative expansion [14]. In particular the photon dispersion in the Double Scaling Limit appears IR stable, though we could only measure it in the commutative plane. If one measures a deviation from the linear behavior in an NC direction, the confrontation with GRB data could establish a robust bound on $\|\Theta\|$ (the norm of the non-commutativity tensor) in Nature. So far we provided evidence that the photon may survive in a NC space, although perturbation theory found an negative IR divergence (that result only holds in the weak coupling phase, which turned out to be irrelevant for the NC continuum limit).

As an outlook on the phenomenological side, we mention that the Fermi Gamma-ray Space Telescope has been in orbit since June 2008. It monitors hundreds of GRBs per year, with sensitivity to photon energies $E \simeq 8$ keV ... 300 GeV [21]. This could further boost the precision of Lorentz Invariance tests on cosmic photons [22], cf. Section 2.

Acknowledgment: I thank Jun Nishimura, Yoshiaki Susaki and Jan Volkholz for their collaboration in the work that I summarized here, and Frank Hofheinz for his assistance.

[1] D. Mattingly, *Living Rev. Rel.* **8**, 5 (2005). T. Jacobson, S. Liberati and D. Mattingly, *Annals Phys.* **321**, 150–196 (2006). W. Bietenholz, arXiv:0806.3713 [hep-ph]; *Fortschr. Phys.* **57**, 505–513 (2009). M. Galaverni and G. Sigl, *Phys. Rev.* **D 78**, 063003 (2008). S. Liberati and L. Maccione, *Ann. Rev. Nucl. Part. Sci.* **59**, 245–267 (2009). L. Shao and B.-Q. Ma, arXiv:1007.2269 [hep-ph].
[2] http://www.nasa.gov/mission_pages/swift/bursts/farthest_grb.html
[3] G. Amelino-Camelia *et al.*, *Nature* **393**, 763–765 (1998).
[4] J.R. Ellis *et al.*, *Astropart. Phys.* **25**, 402–411 (2006); erratum, arXiv:0712.2781 [astro-ph].
[5] V.A. Kostelecký and R. Lehnert, *Phys. Rev.* **D 63**, 065008 (2001).
[6] S. Doplicher, K. Fredenhagen and J.E. Roberts, *Commun. Math. Phys.* **172**, 187–220 (1995).
[7] S. Minwalla, M. Van Raamsdonk and N. Seiberg, *JHEP* **02**, 020 (2000).
[8] R.J. Szabo, *Phys. Rept.* **378**, 207–299 (2003).
[9] A. González-Arroyo and M. Okawa, *Phys. Rev.* **D 27**, 2397–2411 (1983).
[10] H. Aoki, N. Ishibashi, S. Iso, H. Kawai, Y. Kitazawa and T. Tada, *Nucl. Phys.* **B 565**, 176–192 (2000). J. Ambjørn, Y. Makeenko, J. Nishimura and R.J. Szabo, *JHEP* **05**, 023 (2000).
[11] N. Ishibashi, S. Iso, H. Kawai and Y. Kitazawa, *Nucl. Phys.* **B 573**, 573–593 (2000). D.J. Gross, A. Hashimoto and N. Itzhaki, *Adv. Theor. Math. Phys.* **4**, 893–928 (2000).
[12] A. Matusis, L. Susskind and N. Toumbas, *JHEP* **0012**, 002 (2000).
[13] G. Amelino-Camelia, L. Doplicher, S.-K. Nam and Y.-S. Seo, *Phys. Rev.* **D 67**, 085008 (2003).
[14] F. Ruiz Ruiz, *Phys. Lett.* **B 502**, 274–278 (2001). K. Landsteiner, E. Lopez and M.H.G. Tytgat, *JHEP* **0106**, 055 (2001). M. Van Raamsdonk, *JHEP* **0111**, 006 (2001).
[15] T. Mariz, J.R. Nascimento and V.O. Rivelles, *Phys. Rev.* **D 75**, 025020 (2007).
[16] W. Bietenholz, J. Nishimura, Y. Susaki and J. Volkholz, *JHEP* **0610**, 042 (2006).
[17] K. Osterwalder and R. Schrader, *Commun. Math. Phys.* **31**, 83–112 (1973).
[18] W. Bietenholz, F. Hofheinz and J. Nishimura, *JHEP* **0209**, 009 (2002).
[19] W. Bietenholz, A. Bigarini and A. Torrielli, *JHEP* **0708**, 041 (2007).
[20] W. Bietenholz, F. Hofheinz and J. Nishimura, *JHEP* **0406**, 042 (2004).
[21] R. Lamon, *JCAP* **0808**, 022 (2008) 022.
[22] Fermi GBM/LAT Collaborations (A.A. Abdo *et al.*), *Nature* **462**, 331–334 (2009). L. Shao, Z. Xiao and B.-Q. Ma, *Astropart. Phys.* **33**, 312–315 (2010).

Indirect Search of Dark Matter with the ANTARES Neutrino Telescope

V. Bertin[1] on behalf of the ANTARES Collaboration

[1]CPPM, Aix-Marseille Université, CNRS/IN2P3,
163 avenue de Luminy – Case 902, 13288 Marseille cedex 9, France

E-mail : bertin@cppm.in2p3.fr

Abstract. Indirect search for Dark Matter trapped inside celestial bodies is one of the main physics goals of neutrino telescopes. The expected flux coming from supersymmetric Dark Matter annihilations in the Sun and the sensitivity of the ANTARES and KM3NeT detectors to such a signal are presented. The ANTARES detector has been taking data during its construction phase in 2007 with five lines operational. This allowed to set a first limit on the neutrino flux coming from Dark Matter annihilations in the Sun with this experiment. A limit on Dark Matter to proton cross section in the framework of Minimal Universal Extra Dimension model is also presented.

1. Introduction

The most popular paradigm of modern cosmology considers the Dark Matter as a population of stable weakly interacting massive particles (WIMPs) relic from the Big Bang, although not yet discovered. Those particles would gravitationally accumulate in the core of massive celestial bodies such as stars or to a lesser extend planets as the Earth, where they could self-annihilate into ordinary matter and eventually produce significant high energy neutrino fluxes. Indirect search for Dark Matter looking at such neutrino fluxes coming from the core of the Sun, the Earth or the Galactic Centre is thus one of the main physics goals of the current and future neutrino telescopes.

2. The ANTARES neutrino telescope

The ANTARES detector [1] is the first undersea neutrino telescope and the largest one of the Northern hemisphere. It is composed of 12 mooring lines, each holding 75 photomultipliers distributed on 25 storeys (the titanium structure holding a triplet of photodetectors), installed at a depth of about 2500 metres off shore the Provençal coast of France, in order to form a 3D-matrix of ~900 photodetectors. The main goal of the experiment is to look for the Cherenkov light induced by high energy muons during their travel in the sea water throughout the detector. The trajectory of the muon track is reconstructed from the detection time of the Cherenkov photons as well as from the positions of the photodetectors. An indirect search for neutrinos can then be performed by selecting the upward-going muons produced by neutrinos which have passed through the entire planet and interacted in the vicinity of the detector. The direction of the incoming neutrino, being almost collinear with the secondary muon, can then be determined with an accuracy reaching 0.2° for high energy neutrinos above 10 TeV. Due to its size and the spacing of the photomultipliers, the ANTARES detector has a low energy threshold of ~20 GeV for reconstructed neutrinos and an effective area of $\sim 10^{-3}$ m² for neutrinos with an energy of 500 GeV. The effective area increases strongly with the neutrino energy and reaches ~1 m^2 for PeV energy neutrinos. Its location in the Northern hemisphere makes it complementary in sky coverage with the South Pole neutrino telescopes AMANDA and IceCube. In addition, a large fraction of the full sky can be observed with ANTARES thanks to the rotation of the Earth, including the central part of the Galaxy which is believed to be the host of many high energy phenomena.

Data taking with the ANTARES detector started after the undersea connection of the first line in February 2006 followed by the second line in September 2006. Three additional lines were connected in January 2007, and another five connected in December 2007. The apparatus reached its complete configuration with the last two lines being connected in May 2008. Some maintenance operations of the detector have successfully been carried out in 2009 and 2010. Line 12 was repaired and reconnected in 2009. Line 6 was repaired, redeployed and is currently waiting for the next submarine connection operation scheduled at the end of 2010, while Line 9 was recovered in Spring 2010 and is currently under repair.

In ANTARES data, event reconstruction is performed by a χ^2 fit of the photodetector hit times as a function of their positions assuming that the light originates from the Cherenkov cone of a muon track passing through the detector. Although the photomultipliers point at 45° downwards, the vast majority of reconstructed events are due to down-going atmospheric muons. These events may represent a background to the neutrino sample if badly reconstructed and not recognised as such. Within the Collaboration, extensive efforts have been carried out to develop efficient reconstruction methods, as well as to improve the Monte Carlo simulation of the detector. An illustration of the present situation is given in figure 1 showing the elevation angle of the reconstructed events from the data recorded in 2007 with a 5-line detector and in 2008 with a 9-to-12-line detector. A clear separation between the upward-going neutrino events and the down-going muon events is obtained as well as a good

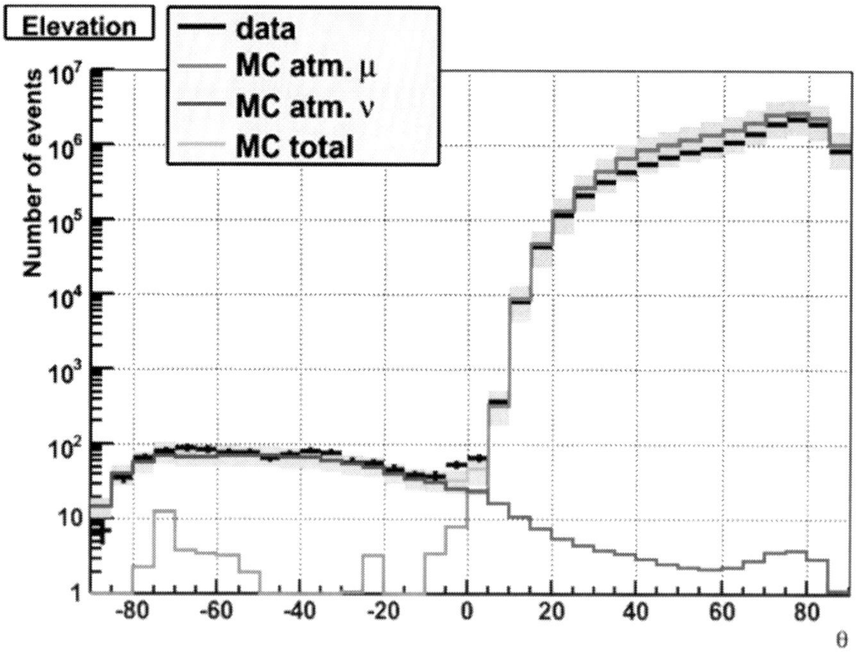

Figure 1. Elevation angle distribution of reconstructed events in data recorded in 2007 with a 5-line detector and in 2008 with a 9-to-12-line detector. The solid lines show the expected distribution for the atmospheric neutrino (blue) and atmospheric muon sample (red), while the green band shows the systematic uncertainty on the simulated total distribution.

agreement between the data sample and the simulated events. A total of more than 1000 neutrino candidate events is selected within this period, corresponding to 3.1 neutrino per day of data taking on average.

3. Sensitivity to neutralino annihilations in the Sun in CMSSM models

The supersymmetric extensions of the Standard Model provide a natural Dark Matter candidate in the form of the lightest neutralino, a particle composed of the supersymmetric partners of the neutral gauge and Higgs bosons. In models with conserved R-parity, the neutralino is a Majorana particle and a stable WIMP which for suitable choice of model parameters can have a relic density in agreement with the one derived from the WMAP measurements [2] of the Cosmic Microwave Background.

The signal of neutralino annihilations into the Sun has been studied within the context of the Constrained Supersymmetric Standard Model (CMSSM) scenario for which the neutralino properties depend on the four parameters m_0, $m_{1/2}$, A_0, $\tan(\beta)$ and $\mathrm{sign}(\mu)$. From those parameters defined at the GUT energy scale, the properties of the supersymmetric particle spectrum including the neutralino at the electroweak scale are calculated using renormalization group equations (RGE). The expected neutrino flux resulting from neutralino annihilations into the Sun was calculated for approximately four million parameter sets with a modified version of DarkSUSY 4.1 [3] using a random walk method to scan the regions of the parameter space allowed by theoretical and experimental constraints and to highlight models predicting a neutralino relic density in agreement with the WMAP constraints. The RGE code ISASUGRA [4] was used for the calculation of the supersymmetric particle spectrum and the halo model of Navarro, Frenk and White [5] was assumed with a local Dark Matter density of

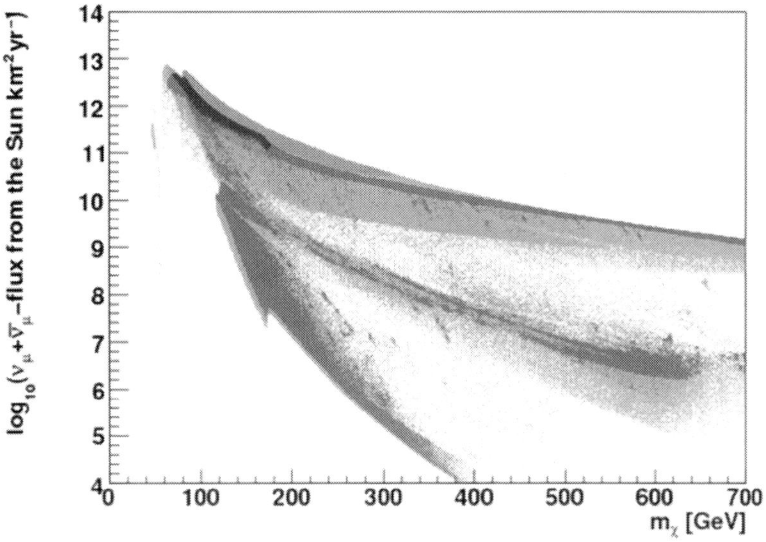

Figure 2. Sensitivity of ANTARES to neutrinos produced by annihilations of neutralinos into the Sun in CMSSM models. The flux of ν_μ and anti-ν_μ integrated for neutrino energies above 10 GeV is given as a function of the neutralino mass. Blue points indicate models within the sensitivity of ANTARES in 3 years, while red points show models outside the reach of this experiment. Brightly coloured points indicate models with a relic density predicted within 2σ of the WMAP region, shaded ones are those outside.

0.3 GeV/c² per cm^3. The flux calculation takes into account the effect of absorption and oscillations of neutrinos inside the Sun as well as during their propagation through vacuum from Sun to Earth.

Knowing the effective area of the ANTARES detector as a function of the neutrino energy, an estimated detection rate can then be calculated from the neutrino flux. Taking into account the irreducible background coming from atmospheric neutrinos as well as an additional background due to misreconstructed atmospheric muon events, a sensitivity is derived considering the signal and background events integrated within a cone of 3° radius around the direction of the Sun. Assuming that only the averaged background rate will be measured, an achievable upper limit for three years of data taking with the complete ANTARES detector can be derived and compared to the detection rate predicted for each individual CMSSM model. Figure 2 highlights the models which can be excluded by three years of data taking with the complete ANTARES detector. The sensitivity of ANTARES will allow to put constraints on part of the CMSSM parameter space, in particular in the so-called "Focus Point" region [6] where the neutralino is mainly of Higgsino type and for which higher neutrino fluxes and harder neutrino spectra are expected.

In order to compare this sensitivity with current limits or expected sensitivities published by other indirect detection experiments, it is more common to present it in term of muon flux at the detector. This quantity becomes however site dependent, being related to the neutrino propagation through the Earth and the target density around the detector. Figure 3 shows the corresponding neutrino induced muon flux integrated above an energy threshold of 1 GeV, which was calculated using neutrino-to-muon conversion rates extracted from DarkSUSY. Figure 3 (left) shows the expected sensitivity of ANTARES to CMSSM models in three years of data taking, while figure 3 (right) highlight the models which could be tested by KM3NeT [7], a next generation of km-scale neutrino telescope in the Mediterranean Sea, in ten years of operation. The expected sensitivity of the IceCube detector with its DeepCore extension [8] is also presented. It can be seen that these new generation neutrino telescope will be able to test almost entirely the "Focus Point" region of the CMSSM parameter space.

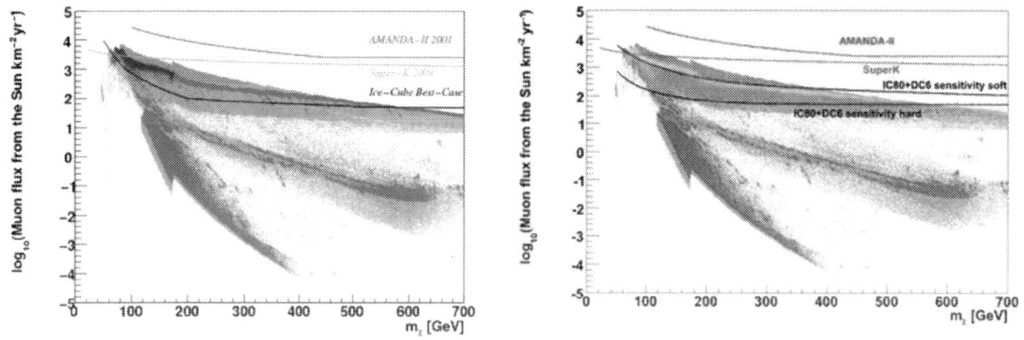

Figure 3. Sensitivity of ANTARES (left) and KM3NeT (right) to neutrino induced muons produced by annihilations of neutralinos into the Sun in CMSSM models. The muon flux integrated for energies above 1 GeV is given as a function of the neutralino mass. Blue and green points indicate models within the sensitivity of ANTARES (in 3 years) and KM3NeT (in 10 years) respectively, while red points show models outside the reach of both experiments. Brightly coloured points indicate models with a relic density predicted within 2σ of the WMAP region, shaded ones are those outside.

Journal of Physics: Conference Series **315** (2011) 012030 doi:10.1088/1742-6596/315/1/012030

4. Limit on neutrino and muon flux from Dark Matter annihilations in the Sun with ANTARES

A search for neutrinos produced by Dark Matter annihilations into the Sun has been carried out in the data sample collected by the ANTARES detector during its five line operation phase in 2007. During this period, corresponding to 167 days of effective lifetime of data taking, more than 15 millions of muon events have been recorded. After reconstruction and selection cuts, essentially based on the quality of the track fit, a sample of about 200 upward-going events representing the neutrino candidates are selected. This total event rate as well as their zenith angle distribution is found to be in good agreement with expectations from the background of atmospheric neutrinos.

This sample of upward-going events has been used to look for a possible excess of neutrinos coming from the direction of the Sun. With the condition that the Sun has to be below the horizon, and taking into account some trigger dead-time, the effective lifetime of this search period reduces to 68.4 days. The analysis is performed by counting the number of observed events inside a search cone centred towards the direction of the Sun. The expected background, mainly due to atmospheric neutrino events, has been estimated as a function of the cone opening angle by Monte Carlo simulation. This has been found in very good agreement with an alternative estimation obtained with the data sample by randomizing the direction of the upward-going events. In the data sample recorded with the ANTARES 5-line detector, the distribution of events observed towards the direction of the Sun as a function of the cone opening angle is found to be in good agreement with background expectation. A limit on a possible excess of events as a function of the cone opening angle has then been derived following the unified approach method of Feldman and Cousins [9].

Using the two extreme cases of a hard and a soft neutrino spectrum resulting from neutralino annihilations into W^+W^-/ZZ gauge bosons and b anti-b quarks respectively, an optimal cone size has been derived for every neutralino mass before analyzing the data. This allowed to set a limit on the flux of neutrinos, integrated above an energy threshold of 10 GeV, produced by neutralino annihilations inside the Sun as a function of the neutralino mass for these two hard and soft spectra, as shown in figure 4 (left). In order to compare this result with the limits published by other indirect detection experiments, the corresponding neutrino induced muon flux integrated above an energy threshold of 1 GeV was calculated. The limits obtained on that quantity are presented in figure 4 (right). Although they are not yet competitive, the limits obtained by ANTARES based on a data sample of about six months recorded with half of its final detector design are already promising.

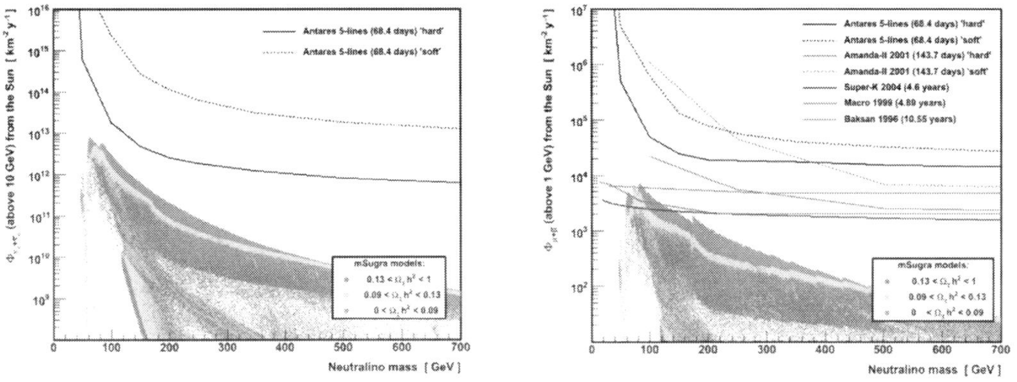

Figure 4. Limit on neutrino flux (Eν > 10 GeV) (left) and neutrino induced muon flux (Eμ > 1 GeV) (right) coming from the Sun obtained by ANTARES with the data of the 5-line period in comparison to the expected flux from neutralino annihilations in CMSSM models. Existing limits from other experiments are also shown. Green points indicate models with a relic density within 2σ of the WMAP region, blue and red points show models predicting respectively a lower or larger relic density.

5. Limit in the Minimal Universal Extra Dimension Framework

The same analysis result can also be interpreted in other alternative Dark Matter scenario in order to put some constraints on their parameter space. Another popular phenomenology is given by the Minimal Universal Extra Dimension (MUED) model with one extra-dimension, in which the first Kaluza-Klein (KK) excitation of the hypercharge gauge boson $B^{(1)}$ is the Lightest KK particle (LKP) and a viable Dark Matter candidate [10]. This model provides a highly predictive phenomenology with mainly two free parameters: the LKP mass M_B^1 and the relative mass splitting between the LKP and the first quark excitation $\Delta = (M_Q^1 - M_B^1)/M_B^1$. Accelerator measurements constraint the lower band for the LKP mass at 300 GeV [11]. The upper band is limited to a few TeV in order not to exceed the observed relic density and overclose the Universe. Figure 5 shows the upper limit on the LKP-to-proton cross section, governing the LKP capture within the Sun, which can be derived from the absence of high energy neutrinos coming from the direction of the Sun within the data recorded by ANTARES in 2007 with a 5-line detector. Although MUED models with a LKP relic density compatible with the WMAP constraints are beyond the reach of existing neutrino telescope, its illustrates the much higher sensitivity of such detectors with respect to direct detection experiments.

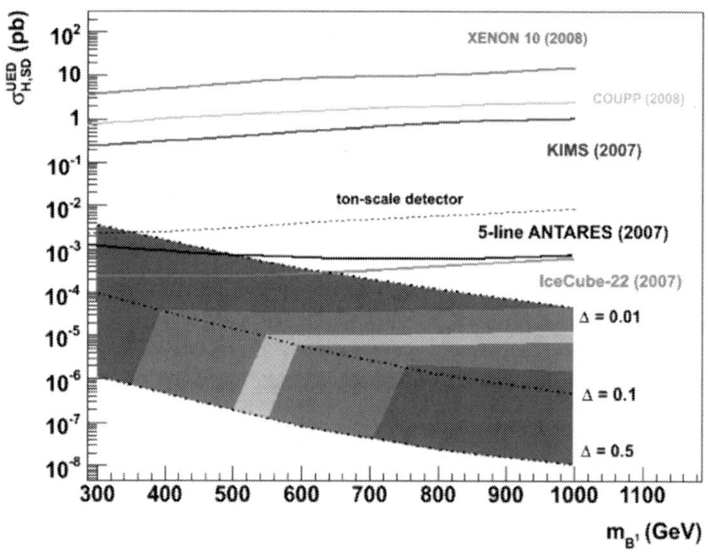

Figure 5. Limit on the LKP-to-proton cross section as a function of the mass of the LKP $B^{(1)}$ particle obtained from the search of neutrino events produced by Dark Matter annihilation inside the Sun in the data recorded by ANTARES in 2007 with a 5-line detector, compared to existing limit from other direct and indirect detection experiments. The light pink (resp. dark pink) region indicates MUED models with a LKP relic density within 1σ of the WMAP region (resp. within $0.05 < \Omega_{LKP} h^2 < 0.20$). The blue area indicates other MUED models.

6. Conclusion and perspectives

The ANTARES detector is the first undersea neutrino telescope and the largest one in the Northern hemisphere. A sample of about 200 neutrino events has been recorded during the building stage of the experiment in 2007, when five lines were operational. This set of events is found to be in good agreement with the expected atmospheric neutrino background. In particular, no excess has been found towards the direction of the Sun allowing to set a promising limit on the flux of neutrinos produced by Dark Matter annihilations inside the Sun. A similar analysis looking for signals of Dark

Matter annihilations in the Earth or towards the centre of the Galaxy is currently under progress, while more than 2000 neutrino candidates have already been recorded by ANTARES up to 2010.

Further studies of supersymmetric models show that the ANTARES experiment and a future km-scale undersea neutrino telescope KM3NeT will be sensitive to neutrino fluxes predicted by an interesting class of models for which the Dark Matter relic density is in agreement with current cosmological constraints.

References

[1] A. Aguilar *et al.* (ANTARES Collaboration), "ANTARES, the first undersea neutrino telescope", to appear in *Nuclear Instruments & Methods*, arXiv:1104.1607v2
[2] J. Dunkley *et al.*, *Astrophys. J. Suppl.* **180**, 306 (2009).
[3] P. Gondolo *et al.*, *JCAP* **0407**, 008 (2004).
[4] F.E. Page, S.D. Protopescu, H. Baer and X. Tata, "Isajet 7.69: A Monte Carlo generator for p p, anti-p p and e+e- reactions", hep/ph0312045 (2003).
[5] J.F. Navarro, C.S. Frenk and S.D.M. White, *Astrophys. J.* **490**, 493 (1997).
[6] J.L. Feng, K.T. Matchev and T. Moroi, *Phys. Rev. D* **61**, 075005 (2000).
[7] "Technical Design Report of the KM3NeT detector", http://www.km3net.org/public.php
[8] E. Resconi for the IceCube Coll., Nucl. *Instr. Meth. A* **602**, 7 (2009).
[9] G.L. Feldman and R.D. Cousins, *Phys. Rev. D* **57**, 3873 (1998).
[10] G. Servant and T.M.P. Tait, *Nucl. Phys. B* **650**, 391 (2003).
[11] J.F. Oliver *et al.*, *Phys. Rev. D* **67**, 056002 (2003).